D1636303

The Moscow Puzzles

359 Mathematical Recreations

BORIS A. KORDEMSKY

Edited and with an introduction by

MARTIN GARDNER

Translated by Albert Parry,
Professor Emeritus of Russian Civilization and Language,
Colgate University

DOVER PUBLICATIONS, INC.
NEW YORK

This Dover edition, first published in 1992, is a slightly altered, slightly corrected republication of the work first published by Charles Scribner's Sons, New York, in 1972.

Library of Congress Cataloging-in-Publication Data

Kordemskiĭ, B. A. (Boris Anastas'evich), 1907–1999
[Matematicheskaia smekalka. English]
The Moscow puzzles : 359 mathematical recreations / Boris A. Kordemsky.
p. cm.
Translation of: Matematicheskaia smekalka.
Originally published: New York : C. Scribner's Sons, 1972.
Includes index.
ISBN-13: 978-0-486-27078-4
ISBN-10: 0-486-27078-5
1. Mathematical recreations. I. Title.
[QA95.K5613 1992]
793.7'4—dc 20 91-39094
CIP

Manufactured in the United States by RR Donnelley
27078524 2015
www.doverpublications.com

Contents

Introduction

The book now in the reader's hands is the first English translation of *Mathematical Know-how,* the best and most popular puzzle book ever published in the Soviet Union. Since its first appearance in 1956 there have been eight editions, as well as translations from the original Russian into Ukrainian, Estonian, Lettish, and Lithuanian. Almost a million copies of the Russian version alone have been sold. Outside the U.S.S.R. the book has been published in Bulgaria, Rumania, Hungary, Czechoslovakia, Poland, Germany, France, China, Japan, and Korea.

The author, Boris A. Kordemsky, who was born in 1907, is a talented high school mathematics teacher in Moscow. His first book on recreational mathematics, *The Wonderful Square,* a delightful discussion of curious properties of the ordinary geometric square, was published in Russian in 1952. In 1958 his *Essays on Challenging Mathematical Problems* appeared. In collaboration with an engineer he produced a picture book for children, *Geometry Aids Arithmetic* (1960), which by lavish use of color overlays, shows how simple diagrams and graphs can be used in solving arithmetic problems. His *Foundations of the Theory of Probabilities* appeared in 1964, and in 1967 he collaborated on a textbook about vector algebra and analytic geometry. But it is for his mammoth puzzle collection that Kordemsky is best known in the Soviet Union, and rightly so, for it is a marvelously varied assortment of brain teasers.

Admittedly many of the book's puzzles will be familiar in one form or another to puzzle buffs who know the Western literature, especially the books of England's Henry Ernest Dudeney and America's Sam Loyd. However, Kordemsky has given the old puzzles new angles and has presented them in such amusing and charming story forms that it is a pleasure to come upon them again, and the story backgrounds incidentally convey a valuable impression of contemporary Russian life and customs. Moreover, mixed with the known puzzles are many that will be new to Western readers, some of them no doubt invented by Kordemsky himself.

The only other Russian writer on recreational mathematics and science who can be compared with Kordemsky is Yakov I. Perelman (1882-1942), who in addition to books on recreational arithmetic, algebra, and geometry, wrote similar books on mechanics, physics, and astronomy. Paperback editions of Perelman's works are still widely sold throughout the U.S.S.R., but Kordemsky's book is now regarded as *the* outstanding puzzle collection in the history of Russian mathematics.

The translation of Kordemsky's book was made by Dr. Albert Parry, former chairman of Russian Studies at Colgate University, and more recently at Case Western Reserve University. Dr. Parry is a distinguished American scholar of Russian origin whose many books range from the early *Garrets and Pretenders* (a colorful history of American bohemianism) and a biography entitled *Whistler's Father* (the father of the painter was a pioneer railroad builder in prerevolutionary Russia) to *The New Class Divided,* a comprehensive, authoritative account of the growing conflict in the Soviet Union between its scientific-technical elite and its ruling bureaucracy.

As editor of this translation I have taken certain necessary liberties with the text. Problems involving Russian currency, for example, have been changed to problems about dollars and cents wherever this could be done without damaging the puzzle. Measurements in the metric system have been altered to miles, yards, feet, pounds, and other units more familiar to readers in a nation where, unfortunately, the metric system is still used only by scientists. Throughout, wherever Kordemsky's original text could be clarified and sometimes simplified, I have not hesitated to rephrase, cut, or add new sentences. Occasionally a passage or footnote referring to a Russian book or article not available in English has been omitted. Toward the end of his volume Kordemsky included some problems in number theory that have been omitted because they seemed so difficult and technical, at least for American readers, as to be out of keeping with the rest of the collection. In a few instances where puzzles were inexplicable without a knowledge of Russian words, I substituted puzzles of a similar nature using English words.

The original illustrations by Yevgeni Konstantinovich Argutinsky have been retained, retouched where necessary and with Russian letters in the diagrams replaced by English letters.

In brief, the book has been edited to make it as easy as possible for an English-reading public to understand and enjoy. More than 90 percent of the original material has been retained, and every effort has been made to convey faithfully its warmth and humor. I hope that the result will provide many weeks or even months of entertainment for all who enjoy such problems.

Martin Gardner

The Moscow Puzzles

I
Amusing Problems

Using Elementary Operations

To see how good your brain is, let's first put it to work on problems that require only perseverance, patience, sharpness of mind, and the ability to add, subtract, multiply, and divide whole numbers.

1. OBSERVANT CHILDREN

A schoolboy and a schoolgirl have just completed some meteorological measurements. They are resting on a knoll. A freight train is passing, its locomotive fiercely fuming and huffing as it pulls the train up a slight incline. Along the railroad bed the wind is wafting evenly, without gusts.

"What wind speed did our measurements show?" the boy asked.

"Twenty miles per hour."

"That is enough to tell me the train's speed."

"Well now." The girl was dubious.

"All you have to do is watch the movement of the train a bit more closely."

The girl thought awhile and also figured it out.

What they saw was precisely what the artist has drawn. What was the train's speed?

2. THE STONE FLOWER

Do you remember the smart craftsman Danila from P. Bazhov's fairy tale, "The Stone Flower"?

They tell in the Urals that Danila, while still an apprentice, took semiprecious Ural stones and chiseled two flowers whose leaves, stems, and petals could be separated. From the parts of these flowers it was possible to make a circular disk.

Take a piece of paper or cardboard, copy Danila's flowers from the diagram, then

cut out the petals, stems, and leaves and see if you can put them together to make a circle.

3. MOVING CHECKERS

Place 6 checkers on a table in a row, alternating them black, white, black, white, and so on, as shown.

Leave a vacant place large enough for 4 checkers on the left.

Move the checkers so that all the white ones will end on the left, followed by all the black ones. The checkers must be moved in pairs, taking 2 adjacent checkers at a time, without disturbing their order, and sliding them to a vacant place. To solve this problem, only three such moves are necessary.

The theme of this problem is further developed in Problems 94–97.

If no checkers are available, use coins, or cut pieces out of paper or cardboard.

4. THREE MOVES

Place three piles of matches on a table, one with 11 matches, the second with 7, and the third with 6. You are to move matches so that each pile holds 8 matches. You may add to any pile only as many matches as it already contains, and all the matches must come from one other pile. For example, if a pile holds 6 matches, you may add 6 to it, no more or less.

You have three moves.

5. COUNT!

How many different triangles are there in the figure?

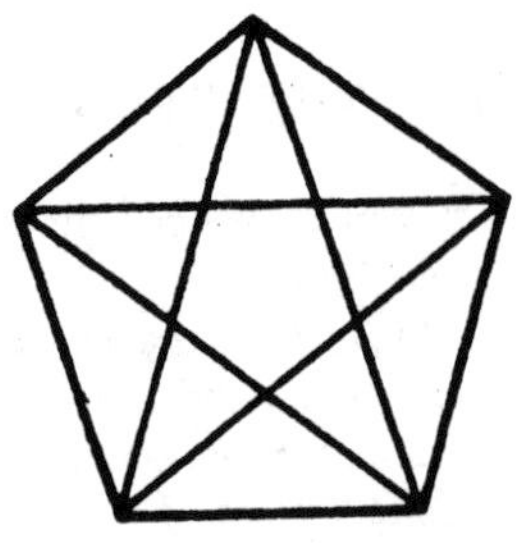

6. THE GARDENER'S ROUTE

The diagram shows the plan of an apple orchard (each dot is an apple tree). The gardener started with the square containing a star, and he worked his way through

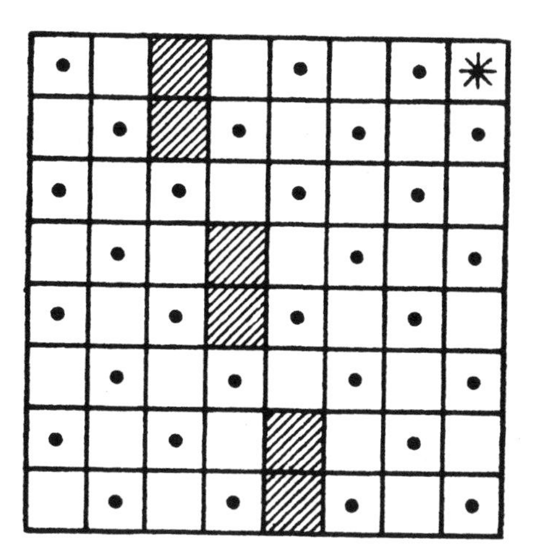

all the squares, with or without apple trees, one after another. He never returned to a square previously occupied. He did not walk diagonally and he did not walk through the six shaded squares (which contain buildings). At the end of his route the gardener found himself on the starred square again.

Copy the diagram and see if you can trace the gardener's route.

7. FIVE APPLES

Five apples are in a basket. How do you divide them among five girls so that each girl gets an apple, but one apple remains in the basket?

8. DON'T THINK TOO LONG

How many cats are in a small room if in each of the four corners a cat is sitting, and opposite each cat there sit 3 cats, and at each cat's tail a cat is sitting?

9. DOWN AND UP

A boy presses a side of a blue pencil to a side of a yellow pencil, holding both pencils vertically. One inch of the pressed side of the blue pencil, measuring from its lower end, is smeared with paint. The yellow pencil is held steady while the boy slides the blue pencil down 1 inch, continuing to press it against the yellow one. He returns the blue pencil to its former position, then again slides it down 1 inch. He continues until he has lowered the blue pencil 5 times and raised it 5 times—10 moves in all.

Suppose that during this time the paint neither dries nor diminishes in quantity. How many inches of each pencil will be smeared with paint after the tenth move?

This problem was thought up by the mathematician Leonid Mikhailovich Rybakov while on his way home after a successful duck hunt. What led him to make up this puzzle is explained in the answer, but don't read it until you have solved the problem.

10. CROSSING A RIVER

A detachment of soldiers must cross a river. The bridge is broken, the river is deep. What to do? Suddenly the officer in charge spots 2 boys playing in a rowboat by the shore. The boat is so tiny, however, that it can only hold 2 boys or 1 soldier. Still, all the soldiers succeed in crossing the river in the boat. How?

Solve this problem either in your mind or practically—that is, by moving checkers, matches, or the like on a table across an imaginary river.

11. WOLF, GOAT, AND CABBAGE

This problem can be found in eighth-century writings.

A man has to take a wolf, a goat, and some cabbage across a river. His rowboat has enough room for the man plus either the wolf or the goat or the cabbage. If he takes the cabbage with him, the wolf will eat the goat. If he takes the wolf, the goat will eat the cabbage. Only when the man is present are the goat and the cabbage safe from their enemies. All the same, the man carries wolf, goat, and cabbage across the river.

How?

12. ROLL THEM OUT

In a long, narrow chute there are 8 balls: 4 black ones on the left, and 4 white ones—slightly larger—on the right. In the middle of the chute there is a small niche

that can hold 1 ball of either color. The chute's right end has an opening large enough for a black but not a white ball.

Roll all the black balls out of the chute. (No, you can't pick them up.)

13. REPAIRING A CHAIN

Do you know why the young craftsman in the picture is so deep in thought? He has 5 short pieces of chain that must be joined into a long chain. Should he open ring 3

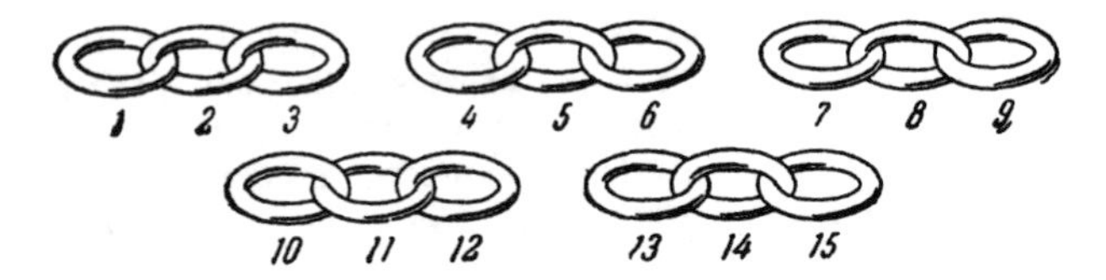

(first operation), link it to ring 4 (second operation), then unfasten ring 6 and link it to ring 7, and so on? He could complete his task in 8 operations, but he wants to do it in 6. How does he do it?

14. CORRECT THE ERROR

With 12 matches form the "equation" shown.

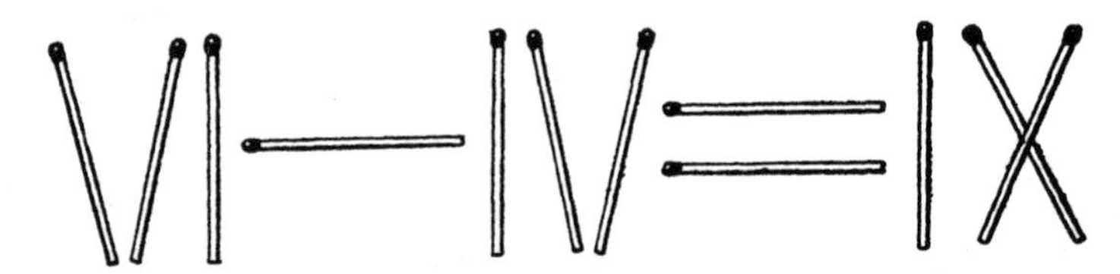

The equation shows that 6 – 4 = 9. Correct it by shifting just 1 match.

15. FOUR OUT OF THREE (A JOKE)

Three matches are on a table. Without adding another, make 4 out of 3. You are not allowed to break the matches.

16. THREE AND TWO IS EIGHT (ANOTHER JOKE)

Place 3 matches on a table. Ask a friend to add 2 more matches to make 8.

17. THREE SQUARES

Take 8 small sticks (or matches), 4 of which are half the length of the other 4. Make three equal squares out of the 8 sticks (or matches).

18. HOW MANY ITEMS?

An item is made from lead blanks in a lathe shop. Each blank suffices for 1 item.

Lead shavings accumulated from making 6 items can be melted and made into a blank. How many items can be made from 36 blanks?

19. ARRANGING FLAGS

Komsomol youths have built a small hydroelectric powerhouse. Preparing for its opening, young Communist boys and girls are decorating the powerhouse on all four sides with garlands, electric bulbs, and small flags. There are 12 flags.

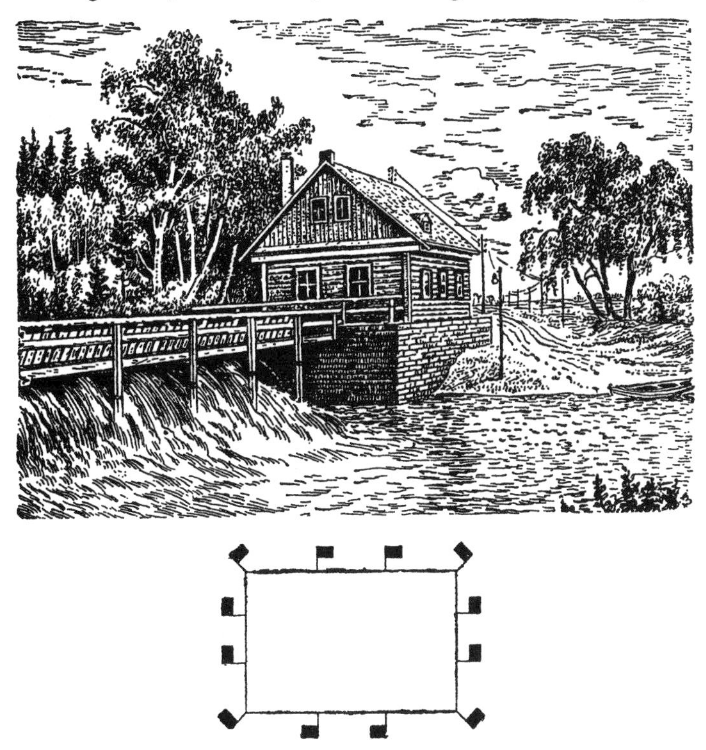

At first they arrange the flags 4 to a side, as shown, but then they see that the flags can be arranged 5 or even 6 to a side. How?

20. TEN CHAIRS

In a rectangular dance hall, how do you place 10 chairs along the walls so that there are an equal number of chairs along each wall?

21. KEEP IT EVEN

Take 16 objects (pieces of paper, coins, plums, checkers) and put them in four rows of 4 each. Remove 6, leaving an even number of objects in each row and each column. (There are many solutions.)

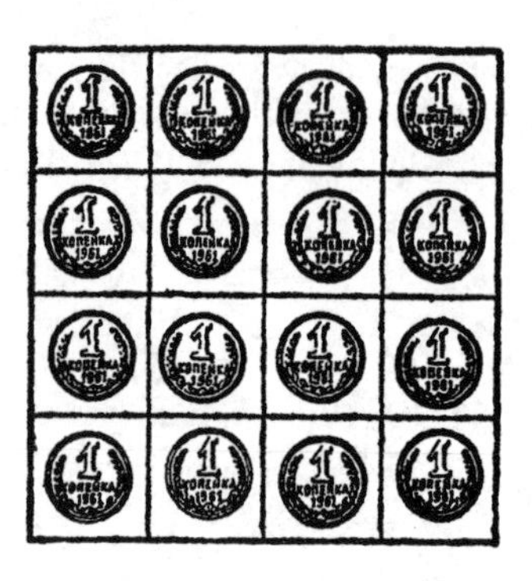

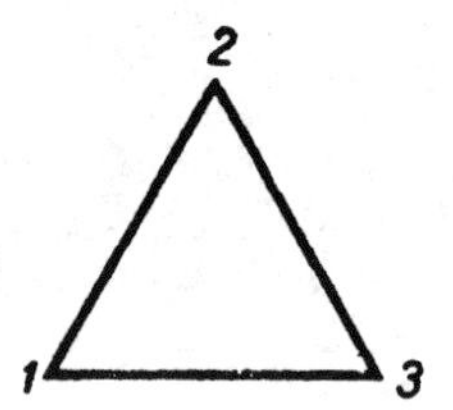

22. A MAGIC TRIANGLE

I have placed the numbers 1, 2, and 3 at the vertices of a triangle. Arrange 4, 5, 6, 7, 8, and 9 along the sides of the triangle so that the numbers along each side add to 17.

This is harder: without being told which numbers to place at the vertices, make a similar arrangement of the numbers from 1 through 9, adding to 20 along each side. (Several solutions are possible.)

23. GIRLS PLAYING BALL

Twelve girls in a circle began to toss a ball, each girl to her neighbor on the left. When the ball completed the circle, it was tossed in the opposite direction.

After a while one of the girls said: "Let's skip 1 girl as we toss the ball."

"But since there are 12 of us, half the girls will not be playing," Natasha objected.

"Well, let's skip 2 girls!"

"This would be even worse—only 4 would be playing. We should skip 4 girls—the fifth would catch it. There is no other combination."

"And if we skip 6?"

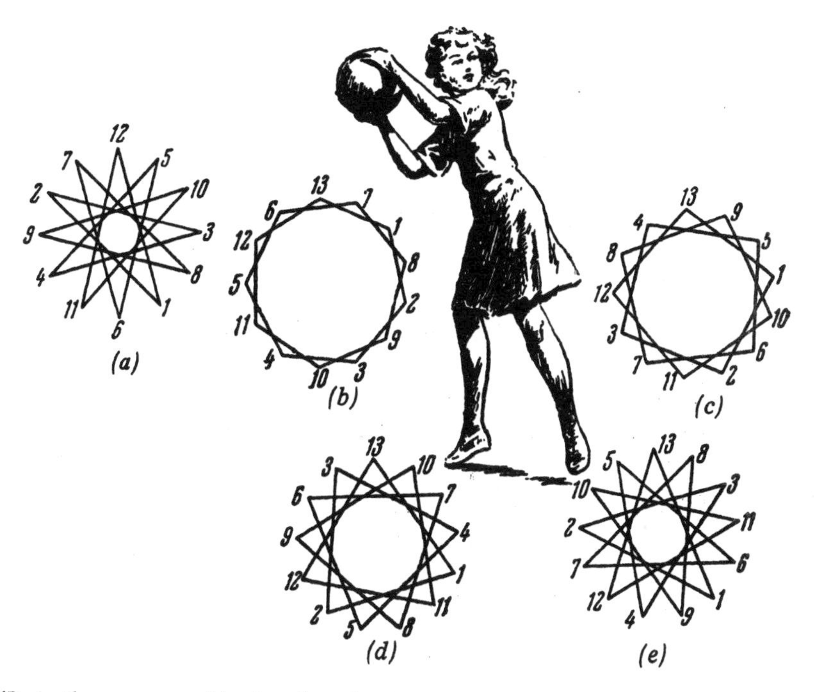

"It is the same as skipping 4, only the ball goes in the opposite direction," Natasha answered.

"And if we skip 10 girls each time, so that the eleventh girl catches it?"

"But we have already played that way," said Natasha.

They began to draw diagrams of every such way to toss the ball, and were soon convinced that Natasha was right. Besides skipping none, only skipping 4 (or its mirror image 6) let all the girls participate (see *a* in the picture).

If there had been 13 girls, the ball could have been tossed skipping 1 girl *(b)*, or 2 *(c)*, or 3 *(d)*, or 4 *(e)*, without leaving any girls out. How about 5 and 6? Draw diagrams.

24. FOUR STRAIGHT LINES

Make a square with 9 dots as shown. Cross all the dots with 4 straight lines without taking your pencil off the paper.

25. GOATS FROM CABBAGE

Now, instead of joining points, separate all the goats from the cabbage in the picture by drawing 3 straight lines.

26. TWO TRAINS

A nonstop train leaves Moscow for Leningrad at 60 miles per hour. Another nonstop train leaves Leningrad for Moscow at 40 miles an hour.

How far apart are the trains 1 hour before they pass each other?

27. THE TIDE COMES IN (A JOKE)

Not far off shore a ship stands with a rope ladder hanging over her side. The rope has 10 rungs. The distance between each rung is 12 inches. The lowest rung touches the water. The ocean is calm. Because of the incoming tide, the surface of the water rises 4 inches per hour. How soon will the water cover the third rung from the top rung of the rope ladder?

28. A WATCH FACE

Can you divide the watch face with 2 straight lines so that the sums of the numbers in each part are equal?

Can you divide it into 6 parts so that each part contains 2 numbers and the six sums of 2 numbers are equal?

29. A BROKEN CLOCK FACE

In a museum I saw an old clock with Roman numerals. Instead of the familiar IV there was an old-fashioned IIII. Cracks had formed on the face and divided it into 4 parts. The picture shows unequal sums of the numbers in each part, ranging from 17 to 21.

Can you change one crack, leaving the others untouched, so that the sum of the numbers in each of 4 parts is 20?

(Hint: The crack, as changed, does not have to run through the center of the clock.)

30. THE WONDROUS CLOCK

A watchmaker was telephoned urgently to make a house call to replace the broken hands of a clock. He was sick, so he sent his apprentice.

The apprentice was thorough. When he finished inspecting the clock it was dark. Assuming his work was done, he hurriedly attached the new hands and set the clock by his pocket watch. It was six o'clock, so he set the big hand at 12 and the little hand at 6.

The apprentice returned, but soon the telephone rang. He picked up the receiver only to hear the client's angry voice:

"You didn't do the job right. The clock shows the wrong time."

Surprised, he hurried back to the client's house. He found the clock showing not much past eight. He handed his watch to the client, saying: "Check the time, please. Your clock is not off even by 1 second."

The client had to agree.

Early the next morning the client telephoned to say that the clock hands, apparently gone berserk, were moving around the clock at will. When the apprentice rushed over, the clock showed a little past seven. After checking with his watch, the apprentice got angry:

"You are making fun of me! Your clock shows the right time!"

Have you figured out what was going on?

31. THREE IN A ROW

On a table, arrange 9 buttons in a 3-by-3 square. When 2 or more buttons are in a

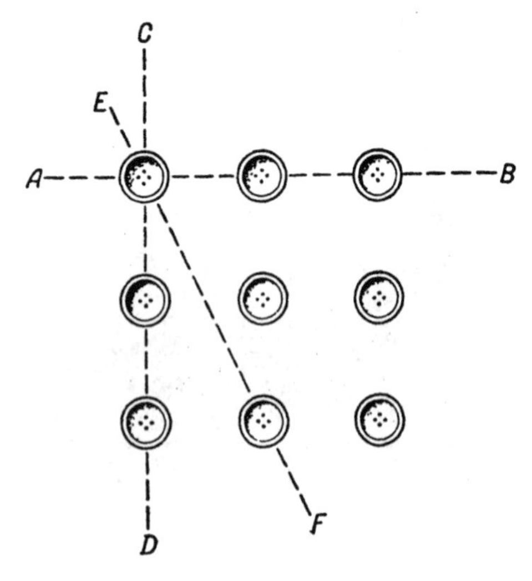

straight line we will call it a row. Thus rows *AB* and *CD* have 3 buttons, and row *EF* has 2.

How many 3- and 2-button rows are there?

Now remove 3 buttons. Arrange the remaining 6 buttons in 3 rows so that each row contains 3 buttons. (Ignore the subsidiary 2-button rows this time.)

32. TEN ROWS

It is easy to arrange 16 checkers in 10 rows of 4 checkers each, but harder to arrange 9 checkers in 6 rows of 3 checkers each. Do both.

33. PATTERN OF COINS

Take a sheet of paper, copy the diagram on it, enlarging it two or three times, and have ready 17 coins:

20 kopeks	5
15 kopeks	3
10 kopeks	3
5 kopeks	6

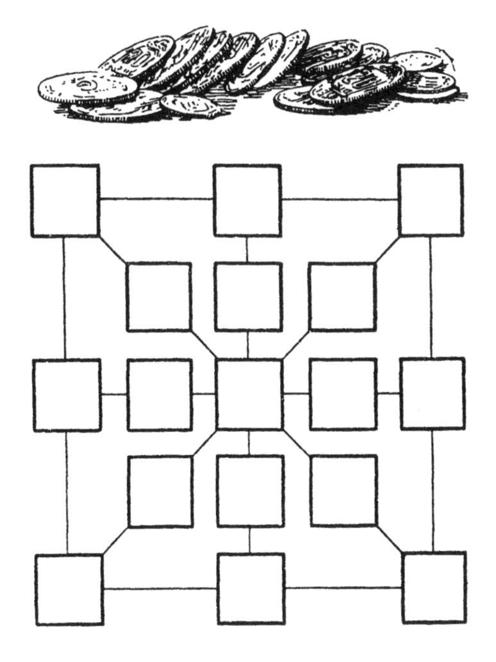

Place a coin in each square so that the number of kopeks along each straight line is 55.

[This problem cannot be translated into United States coinage, but you can work on it by writing the kopek values on pieces of paper–*M.G.*]

34. FROM 1 THROUGH 19

Write the numbers from 1 through 19 in the circles so that the numbers in every 3 circles on a straight line total 30.

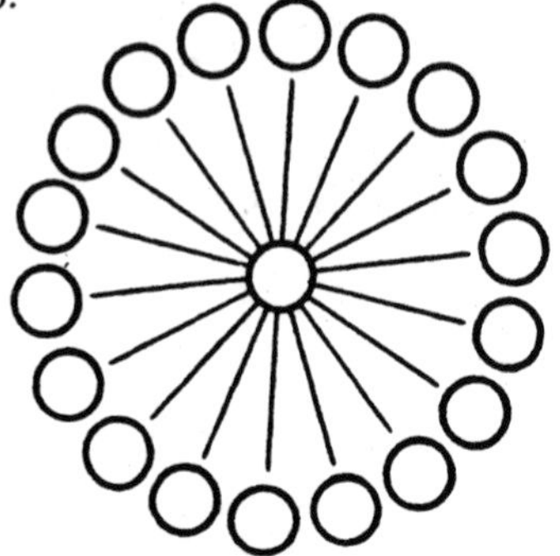

35. SPEEDILY YET CAUTIOUSLY

The title of the problem tells you how to approach these four questions.

(A) A bus leaves Moscow for Tula at noon. An hour later a cyclist leaves Tula for Moscow, moving, of course, slower than the bus. When bus and bicycle meet, which of the two will be farther from Moscow?

(B) Which is worth more: a pound of $10 gold pieces or half a pound of $20 gold pieces?

(C) At six o'clock the wall clock struck 6 times. Checking with my watch, I noticed that the time between the first and last strokes was 30 seconds. How long will the clock take to strike 12 at midnight?

(D) Three swallows fly outward from a point. When will they all be on the same plane in space?

Now check the Answers. Did you fall into any of the traps which lurk in these simple problems?

The attraction of such problems is that they keep you on your toes and teach you to think cautiously.

36. A CRAYFISH FULL OF FIGURES

The crayfish is made of 17 numbered pieces. Copy them on a sheet of paper and cut them out.

Using all the pieces, make a circle and, by its side, a square.

37. THE PRICE OF A BOOK

A book costs $1 plus half its price. How much does it cost?

38. THE RESTLESS FLY

Two cyclists began a training run simultaneously, one starting from Moscow, the other from Simferopol.

When the riders were 180 miles apart, a fly took an interest. Starting on one cyclist's shoulder, the fly flew ahead to meet the other cyclist. On reaching the latter, the fly at once turned back.

The restless fly continued to shuttle back and forth until the pair met; then it settled on the nose of one of the cyclists.

The fly's speed was 30 miles per hour. Each cyclist's speed was 15 miles per hour. How many miles did the fly travel?

39. UPSIDE-DOWN YEAR

When was the latest year that is the same upside down?

40. TWO JOKES

(A) A man phoned his daughter to ask her to buy a few things he needed for a trip. He told her she would find enough dollar bills for the purchases in an envelope on his desk. She found the envelope with 98 written on it.

In a store she bought $90 worth of things, but when it was time to pay she not only didn't have $8 left over but she was short.

By how much, and why?

(B) Mark 1, 2, 3, 4, 5, 7, 8, and 9 on 8 pieces of paper and place them in 2 rows as shown.

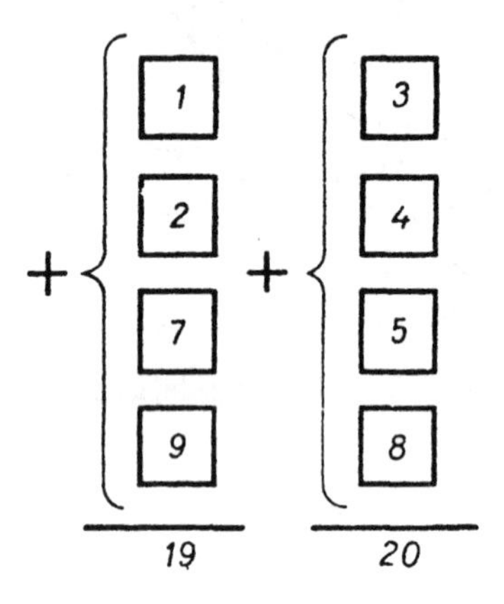

Move 2 pieces so as to make the sums of the two columns equal.

41. HOW OLD AM I?

When my father was 31 I was 8. Now he is twice as old as I am. How old am I?

42. TELL "AT A GLANCE"

Here are two columns of numbers:

123456789	1
12345678	21
1234567	321
123456	4321
12345	54321
1234	654321
123	7654321
12	87654321
1	987654321

Look closely: the numbers on the right are the same as on the left, but reversed and in reverse order. Which column has the larger total? (First answer "at a glance"; then check by adding.)

43. A QUICK ADDITION

(A) These six-digit numbers:

328,645
491,221
816,304
117,586
671,355
508,779
183,696
882,414

can be grouped mentally and added in 8 seconds. How?

(B) Say to a friend: "Write down as many four-digit numbers as you like. Then I will jot down just as many numbers and add them all up, yours and mine, in an instant."

Suppose he writes:

7,621
3,057
2,794
4,518

For your first number, match his fourth number: his 4 with a 5, his 5 with a 4, his 1 with an 8, and his 8 with a 1. His 4,518 plus your 5,481 equals 9,999. Match his other numbers the same way. The complete list is:

7,621
3,057
2,794
4,518
5,481
7,205
6,942
2,378

How can you know, in just a few seconds, that the correct sum is 39,996?

(C) Say: "Write down any two numbers. I will write a third and at once write (from left to right) the sum of the three numbers."

If he writes:

72,603,294
51,273,081

what number should you write, and how do you find the total so quickly?

44. WHICH HAND?

Give a friend an "even" coin (say, a dime—ten is an even number) and an "odd" coin (say, a nickel). Ask him to hold one coin in his right hand and the other in his left.

Tell him to triple the value of the coin in his right hand and double the value of the coin in his left, then add the two.

If the sum is even, the dime is in his right hand; if odd, in his left.

Explain, and think up some variations.

45. HOW MANY?

A boy has as many sisters as brothers, but each sister has only half as many sisters as brothers.

How many brothers and sisters are there in the family?

46. WITH THE SAME FIGURES

Combine plus signs and five 2s to get 28. Combine plus signs and eight 8s to get 1,000.

47. ONE HUNDRED

Express 100 with five 1s. Express 100 three ways with five 5s. You can use brackets, parentheses, and these signs: +, −, ×, ÷ .

48. A DUEL IN ARITHMETIC

The Mathematics Circle in our school had this custom: Each applicant was given a simple problem to solve—a little mathematical nut to crack, so to speak. You became a full member only if you solved the problem.

An applicant named Vitia was given this array:

$$\begin{array}{ccc} 1 & 1 & 1 \\ 3 & 3 & 3 \\ 5 & 5 & 5 \\ 7 & 7 & 7 \\ 9 & 9 & 9 \end{array}$$

He was asked to replace 12 digits with zeros so that the sum would be 20.

Vitia thought a little, then wrote rapidly:

```
0 1 1      0 1 0
0 0 0      0 0 3
0 0 0      0 0 0
0 0 0      0 0 7
0 0 9      0 0 0
-----      -----
  2 0        2 0
```

He smiled and said: "If you substitute just ten zeros for digits, the sum will be 1,111. Try it!"

The Circle's president was taken aback briefly. But he not only solved Vitia's problem, he improved on it:

"Why not replace only 9 digits with zeros—and still get 1,111?"

As the debate continued, ways of getting 1,111 by replacing 8, 7, 6, and 5 digits with zeros were found.

Solve the six forms of this problem.

49. TWENTY

There are three ways to add four odd numbers and get 10:

$$1 + 1 + 3 + 5 = 10;$$
$$1 + 1 + 1 + 7 = 10;$$
$$1 + 3 + 3 + 3 = 10.$$

Changes in the order of numbers do not count as new solutions.

Now add eight odd numbers to get 20. To find all eleven solutions you will need to be systematic.

50. HOW MANY ROUTES?

"In our Mathematics Circle we diagramed 16 blocks of our city. How many different routes can we draw from A to C moving only upward and to the right?

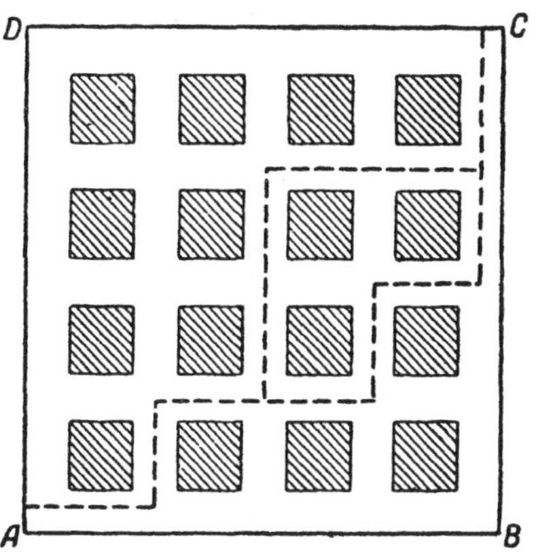

Different routes may, of course, have portions that coincide (as in the diagram).
"This problem is not easy. Have we solved it by counting 70 different routes?"
What answer should we give these students?

51. ORDER THE NUMBERS

The diagram shows 1 through 10 (in order) at the tips of five diameters. Only once does the sum of two adjacent numbers equal the sum of the opposite two numbers:

$$10 + 1 = 5 + 6.$$

Elsewhere, for example:

$$1 + 2 \neq 6 + 7;$$
$$2 + 3 \neq 7 + 8.$$

Rearrange the numbers so that all such sums are equal. You can expect more than one solution to this problem. How many basic solutions are there? How many variants (not including simple rotations of variants)?

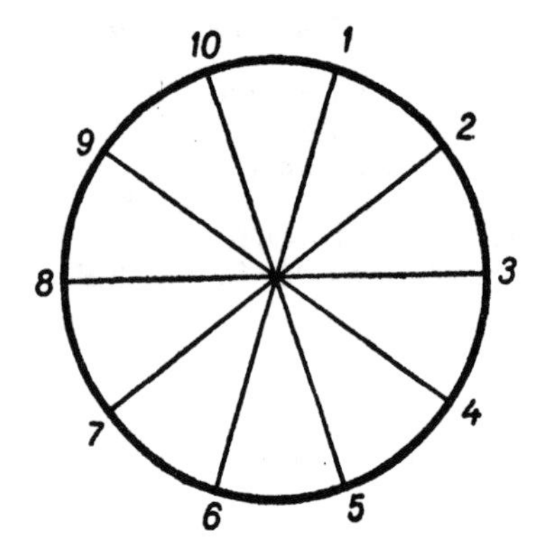

52. DIFFERENT ACTIONS, SAME RESULT

Given two 2s, "plus" can be changed to "times" without changing the result: $2 + 2 = 2 \times 2$. The solution with 3 numbers is easy too: $1 + 2 + 3 = 1 \times 2 \times 3$.
Now find the answer for 4 numbers and the answer(s) for 5 numbers.

53. NINETY-NINE AND ONE HUNDRED

How many pluses should we put between the digits of 987,654,321 to get a total of 99, and where?

There are two solutions. To find even one is not easy. But the experience will help you put pluses between 1, 2, 3, 4, 5, 6, and 7, in order to get a total of 100. (A schoolgirl from Kemerovo, central Siberia, has found two solutions.)

54. A CUT-UP CHESSBOARD

A merry chess player cut his cardboard chessboard into 14 parts, as shown. Friends who wanted to play chess with him had to put the parts back together again first.

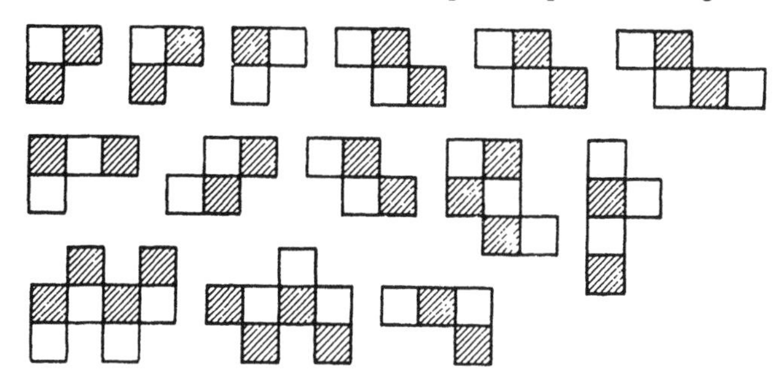

55. LOOKING FOR A LAND MINE

A colonel gave a group of military school cadets a puzzle to solve. He pointed to a field map and said:

"Two sappers with mine detectors must search this area to find enemy mines and defuse them. They have to examine every square on the diagram except the central

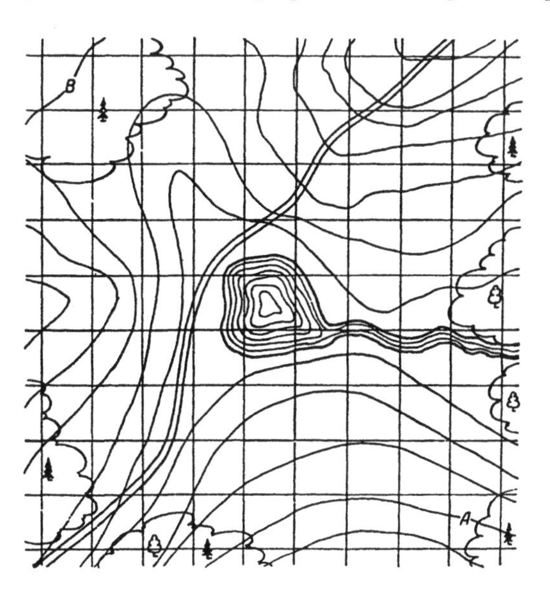

square, which is a small pond. They can proceed horizontally and vertically, but not diagonally, and only one sapper can visit each square, once. The first soldier goes from *A* to *B*, the other from *B* to *A*. Draw their paths so that each one passes through the same number of squares."

Can you, too, solve the colonel's puzzle?

56. GROUPS OF TWO

Ten matches are in a row. I can group them in five pairs, each time moving a match across 2 matches, and placing it on top of a third one, as shown.

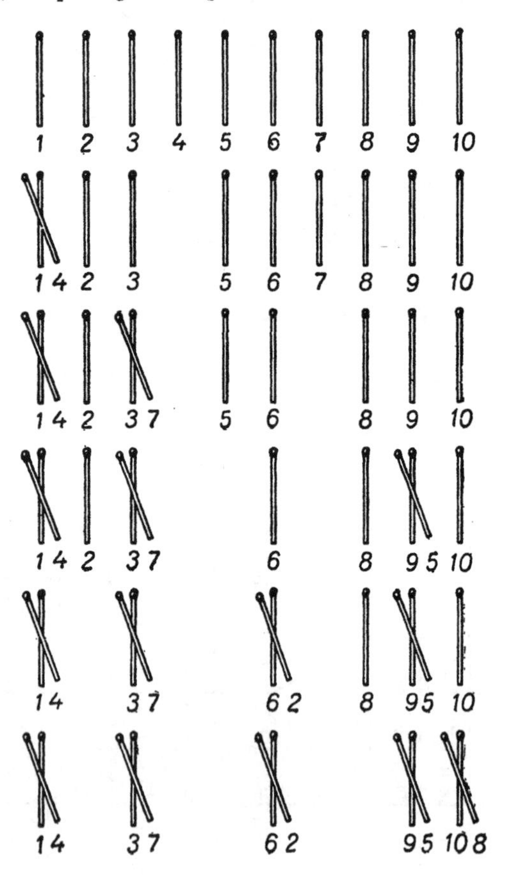

Do it in 10 moves, writing them down.
Find another solution.

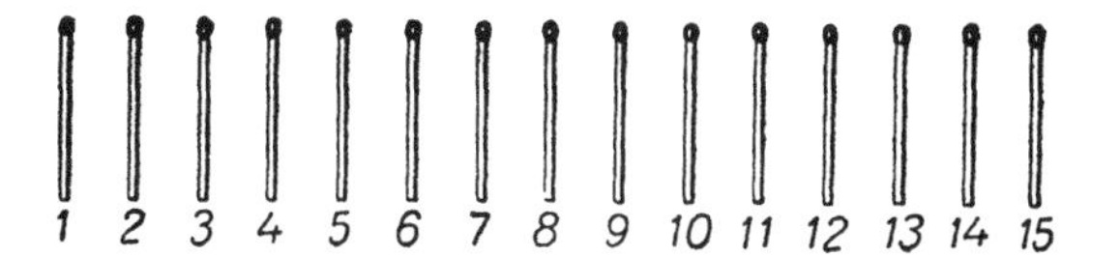

57. GROUPS OF THREE

There are 15 matches in a row. Put them in five groups of 3. In each move a match jumps over 3 matches.

58. THE STOPPED CLOCK

My only timepiece is a wall clock. One day I forgot to wind it and it stopped. I went to visit a friend whose watch is always correct, stayed awhile, and returned home. There I made a simple calculation and set the clock right.

How did I do this when I had no watch on me to tell how long it took me to return from my friend's house?

59. PLUS AND MINUS SIGNS

1 2 3 4 5 6 7 8 9 = 100

Here is the only way to insert 7 plus and minus signs between the digits on the left side to make the equation correct:

$$1 + 2 + 3 - 4 + 5 + 6 + 78 + 9 = 100.$$

Can you do it with only three plus or minus signs?

60. THE PUZZLED DRIVER

The odometer of the family car shows 15,951 miles. The driver noticed that this number is palindromic: it reads the same backward as forward.

"Curious," the driver said to himself. "It will be a long time before that happens again."

But 2 hours later, the odometer showed a new palindromic number.

How fast was the car traveling in those 2 hours?

61. FOR THE TSIMLYANSK POWER INSTALLATION

A factory making measuring equipment urgently needed by the famous Tsimlyansk power installation has a brigade of ten excellent workers: the chief (an older, experienced man) and 9 recent graduates of a manual training school.

Each of the nine young workers produces 15 sets of equipment per day, and their chief turns out 9 more sets than the average of all ten workers.

How many sets does the brigade produce in a day?

62. DELIVERING GRAIN ON TIME

A collective farm was due to deliver its quota of grain to the state authorities. The management of the kolkhoz decided the trucks should arrive in the city at exactly 11:00 A.M. If the trucks traveled at 30 miles per hour they would reach the city at ten, an hour early; at 20 miles an hour they would arrive at noon, an hour late.

How far is the kolkhoz from the city, and how fast should the trucks travel to arrive at 11:00 A.M.?

63. RIDING THE TRAIN TO A DACHA

Two schoolgirls were traveling from the city to a dacha (summer cottage) on an electric train.

"I notice," one of the girls said, "that the dacha trains coming in the opposite direction pass us every 5 minutes. What do you think—how many dacha trains arrive in the city in an hour, given equal speeds in both directions?"

"Twelve, of course," the other girl answered, "because 60 divided by 5 equals 12."

The first girl did not agree. What do you think?

64. FROM 1 TO 1,000,000,000

When the celebrated German mathematician Karl Friedrich Gauss (1777-1855) was nine he was asked to add all the integers from 1 through 100. He quickly added 1 to 100, 2 to 99, and so on for 50 pairs of numbers each adding to 101. Answer: 50 × 101 = 5,050.

Now find the sum of all the digits in the integers from 1 through 1,000,000,000. That's all the *digits* in all the numbers, not all the numbers themselves.

65. A SOCCER FAN'S NIGHTMARE

A soccer fan, upset by the defeat of his favorite team, slept restlessly. In his dream a goalkeeper was practicing in a large unfurnished room, tossing a soccer ball against a wall, then catching it.

But the goalkeeper grew smaller and smaller and then changed into a ping-pong ball while the soccer ball swelled up into a huge cast-iron ball. The iron ball circled around madly, trying to crush the ping-pong ball which darted desperately about.

Could the ping-pong ball find safety without leaving the floor?

Using Fractions and Decimals

To solve the following problems you must know how to use fractions and decimals.

If you have not studied fractions and decimals, skip this section and go on to Chapter II.

66. MY WATCH

As I traveled up and down our great and glorious country, I found myself in a place where the temperature goes up sharply in the day and down at night. This had an effect on my watch. I noticed it was 1/2 minute fast at nightfall, but at dawn it had lost 1/3 minute, making it only 1/6 minute fast.

One morning–May 1–my watch showed the right time. By what date was it 5 minutes fast?

67. STAIRS

A house has 6 stories, each the same height. How many times as long is the ascent to the sixth floor as the ascent to the third?

68. A DIGITAL PUZZLE

What arithmetic symbol can we place between 2 and 3 to make a number greater than 2 but less than 3?

69. INTERESTING FRACTIONS

If to the numerator and denominator of the fraction 1/3 you add its denominator, 3, the fraction will double.

Find a fraction which will triple when its denominator is added to its numerator and to its denominator; find one that will quadruple.

70. WHAT IS IT?

A half is a third of it. What is it?

71. THE SCHOOLBOY'S ROUTE

Each morning Boris walks to school. At one-fourth of the way he passes the machine and tractor station; at one-third of the way, the railroad station. At the machine and tractor station its clock shows 7:30, and at the railroad station its clock shows 7:35.

When does Boris leave his house, when does he reach school?

72. AT THE STADIUM

Twelve flags stand equidistant along the track at the stadium. The runners start at the first flag.

A runner reaches the eighth flag 8 seconds after he starts. If he runs at an even speed, how many seconds does he need altogether to reach the twelfth flag?

73. WOULD HE HAVE SAVED TIME?

Our man Ostap was going home from Kiev. He rode halfway—fifteen times as fast as he goes on foot. The second half he went by ox team. He can walk twice as fast as that.

Would he have saved time if he had gone all the way on foot? How much?

74. THE ALARM CLOCK

An alarm clock runs 4 minutes slow every hour. It was set right 3½ hours ago. Now another clock, which is correct, shows noon.

In how many minutes, to the nearest minute, will the alarm clock show noon?

75. LARGE SEGMENTS INSTEAD OF SMALL

In the Soviet machine industry a marker is a man who draws lines on a metal blank. The blank is cut along the lines to produce the desired shape.

A marker was asked to distribute 7 equal-sized sheets of metal among 12 workers, each worker to get the same amount of metal. He could not use the simple solution of dividing each sheet into 12 equal parts, for this would result in too many tiny pieces. What was he to do?

He thought awhile and found a more convenient method.

Later, he easily divided 5 sheets for 6 workers, 13 for 12, 13 for 36, 26 for 21, and so on.

What was his method?

76. A CAKE OF SOAP

If you place 1 cake of soap on a pan of a scale and 3/4 cake of soap and a 3/4-pound weight on the other, the pans balance.

How much does a cake of soap weigh?

77. ARITHMETICAL NUTS TO CRACK

(A) Use two digits to make the smallest possible positive integer.

(B) Five 3s can express 37:

$$37 = 33 + 3 + 3/3.$$

Find another way to do it.

(C) Use six identical digits to make 100. (Several solutions are possible.)

(D) Use five 4s to make 55.

(E) Use four 9s to make 20.

(F) Seven matches are shown that represent 1/7. Can you get a fraction that equals 1/3 without removing or adding any matches?

(G) Express 20 with plus signs and 1, 3, 5, and 7, using each digit three times.

(H) The sum of two numbers formed with plus signs and the digits 1, 3, 5, 7, and 9 equals the sum of two numbers formed with plus signs and the digits 2, 4, 6, and 8. Find these numbers, using each digit only once, and not using improper fractions.

(I) Name two numbers that have the same product and difference.

Such pairs are uncountably many. How are they formed?

(J) From the digits 0 through 9, each used once, form two equal fractions whose sum equals 1. (Several solutions are possible.)

(K) Using 0 through 9 once each, form two numbers–each an integer with a proper fraction–that add to 100. (Several solutions are possible.)

78. DOMINO FRACTIONS

From a box of dominoes remove the doubles (tiles with the same number at both ends), and the tiles that contain a blank. The remaining 15 tiles, regarded as fractions, are shown in three rows such that the sum of each row is 2½.

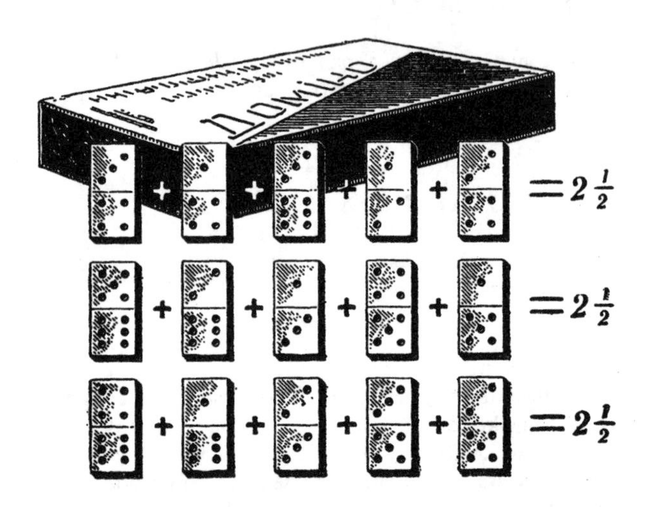

Arrange the 15 tiles in three rows of 5 tiles each so that the sum of the fractions in each row is 10. (You can use improper fractions, such as 4/3, 6/1, 3/2.)

79. MISHA'S KITTENS

Every time young Misha sees a stray kitten he picks up the animal and brings it home. He is always raising several kittens, but he won't tell you how many because he is afraid you may laugh at him.

Someone will ask: "How many kittens do you have now?"

"Not many," he answers. "Three-quarters of their number plus three-quarters of a kitten."

His pals think he is joking. But he is really posing a problem—an easy one.

80. AVERAGE SPEED

A horse travels half his route, with no load, at 12 miles per hour. The rest of the way a load slows him to 4 miles per hour.

What is his average speed?

81. THE SLEEPING PASSENGER

A passenger fell asleep on a train halfway to his destination. He slept till he had half as far to go as he went while he slept. How much of the whole trip was he sleeping?

82. HOW LONG IS THE TRAIN?

A train moving 45 miles per hour meets and is passed by a train moving 36 miles per hour. A passenger in the first train sees the second train take 6 seconds to pass him. How long is the second train?

83. A CYCLIST

After a cyclist has gone two-thirds of his route, he gets a puncture. Finishing on foot, he spends twice as long walking as he did riding.

How many times as fast does he ride as walk?

84. A CONTEST

Volody A. and Kostya B., students in a metal-trade school, are doing lathework. Their foreman-teacher assigns them to make batches of metal parts. They want to finish simultaneously and ahead of the deadline, but after a while Kostya has only done half of what Volodya has left to do, and this is half what Volodya has already done.

How much faster than Volodya does Kostya have to work so they finish at the same time?

85. WHO IS RIGHT?

Masha had to find the product of three numbers in order to calculate the volume of some soil.

She multiplied the first number by the second correctly and was about to multiply the result by the third number when she noticed that the second number had been written incorrectly. It was one-third larger than it should be.

To avoid recalculating, Masha decided it would be safe to merely lower the third number by one-third of itself—particularly since it equaled the second number.

"But you shouldn't do that," a girl friend said to Masha. "If you do, you will be wrong by 20 cubic yards."

"Why?" said Masha.

Why indeed? And what is the correct soil volume?

86. THREE SLICES OF TOAST

Mother makes tasty toast in a small pan. After toasting one side of a slice, she turns it over. Each side takes 30 seconds.

The pan can only hold two slices. How can she toast both sides of three slices in 1½ instead of 2 minutes?

II
Difficult Problems

87. BLACKSMITH KHECHO'S INGENUITY

Last summer, as we traveled through the Georgian Republic, we would make up all sorts of unusual stories. Seeing a relic of old times often inspired us.

One day we came across an old and isolated tower. One of us, a student mathematician, invented an amusing puzzle story:

"Some three hundred years ago a prince lived here, a man of ill heart and much pride. His daughter, who was ripe for marriage, was named Daridjan. He had promised her to a rich neighbor, but she had a different plan: she fell in love with a plain lad, the blacksmith Khecho. The lovers tried to run off to the mountains, but were caught.

"Angered, the prince decided to execute them both the next day. He had them locked up in this tower—a somber structure, unfinished and abandoned. A young girl, a servant who had helped the lovers in their unsuccessful flight, was locked up with them.

"Khecho, calmly looking around, climbed the steps to the tower's upper part and glanced out the window. He realized it would be impossible to jump out and survive. But he saw a rope, forgotten by the masons, hanging near the window. The rope was thrown over a rusty tackle fastened to the tower wall above the window. Empty baskets were tied to each end of the rope. These baskets had been used by the masons to lift bricks and lower rubble. Khecho knew that if one load was 10 pounds more than the other, the heavier basket would descend smoothly to the ground while the other rose to the window.

"Looking at the two girls, Khecho guessed Daridjan's weight at 100 pounds and the servant's at 80. He himself weighed nearly 180. In the tower he found 13 separated pieces of chain, each weighing 10 pounds. Now all three prisoners succeeded in reaching the ground. At no time did a descending basket weigh over 10 pounds more than the ascending basket.

"How did they escape?"

88. CAT AND MICE

Purrer has decided to take a nap. He dreams he is encircled by 13 mice: 12 gray and 1 white. He hears his owner saying: "Purrer, you are to eat each thirteenth mouse,

keeping the same direction. The last mouse you eat must be the white one."
Which mouse should he start from?

89. SISKIN AND THRUSH

At the end of summer camp, the children decided to free the 20 birds they had caught. The Counselor suggested:

"Line up the cages in a row. Counting from left to right, open each fifth cage with a bird in it. When you reach the end of the row, start over. You can take the last 2 birds to the city."

Most of the children did not care which birds would be taken to the city, but Tanya and Alik set their hearts on a siskin and a thrush. As they helped line up the cages, they remembered about the cat and the mice (Problem 88). Which cages did they put the 2 birds in?

90. MATCHES AND COINS

Get 7 matches and 6 coins. Place on a table to form a star, as shown. Count clockwise starting with any match and place a coin at the head of the third match.

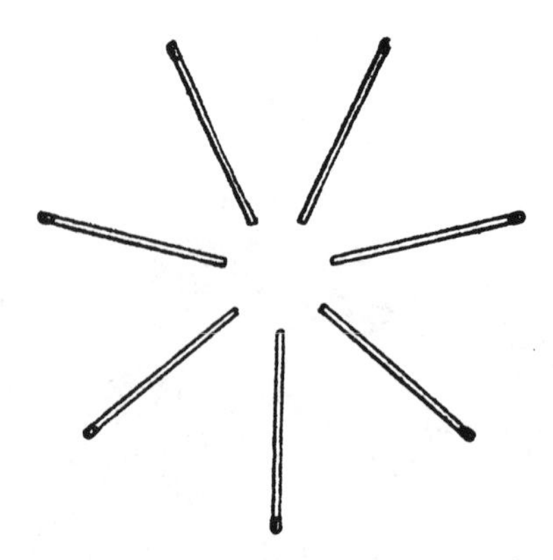

Continue clockwise, beginning with any match that has no coin at its head and placing a coin at the head of the third match. Do not skip matches that already have coins at their heads.

Can you place the 6 coins without placing 2 at the head of any match?

91. LET THE PASSENGER TRAIN THROUGH!

A work train, made up of a locomotive and 5 cars, stops at a small station. The station has a small siding that can hold an engine and 2 cars.

A passenger train is due. How do they let it through?

92. THE WHIM OF THREE GIRLS

The theme of this problem goes back many centuries. Three girls, each with her father, go for a stroll. They come to a small river. One boat, able to carry two persons at a time, is at their disposal. Crossing would be simple, except for the girls' whim: none is willing to be in the boat or ashore with one or two strange fathers unless her own father is present too. The girls, of course, can row.

How do they all get across?

93. AN EXPANSION OF PROBLEM 92

(A) After the six get across, they wonder if, given the conditions, four pairs can cross. Well, they can—if the boat will hold three.

(B) What's more, a boat holding only two can take four girls and their fathers from one shore to the other—if there is an island in the middle which can be used for intermediate loading and unloading.

Show how, for both.

94. JUMPING CHECKERS

Place 3 white checkers in squares 1, 2, and 3 of the figure, and 3 black ones on squares 5, 6, and 7. Shift the white checkers to the squares occupied by the black

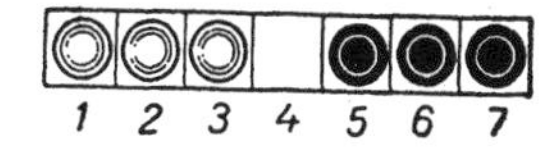

ones, and vice versa. You may move a checker forward to the adjacent unoccupied square, if any. You may jump a checker forward over an adjacent checker into the vacant square. The solution requires 15 moves.

95. WHITE AND BLACK

Take 4 black and 4 white checkers (or 4 pennies and 4 other coins) and put them on a table in a row, white, black, white, black, and so on. Leave a vacant place at one end which can hold 2 checkers. After 4 moves, all the black checkers should be on one side and the white ones on the other.

A move consists of shifting 2 adjacent checkers, keeping their order, into any vacant space.

96. COMPLICATING THE PROBLEM

In the last problem, 8 checkers took 4 moves. Show how 10 checkers take 5 moves, 12 take 6, and 14 take 7.

97. THE GENERAL PROBLEM

From Problems 95–96, can you derive a general procedure for arranging $2n$ checkers in n moves?

98. SMALL CARDS PLACED IN ORDER

Take the ace through 10 from a deck of cards. Deal the ace face down on the table, put the 2 at the bottom of the pile you are holding, deal the 3, put the 4 on the bottom, and so on until all the cards are dealt.

Naturally the cards on the table are not in numerical order.

What order should you start with, top to bottom, to end with cards on the table from ace to 10 with the 10 on top?

99. TWO ARRANGEMENT PUZZLES

(A) Twelve checkers (coins, pieces of paper, etc.) are in a square frame with 4 checkers on each side. Try placing them so there are 5 on each side.

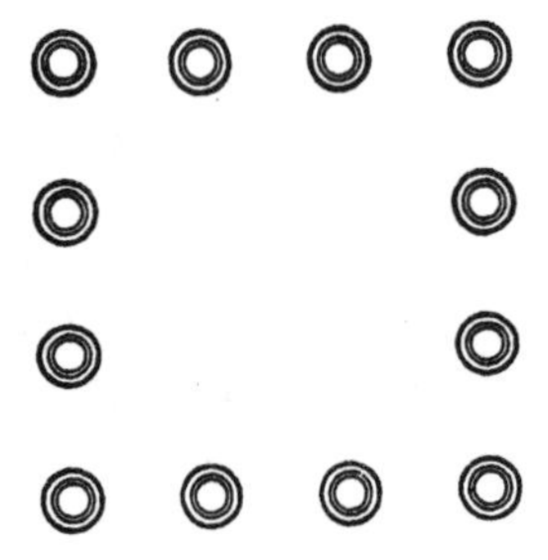

(B) Arrange 12 checkers to form three horizontal and three vertical rows, with 4 checkers in each row.

100. A MYSTERIOUS BOX

Misha brought a pretty little box for his sister Irochka from his Crimean summer camp. She was not of school age yet, but could count to 10. She liked the box because she could count 10 sea shells along each side, as shown.

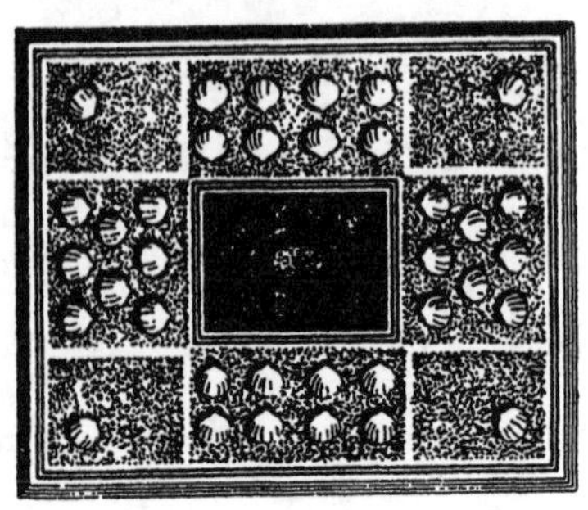

One day Irochka's mother, while cleaning the box, accidentally broke 4 shells.
"No great trouble," Misha said.
He unstuck some of the remaining 32 shells, then pasted them on so that there were again 10 shells along each side of the cover.
A few days later, when the box fell on the floor and 6 more shells were crushed, Misha again redistributed the shells—though not quite so symmetrically—so Irochka could count 10 on each side, as before.
Find both arrangements.

101. THE COURAGEOUS GARRISON

A courageous garrison was defending a snow fort. The commander arranged his forces as shown in the square frame (the inner square showing the garrison's total strength of 40 boys): 11 boys defending each side of the fort.

The garrison "lost" 4 boys during each of the first, second, third, and fourth assaults and 2 during the fifth and last. But after each charge 11 boys defended each side of the snow fort. How?

102. DAYLIGHT LAMPS

A technician was lighting a room for a TV broadcast with tubular neon lamps. At first he put 3 lamps in each corner and 3 lamps along each of the room's four sides, a total of 24 lamps, as shown. He added 4 lamps and again 4 lamps. Then he tried

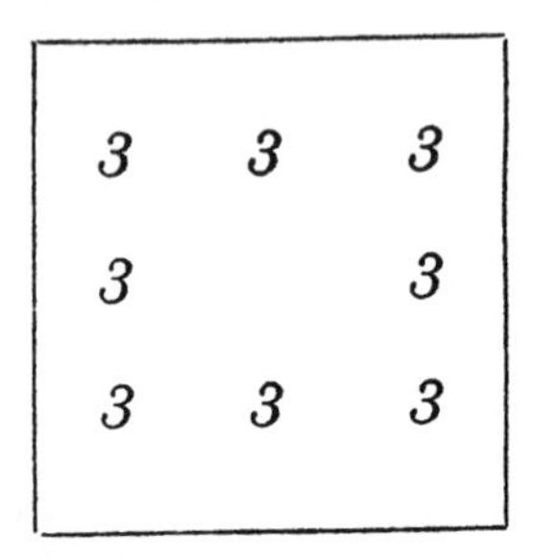

20 lamps, and 18. Always each wall had 9 lamps. How? Could he do it with other numbers of lamps?

103. ARRANGEMENT OF EXPERIMENTAL RABBITS

A special two-level cage has been prepared for experiments with rabbits at a research institute, each level having 9 sections. The rabbits are to occupy 16 sections, 8 on the upper level and 8 on the lower. (The 2 central sections are set aside for equipment.)

There are four conditions for the experiments:

1. All 16 sections must be occupied.
2. No section can hold more than 3 rabbits.
3. Each of the four outer sides (total of both levels) must hold 11 rabbits.
4. The whole upper level must hold twice as many rabbits as the whole lower level.

Although the institute received 3 fewer rabbits than expected, it housed the rabbits in conformity with the four conditions.

How many rabbits were expected, and how many arrived? How were they housed?

104. PREPARING FOR A FESTIVAL

The preceding five problems involved arranging objects along the sides of a rectangle or a square so that their number along each side remained the same when their total number changed. Besides regarding an object in a corner as belonging to two sides, we can regard the intersection of two lines, in general, as belonging to both lines.

For example, in preparing a festive illumination, can you arrange 10 light bulbs in 5 rows with 4 bulbs in each row? The answer is the 5-pointed star shown.

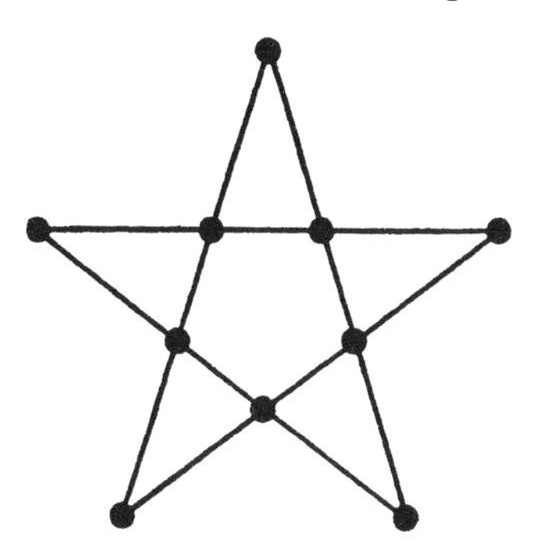

Here are some similar problems. Try to make your solutions symmetrical.

(A) Place 12 light bulbs in 6 rows with 4 bulbs in each row. (There is more than one solution.)

(B) Plant 13 bushes in 12 rows with 3 bushes per row.

(C) On the triangular terrace shown, a gardener raises 16 roses in 12 straight-line rows with 4 roses in each row. Then he prepares a flower bed and transplants to it the 16 roses in 15 rows with 4 roses in each.

How?

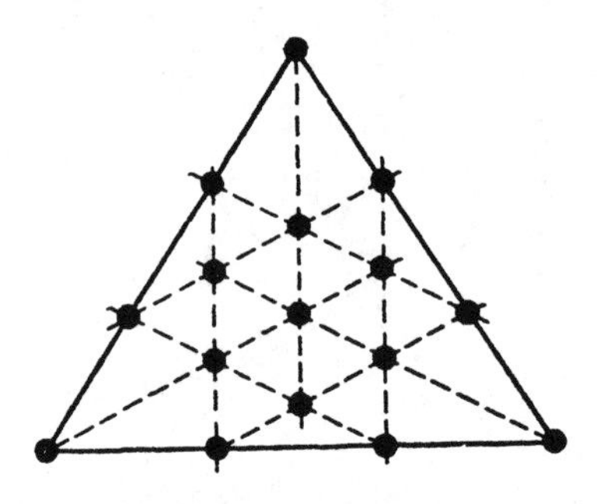

(D) Now arrange 25 trees in 12 rows with 5 trees in each row.

105. PLANTING OAKS

A pretty sight, these 27 oaks in a six-pointed star—9 rows with 6 oaks in each—but a true forester might object to the three isolated trees. An oak loves sunshine from above, but on its sides it prefers greenery. As the saying goes, it likes to wear a coat but no hat.

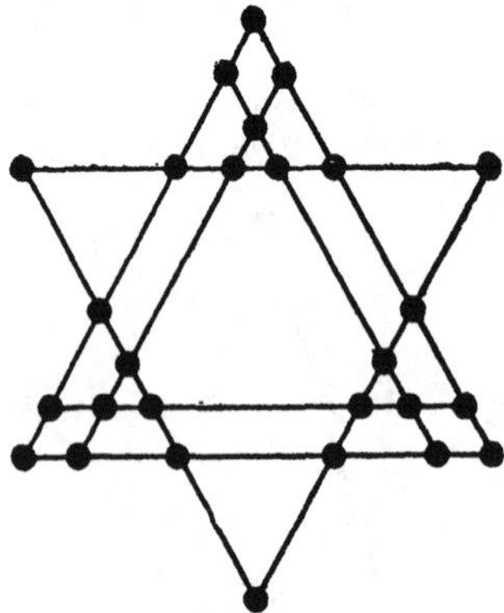

Diagram the 27 oaks in 9 rows with 6 oaks per row, preserving symmetry, but with all the oaks in three clustered groups.

106. GEOMETRICAL GAMES

(A) Place 10 checkers (or coins, buttons, etc.) in 2 rows of 5 each on a table, as shown. Shift 3 checkers from one row and 1 checker from the other (without moving the other checkers and without putting one checker on top of another) so that 5 straight rows with 4 checkers each are formed. Don't move the other checkers and don't pile checkers vertically–but a symmetrical pattern is not required.

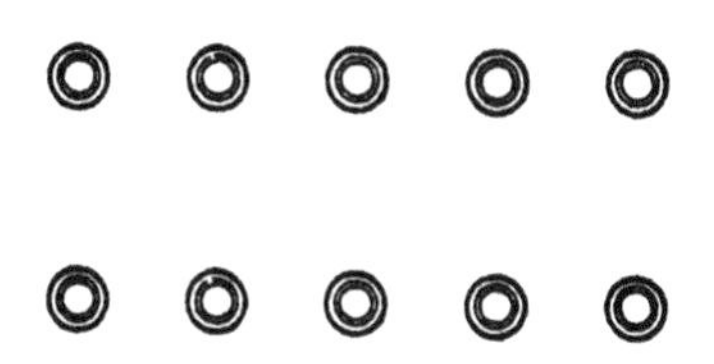

The five solutions shown all have different shapes. And, there are many more solutions. The same checkers can be selected and moved in different ways (*a* and *d*), or different sets of 4 checkers can be selected. Simply selecting the checkers gives fifty solutions: ten ways to select 3 checkers from the top 5, times five ways to select 1 checker from the bottom 5.

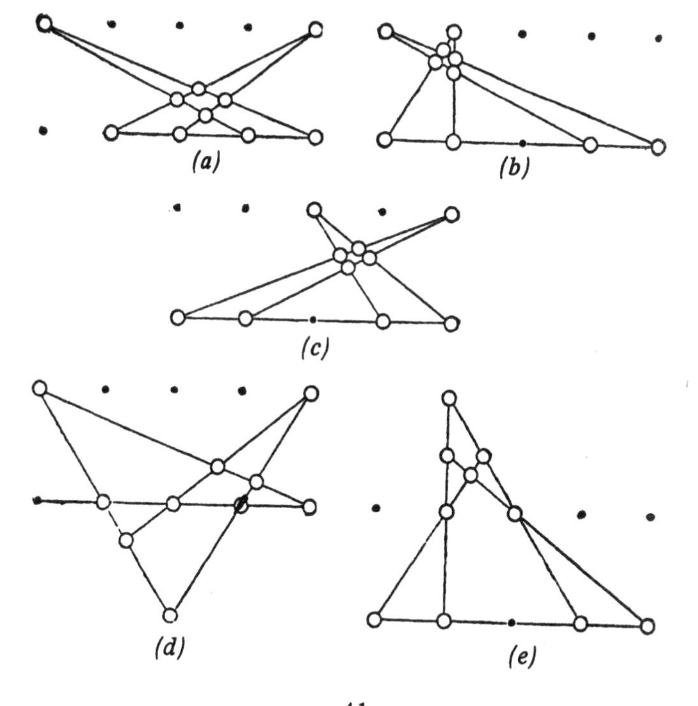

Here is an expansion of the game: In front of each player put 10 checkers in 2 rows. Each player, while alone, shifts 4 checkers (3 from one row and 1 from the other) to form 5 rows with 4 checkers per row. Then they compare solutions. Players with identical configurations get 1 point. A player with a unique pattern gets 2 points. Those not finishing within the time limit get no points.

The game can also be played on paper. Also, you can permit taking 2 checkers from each row, and permit putting checkers on top of each other. This makes many solutions possible like the two shown in diagrams *e* and *f*.

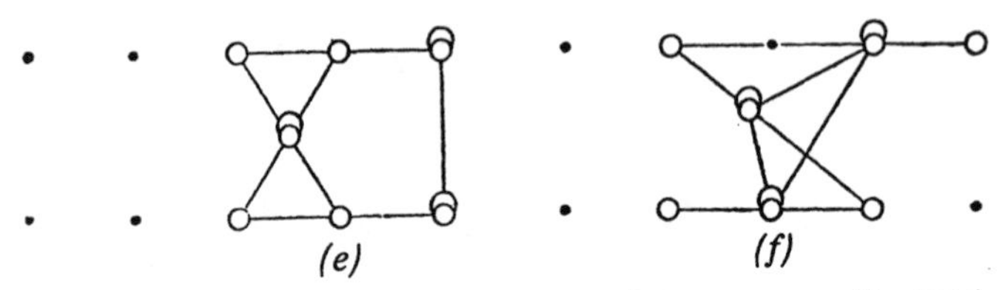

(B) Punch 49 small holes in a piece of cardboard, in a square grille. Stick matches in 10 holes, then try to solve problems such as this:

Take 3 matches and stick them in other holes so as to form 5 rows of 4 matches each.

First solve the problem set up in the diagram below, then vary it by changing the starting pattern and number of rows to be formed.

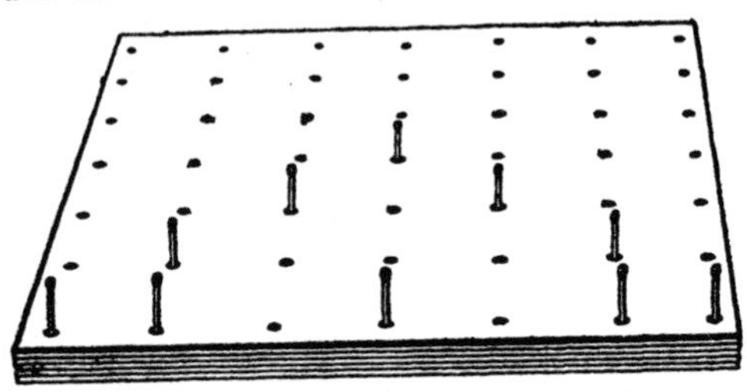

107. EVEN AND ODD

Number 8 checkers and pile them up as shown. Use a minimum of moves to shift checkers 1, 3, 5, and 7 from the center to the "Odd" side circles and checkers 2, 4,

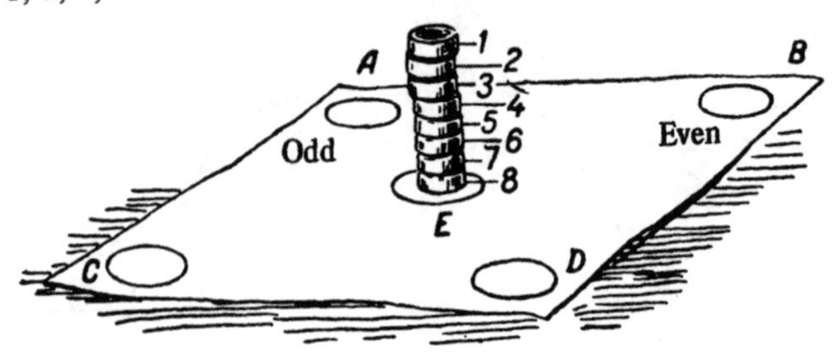

6, and 8 to the "Even" side circles. To move, shift the top checker from one pile to the top of another. It is against the rules to put a checker with a higher number on a checker with a lower one, or to place an odd-numbered checker on an even-numbered checker or vice versa.

Thus you can put checker 1 on 3, 3 on 7, or 2 on 6—not 3 on 1 or 1 on 2.

108. A PATTERN

Place 25 numbered checkers in 25 cells, as shown. By exchanging pairs of checkers put them in numerical order: checkers 1, 2, 3, 4, and 5 in the first row, left to right; 6, 7, 8, 9, and 10 in the second row, and so on.

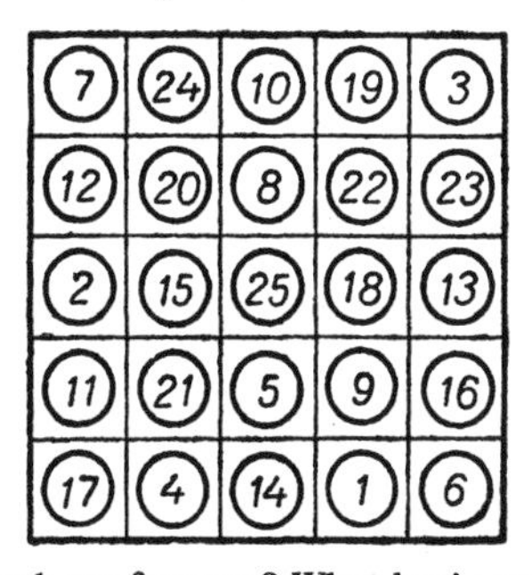

What is the minimum number of moves? What basic method should be used?

109. A PUZZLE GIFT

The well-known Chinese box has a smaller box inside; inside, there is a still smaller box and so on for many boxes.

Make a toy out of four boxes. Put 4 pieces of candy in each of the three smallest boxes and 9 candies in the largest.

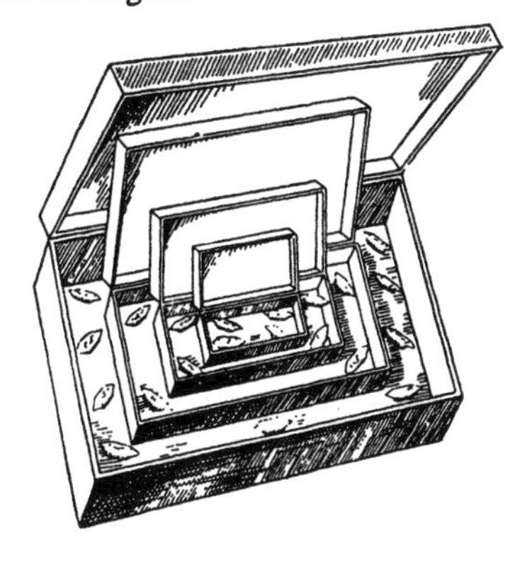

Give this collection of 21 candies as a birthday present and tell your friend not to eat any until he redistributes them so as to have an even number of pairs of candies plus 1 in each box.

Naturally, you should first solve the puzzle yourself.

110. KNIGHT'S MOVE

To solve this problem you need not be a chess player. You need only know the way a knight moves on the chessboard: two squares in one direction and one square at right angles to the first direction. The diagram shows 16 black pawns on a board.

Can a knight capture all 16 pawns in 16 moves?

111. SHIFTING THE CHECKERS

(A) Number 9 checkers, and place them as shown in the first diagram.

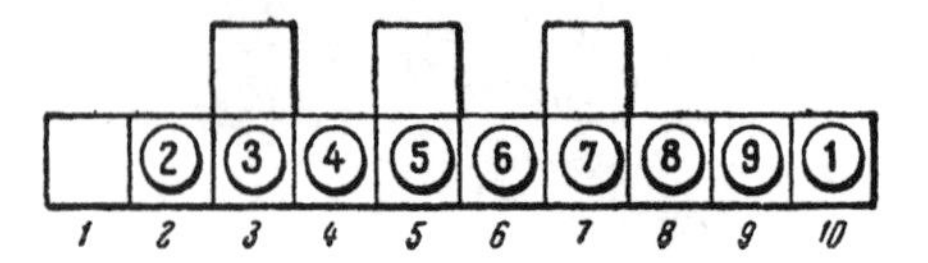

Can you shift checker 1 to cell 1 in 75 moves, with checkers 2 to 9 again in cells 2 to 9?

Checkers move vertically and horizontally into empty cells. No jumps.

(B) In the second diagram, exchange the black and white checkers in 46 moves.

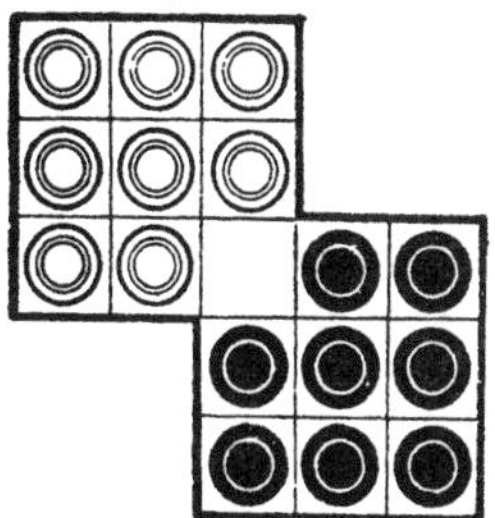

Checkers move horizontally and vertically into the empty cell. You can jump one checker over another and you need not alternate white and black moves.

112. A GROUPING OF INTEGERS 1 THROUGH 15

See how elegantly the integers 1 through 15 can be arranged in five arithmetic progressions of 3 integers:

$$\left.\begin{matrix}1\\8\\15\end{matrix}\right\} d=7; \quad \left.\begin{matrix}4\\9\\14\end{matrix}\right\} d=5; \quad \left.\begin{matrix}2\\6\\10\end{matrix}\right\} \mathrm{d}=4; \quad \left.\begin{matrix}3\\5\\7\end{matrix}\right\} d=2; \quad \left.\begin{matrix}11\\12\\13\end{matrix}\right\} d=1.$$

For example, $8-1=15-8=7$, so d (difference) is 7 for the first triplet.

Now, keeping the first triplet, make four new triplets, still with $d=5$, 4, 2, and 1.

On your own, try arranging the integers from 1 through 15 with other values of d.

113. EIGHT STARS

I have placed a star in one of the white squares of the board shown.

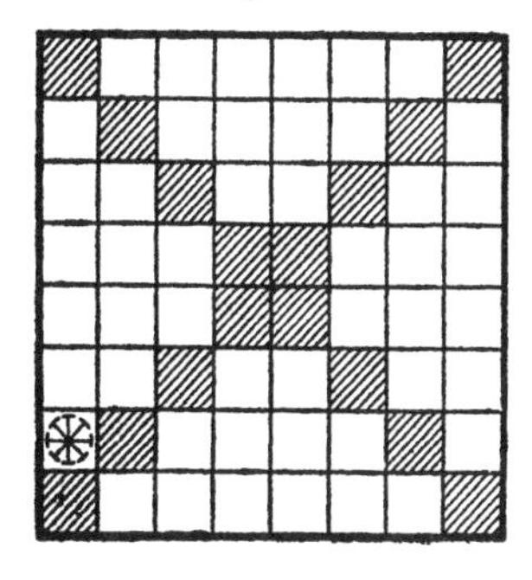

Place 7 more stars in white squares so that no 2 of the 8 stars are in line horizontally, vertically, or diagonally.

114. TWO PROBLEMS IN PLACING LETTERS

(A) Place 4 letters in a 4-by-4 square so that there is only 1 letter in each row, each column, and each of the two main diagonals. How many solutions are there if the 4 letters are identical? If they are different?

(B) Now place 4 *a*'s, 4 *b*'s, 4 *c*'s, and 4 *d*'s in the 4-by-4 square so that there are no identical letters in any row, column, or main diagonal. How many solutions are there?

115. CELLS OF DIFFERENT COLORS

Make 16 square cells of four different colors, say, white, black, red, and green. Write 1, 2, 3, 4 on the 4 white cells, and on the 4 black cells, and so on.

Arrange the cells in a 4-by-4 square so that in each row, column, and main diagonal you will find all the colors and all the numbers. There are many solutions. How many?

116. THE LAST PIECE

Take a piece of cardboard, cut 32 little pieces from it, and place the pieces in circles 2 to 33 of an enlarged copy of the diagram. Circle 1 remains vacant.

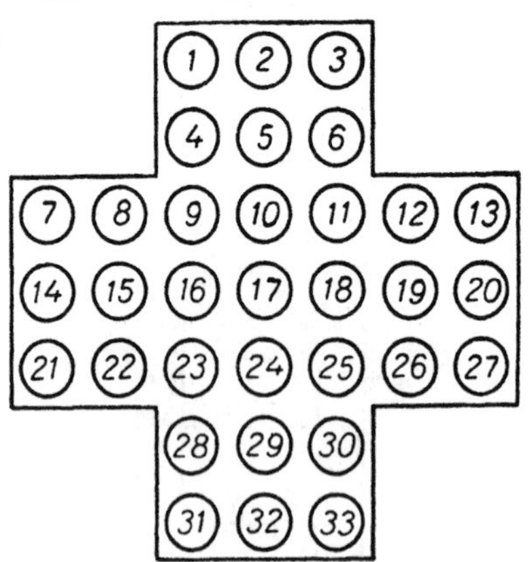

In each move, a piece jumps horizontally or vertically over a second piece into a vacant circle, and the second piece is removed. Make 31 moves so that the last piece jumps into circle 1. (There are many solutions.)

117. A RING OF DISKS

Take 6 equal-sized coins or disks and place them as shown in *a*.

In 4 moves, form them into a new position, ring *(b)*. To move, slide a disk to a new position where it touches at least 2 other disks.

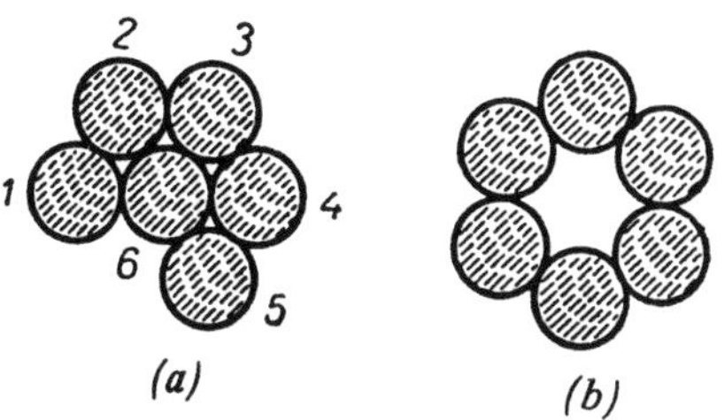

(a) *(b)*

Can you discover the 24 basic solutions of the problem? (The same moves in a different order is not a basic solution.) Here is one, showing the notation 1 to 2, 3, which means slide disk 1 until it touches 2 and 3; 2 to 6, 5, which means slide disk 2 until it touches 6 and 5; 6 to 1, 3; 1 to 6, 2.

118. FIGURE SKATERS

Pupils of a "ballet school on ice" are rehearsing at a Moscow rink. One area is decorated with a square of 64 flowers *(a)*, another with a chessboard *(b)*.

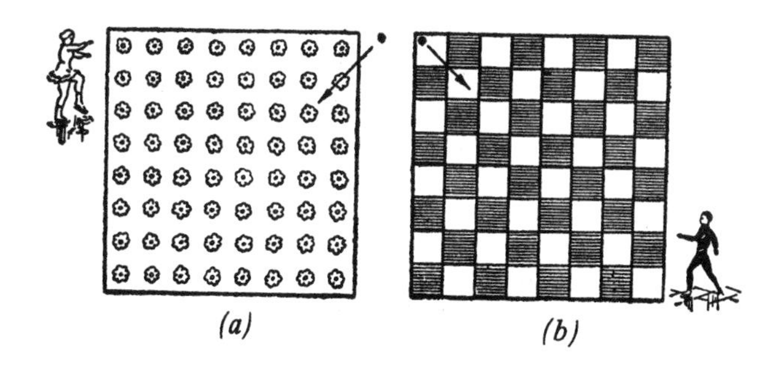
(a) *(b)*

A girl skater enters *(a)* in the direction (but not necessarily the distance) shown by the arrow, starting from the black dot outside the flower square. She skates along 14 straight lines through all the flowers (some several times), returning to the black dot.

Draw her route.

A boy, feeling his skill challenged, sets out to skate along 17 straight lines through all the white squares (some several times, but the top 4 only once, without crossing any black square. Draw his route from the black dot (upper left of *b*) to the square in the lower right corner.

Several routes are possible for each skater.

119. A KNIGHT PROBLEM

Fourth-grade pupil Kolya Sinichkin wants to move a knight from the lower left corner of the chessboard (*a* 1) to the upper right corner (*h* 8), visiting every square en route once. Can he? (See problem 110 for the knight's move if you do not know it.)

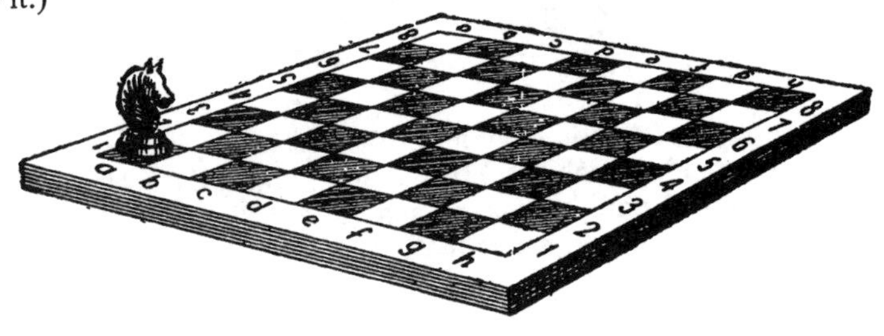

120. ONE HUNDRED AND FORTY-FIVE DOORS

A prisoner was thrown into a medieval dungeon with 145 doors. Nine, shown by black bars, are locked, but each one will open if before you reach it you pass

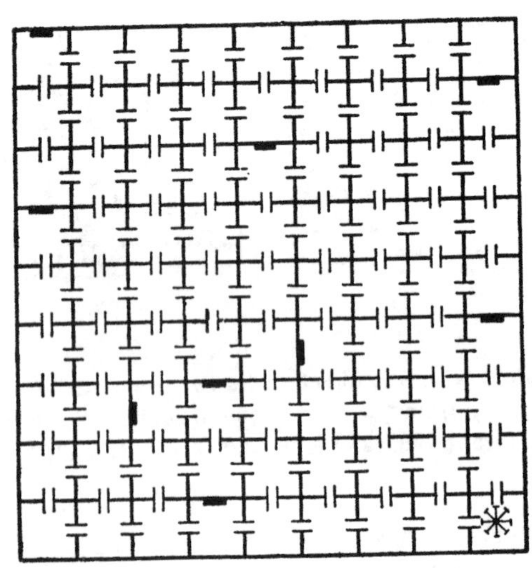

through exactly 8 open doors. You don't have to go through every open door but you do have to go through every cell and all 9 locked doors. If you enter a cell or go through a door a second time, the doors clang shut, trapping you.

The prisoner (in the lower right corner cell) had a drawing of the dungeon. He thought a long time before he set out. He went through all the locked doors and escaped through the last, upper left corner one.

What was his route?

121. HOW DOES THE PRISONER ESCAPE?

This dungeon has 49 cells. In 7 cells (*A* to *G* in the diagram) there is a locked door (black bar). The keys are in cells *a* to *g* respectively. The other doors open only from one side, as shown.

How does the prisoner in cell *O* escape? He can pass through any door any number of times and need not unlock the doors in any special order. His aim is to get the key from cell *g* and use it to escape through cell *G*.

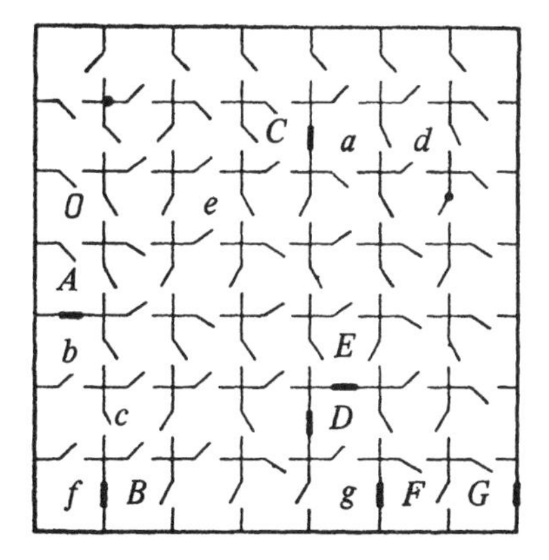

III
Geometry with Matches

Matches or toothpicks of equal length are good tools for geometrical amusements that sharpen your mind. For example: How many identical squares can you form with 24 matches (without breaking them)?

With 6 matches to a side you get 1 square. Five and 4 matches won't work.

With 3, you get 2 squares, as shown.

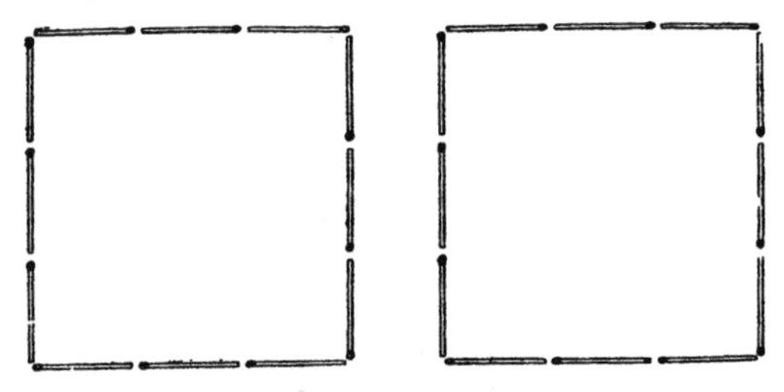

With 2, you get 3 squares, as shown at left, below.

But actually you can get not 2 but 3 squares with 3 matches on a side (see examples below).

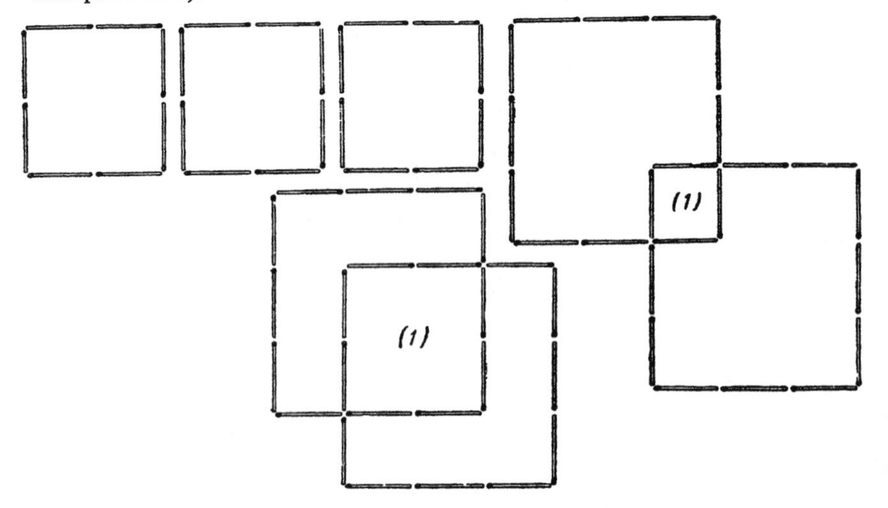

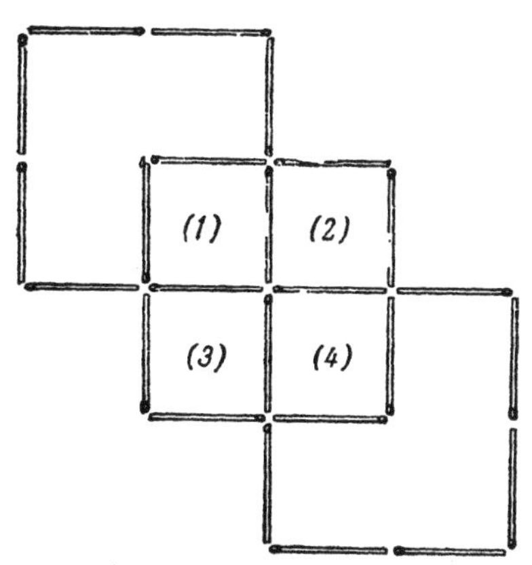

And you can get 4 extra squares (7 in all) with 2 matches on a side (above). The extra squares in both figures, however, are smaller.

With 1 match on a side you can make 6 identical squares (*a*, below), or 7 *(b)*, or 8 *(c, d)*, or 9 *(e)*. And there are large extra squares in the last three: 1 in *c*, 2 in *d*, and 5 in *e*. Do you see them?

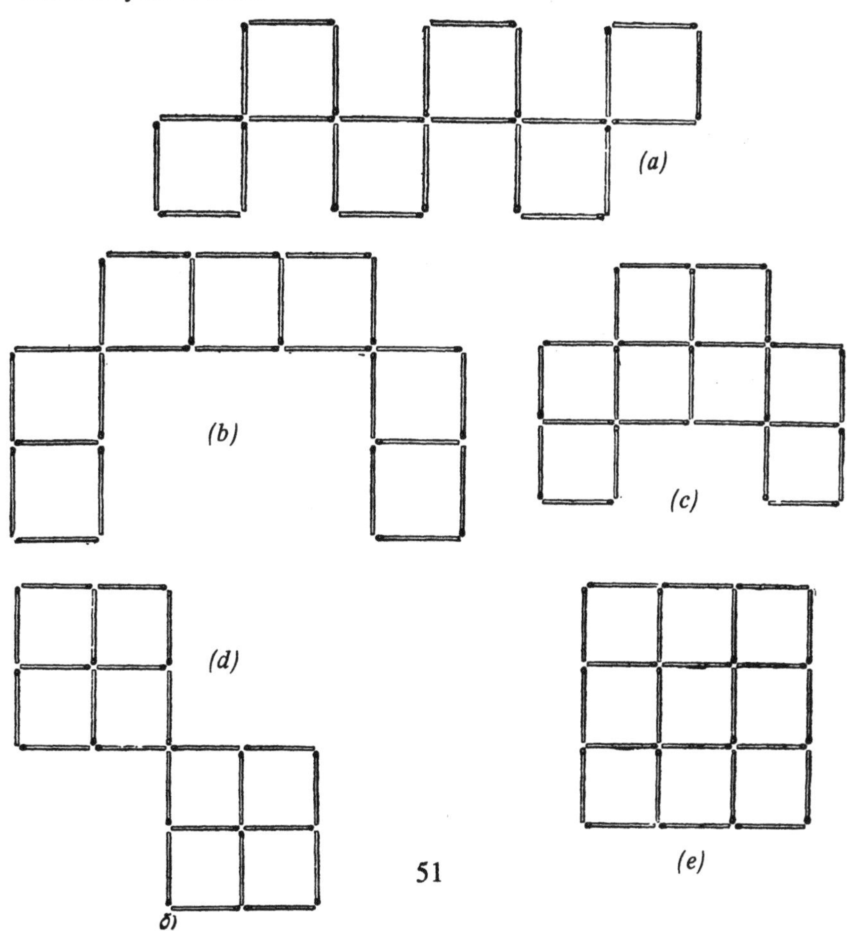

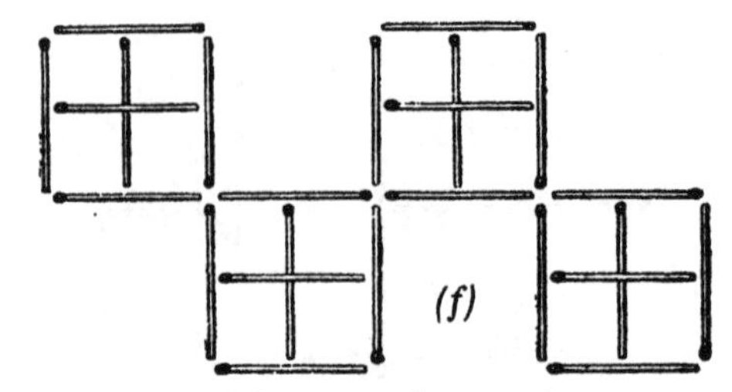

Given half a match as the side of a square (by crossing one match over another, *f*), you can get 16 small squares and 4 larger ones, as shown.

One-third of a match on a side gives 27 small squares and 15 larger ones (see *g* below); one-fifth gives 50 small squares and 60 larger ones *(h)*.

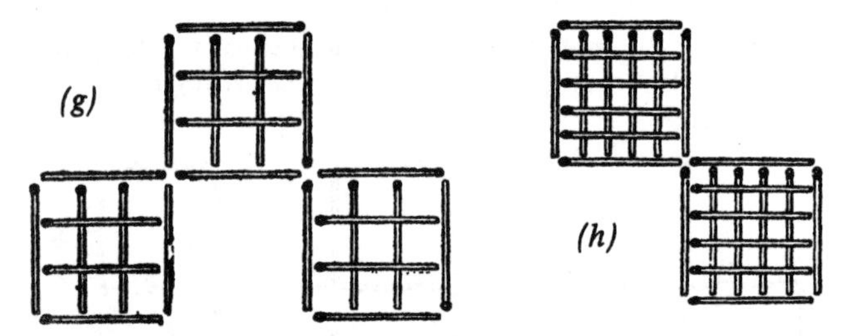

Now try to solve the following match puzzles.

122. FIVE PUZZLES

Start with 4 unit squares (and a large square) made of 12 matches, as shown.

(A) Remove 2 matches leaving 2 squares of different sizes.

(B) Move 3 matches to form 3 identical squares.

(C) Move 4 matches to form 3 identical squares.

(D) Move 2 matches to form 7 squares, not all identical. You can cross one match over another.

(E) Move 4 matches to form 10 squares, not all identical. You can cross one match over another.

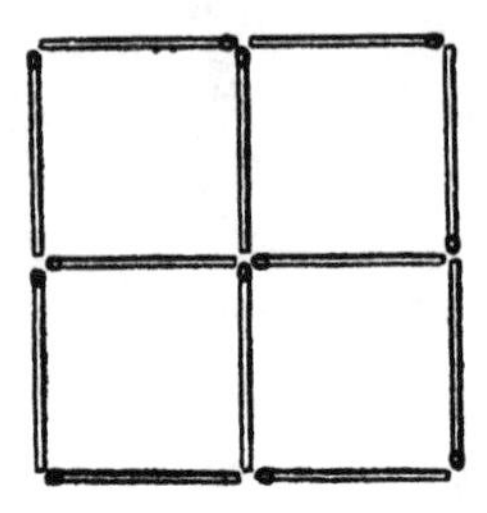

123. EIGHT MORE PUZZLES

Start with 9 unit squares (and 5 larger squares) made out of 24 matches, as shown.

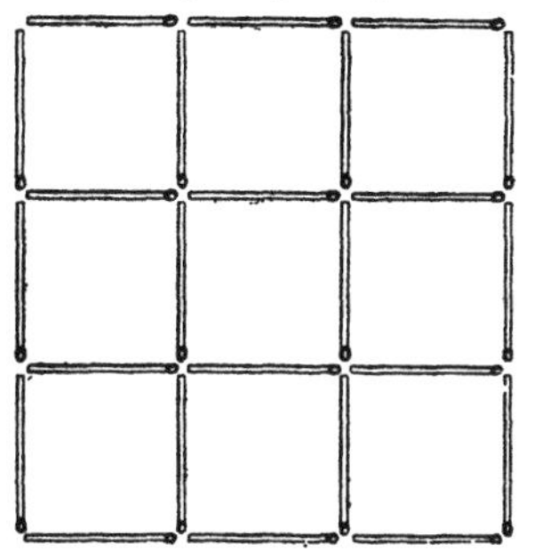

(A) Move 12 matches to make 2 identical squares.
(B) Remove 4 matches leaving 1 large and 4 small squares.
(C) Form 5 unit squares by removing 4, 6, or 8 matches.
(D) Remove 8 matches leaving 4 unit squares (two solutions).
(E) Remove 6 matches leaving 3 squares.
(F) Remove 8 matches leaving 2 squares (two solutions).
(G) Remove 8 matches leaving 3 squares.
(H) Remove 6 matches leaving 2 squares and 2 identical irregular hexagons.

124. NINE MATCHES

Form 6 squares with 9 matches (you can cross one match over another).

125. A SPIRAL

A figure resembling a spiral is shown with 35 matches. Move 4 matches to form 3 squares.

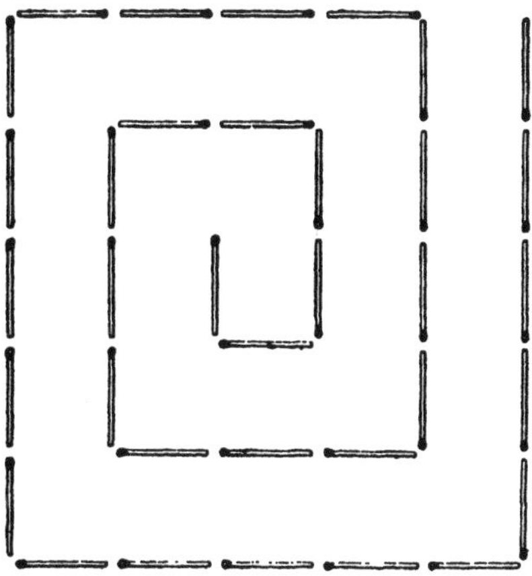

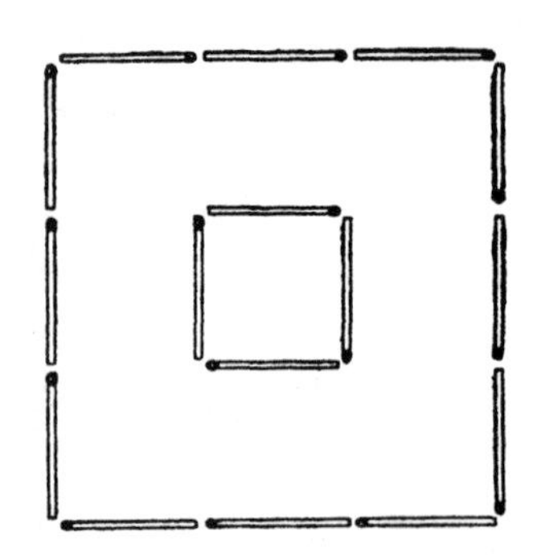

126. CROSSING THE MOAT

A fort surrounded by a deep moat is represented by 16 matches. Add 2 "planks" (matches) so as to cross the moat and enter the fort.

127. REMOVE TWO MATCHES

The figure has 14 squares made out of 8 matches. Remove 2 matches leaving 3 squares.

128. THE FAÇADE OF A HOUSE

An 11-match façade of a house is shown. Move 2 matches and get 11 squares. Move 4 matches and get 15 squares.

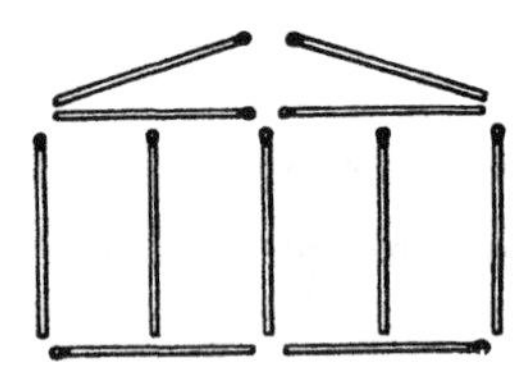

129. A JOKE

Form 1 square from 6 matches, by breaking the rules.

130. TRIANGLES

It takes 3 matches to make an equilateral triangle. Make 6 unit equilateral triangles with 12 matches. Move 4 matches in the resulting figure to make 3 equilateral triangles, not all of the same size.

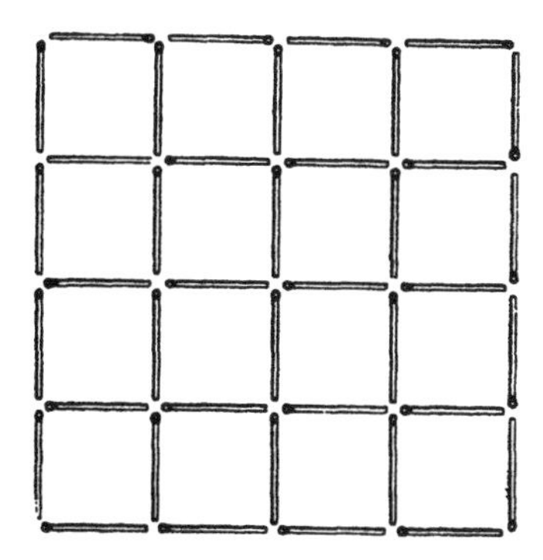

131. HOW MANY MATCHES MUST WE REMOVE?

Including the 16 unit squares shown, how many squares are there? How many matches do you have to remove to leave not one square of any size?

132. ANOTHER JOKE

Thirteen matches are each 2 inches long. Put them together and you get 1 yard.

133. A FENCE

In the fence shown, move 14 matches to make 3 squares.

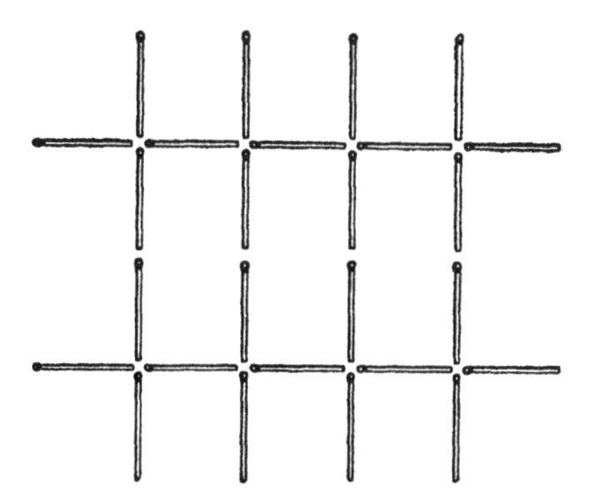

134. A THIRD JOKE

Form a square with 2 matches, without breaking or cutting them.

135. AN ARROW

The figure shows an arrow formed of 16 matches.

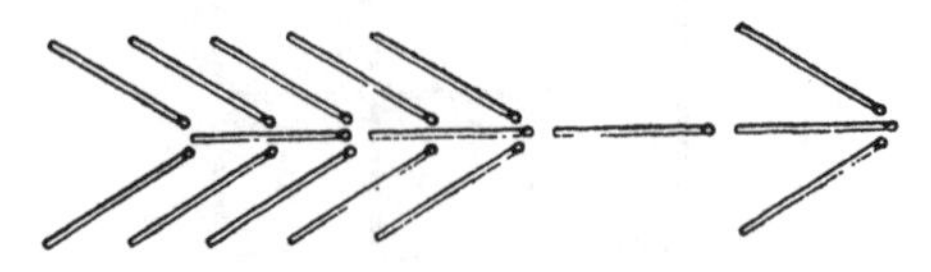

(A) Move 8 matches to make 8 equal triangles.
(B) Move 7 matches to make 5 equal quadrilaterals.

136. SQUARES AND RHOMBUSES

Form 3 squares with 10 matches. Remove 1 match, and use the remaining matches to form 1 square and 2 rhombuses.

137. ASSORTED POLYGONS

Lay out 8 matches to form 2 squares, 8 triangles, and an 8-pointed star. The matches may overlap.

138. PLANNING A GARDEN

Sixteen matches arranged in the form of a square represent the fence around a house (4-match square) and a garden. Using 10 more matches, divide the garden into 5 sectors identical in size and shape.

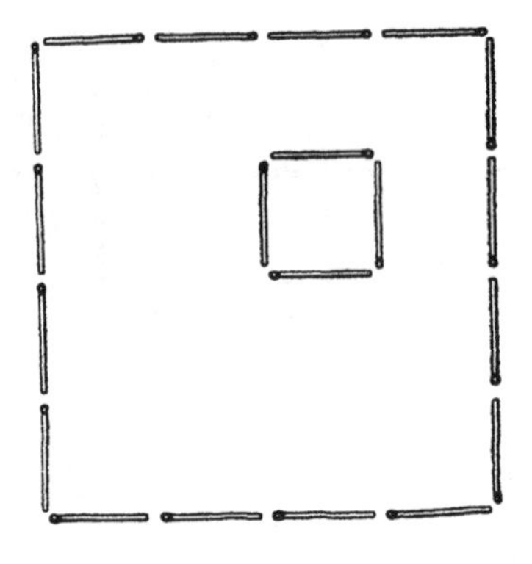

139. PARTS OF EQUAL AREAS

Add 11 matches to the square of 16 matches to form 4 parts of equal areas such that each part borders on the other 3.

140. GARDEN AND WELL

Here we have a garden, formed with 20 matches, in the center of which there is a square well.

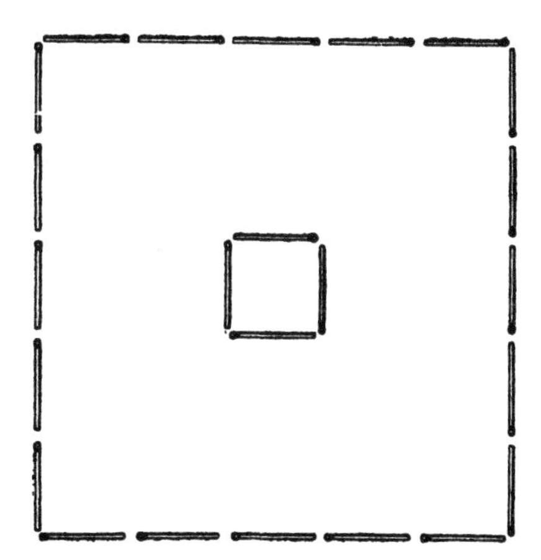

(A) Divide the garden, using 18 more matches, into 6 parts of identical shape and size.

(B) Divide the garden, using 20 more matches, into 8 parts of identical shape and size.

141. PARQUET

How many 2-inch matches do you need to pave 1 square yard with equal squares of 2-inch sides?

142. RATIO OF AREAS

With 20 matches 2 rectangles are formed, one with 6 matches, the other with 14.

Dotted lines divide the first rectangle into 2 squares, the second into 6. Thus the area of the second rectangle is three times the area of the first.

Divide 20 matches into 7 and 13 matches. Form 2 polygons (they need not have the same shape) so that the area of the second is three times the area of the first. (More than one solution is possible.)

143. OUTLINE OF A POLYGON

You have 12 matches of unit length. Make a polygon whose area is 3 square units. (Several ingenious solutions are possible.)

144. FINDING A PROOF

Place 2 matches side by side so they form a straight line. Prove they do so.

You may use extra matches for the proof.

IV
Measure Seven Times Before You Cut

145. IDENTICAL PARTS

Copy the figure on a sheet of paper, then solve the following:

(A) Cut *a* into 4 congruent quadrilaterals.

(B) The upper half of *b* shows how to cut an equilateral triangle into 4 parts. Without the upper triangle, the remaining 3 triangles form a trapezoid (lower half). Cut it into 4 congruent parts.

(C) Cut *c* into 6 congruent parts.

(D) A polygon whose interior angles are equal and whose sides are equal is a regular polygon. Cut the regular hexagon in *d* into 12 congruent quadrilaterals. (Two solutions.)

(E) Not every trapezoid can be cut into 4 congruent smaller trapezoids. But try to do it with *e*, which is made up of 3 congruent isosceles right triangles.

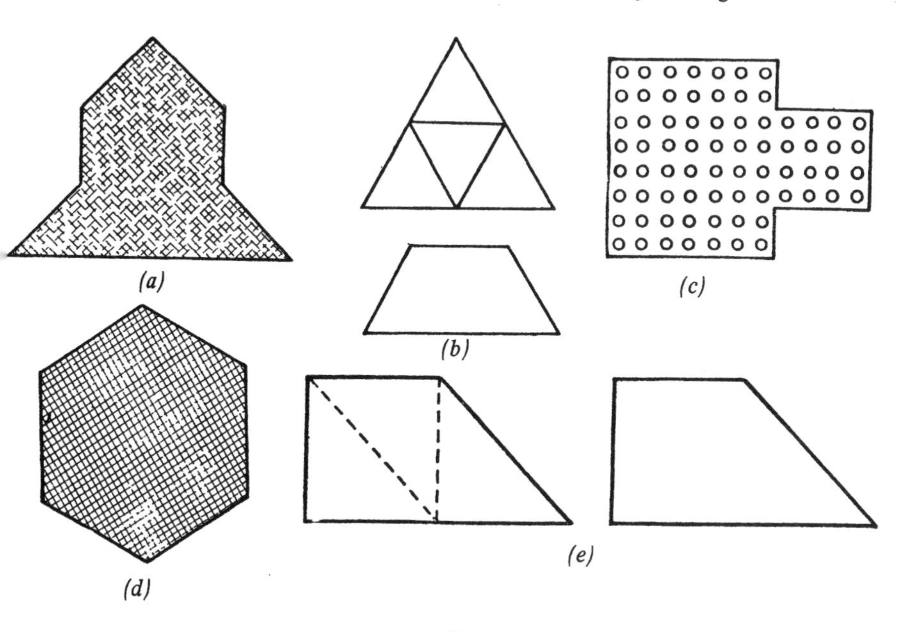

146. SEVEN ROSES ON A CAKE

Cut the cake along 3 straight lines into 7 parts, each with a rose on top.

147. LOST CUTTING LINES

The square, which contains four of each integer from 1 through 4, was marked along the sides of its cells so it could be cut into 4 congruent, symmetrical configurations each of which can be rotated 90° onto the next.

Unfortunately, someone has erased the cutting lines. You can restore them knowing that each configuration contains one 1, one 2, one 3, and one 4. Do you have a method of restoring them, other than trial and error?

			3		1	1	
			3	4			
				2			
	1		4	2			
	1						
		3	3				
					4	2	2
					4		

148. GIVE US ADVICE

The figure is the plan of part of a device. Cut it into 4 congruent chambers with 2 brads (dots) and 1 aperture (small square) in each chamber.

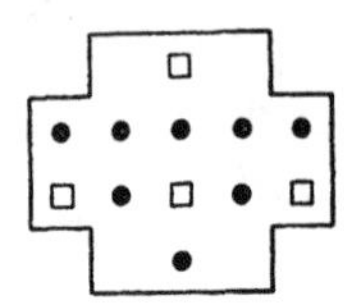

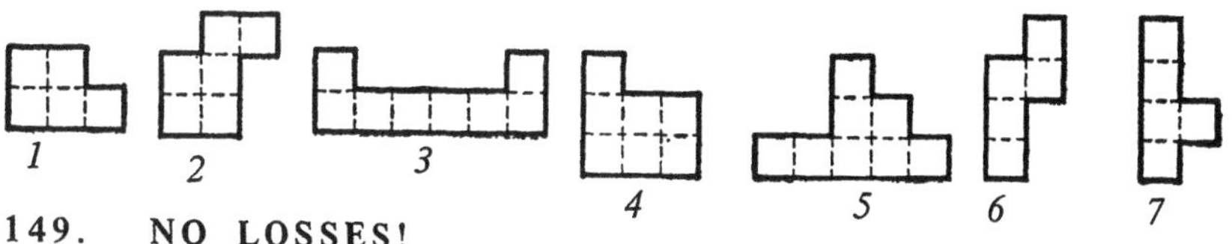

149. NO LOSSES!

In factories, blanks are not sent directly to the lathes for processing, but go first to a marker who puts on each blank the lines and dots it needs.

A factory needed a large quantity of polygonal brass plates in seven shapes, as shown at the top of the page. The marker noticed that one plate fitted exactly six times in a small rectangular sheet of brass. (Which one?)

The marker found that the six patterns shown in the figures below could be cut into plates without wasting a single square of brass.

Three no. 4 plates came from pattern I, 5 no. 7 plates from pattern II, and so on. Show the cutting lines on the patterns. Pattern I is cut into 3 congruent plates; pattern IV into 4; pattern V into 6; pattern VI into 4.

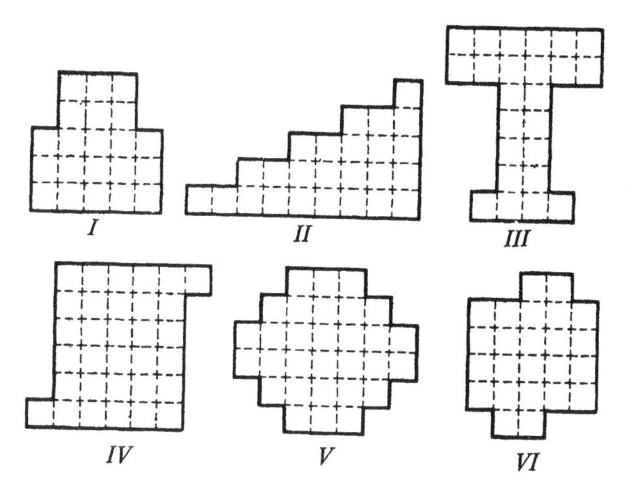

150. WHEN THE FASCISTS ATTACKED OUR COUNTRY

During World War II, Russian cities near the front were blacked out. Once when it was time to darken the windows, schoolboy Vasya's parents could not find a shade for a window 120 by 120 units. All that was available was a rectangular sheet of plywood. Its area was correct, but it was 90 by 160.

Vasya picked up a ruler and drew quick lines on the plywood. He cut it into two parts along the lines he had drawn. With these parts he made a square covering for the window. How?

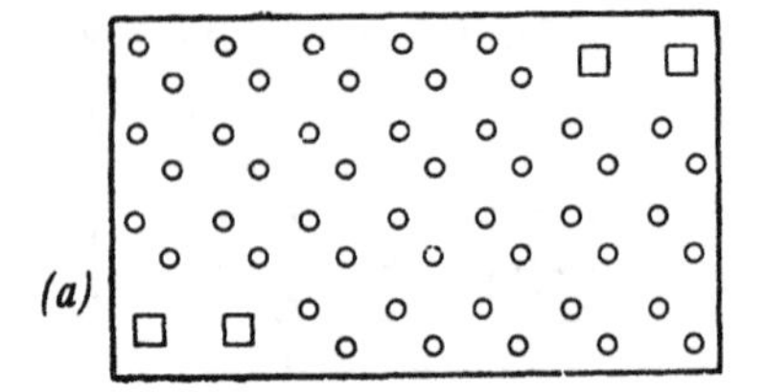

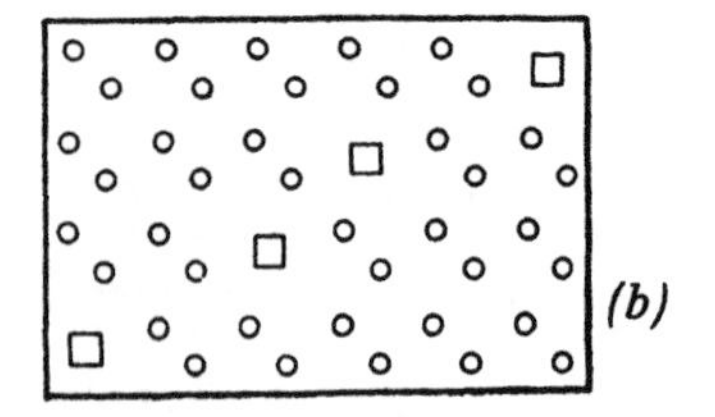

151. AN ELECTRICIAN'S REMINISCENCES

Every apartment house has a fuseboard and every factory has a large number of them. They are usually rectangular or square, but during World War II, to economize, electricians sometimes used unorthodox shapes.

Once we had two large boards in which round and square apertures had been drilled. From these boards we had to cut 8 small panels. Our brigade leader cut the first board *(a)* into 4 congruent panels, each with 1 square and 12 round apertures, and the second board *(b)* into 4 congruent panels, each with 1 square and 10 round apertures.

How did he cut the boards?

152. NOTHING WASTED

"Can I make a chessboard out of this wooden board without any waste?" I asked myself.

When I drew lines on it, making 64 equal squares, each protrusion made 2 squares. Find the lines I sawed along to get 2 congruent parts that would go together to make a chessboard.

153. A DISSECTION PUZZLE

Cut *ABCDEF* into 2 parts that can be fitted together as a square frame. The opening in the frame should be a square congruent to each of the 3 squares composing the original figure.

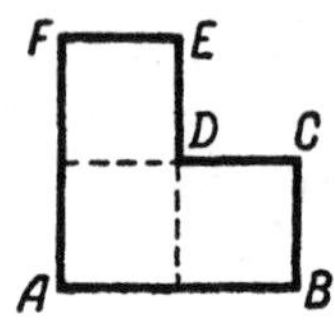

154. HOW TO CUT A HORSESHOE

How do you cut a horseshoe into 6 parts with 2 straight lines? No rearranging parts after the first cut, in this problem.

155. A HOLE IN EACH PART

The horseshoe shown has 6 holes for nails. Using 2 straight-line blows, can you chop it into 6 parts with a hole in each?

156. MAKE A SQUARE OUT OF A JUG

Copy the jug, cut it into 3 parts with 2 straight cuts, and form a square out of the parts.

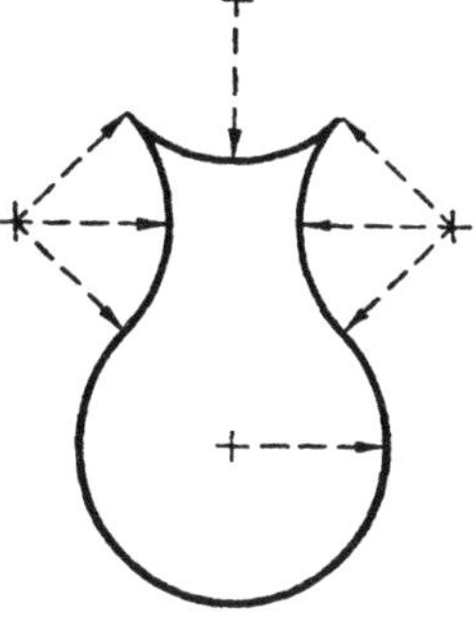

157. SQUARE THE LETTER E

Copy the figure whose outline resembles the letter E, cut it into 7 parts with 4 straight cuts, and form a square out of the parts.

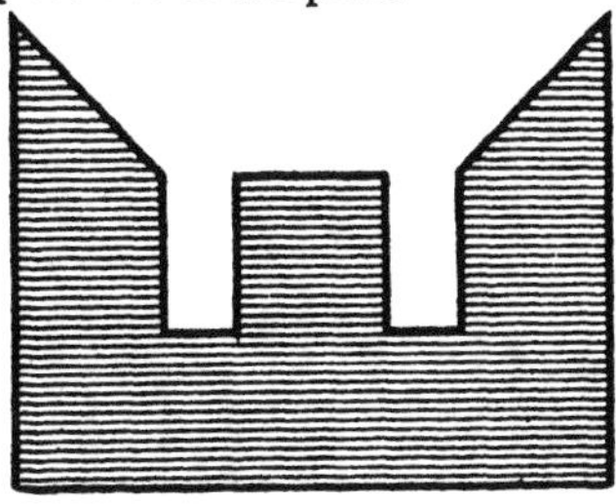

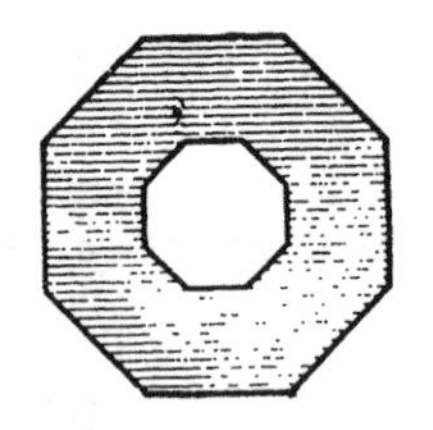

158. AN OCTAGON TRANSFORMATION

Cut a copy of the octagonal frame into 8 congruent parts and with them form an 8-pointed star which also has an octagonal hole at its center.

159. RESTORING A RUG

Two small triangular pieces (shaded in the picture) had been removed from a valuable old rug.

Students at an arts and crafts school decided to restore the rectangular shape of the rug without any waste. They cut the mosaic into 2 parts along straight lines and from these they composed a new rectangle (which turned out to be a square). The rug's pattern was preserved.

How did they do it?

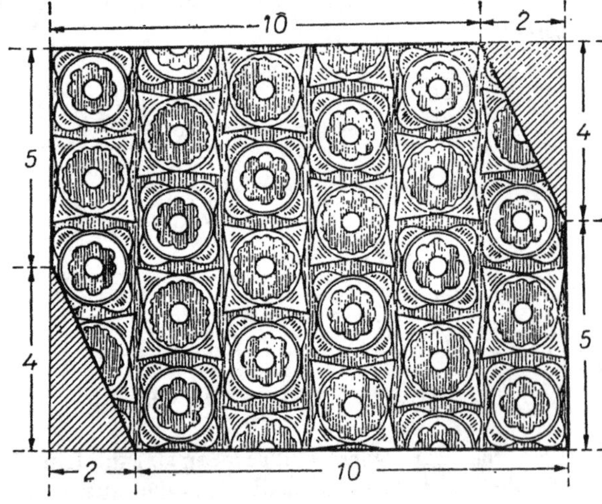

160. A CHERISHED REWARD

When Nuriya Saradzheva was an adolescent, she was awarded a beautiful Turkmenian rug for being first on her collective farm to use an improved method of picking cotton.

Now Nuriya works as an agronomist. Once, while doing some research, she spilled acid on the rug. After the damaged part was cut out, there was a large rectangular hole, 1 by 8 feet.

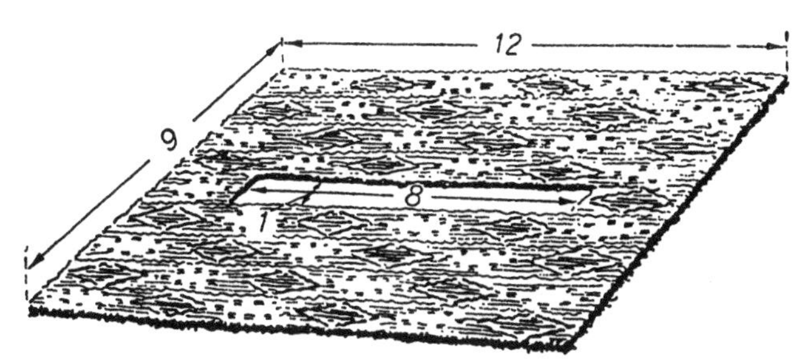

Nuriya decided to repair the rug. Using straight-line cuts, she cut the undamaged part of the rug into 2 parts that, when sewn together, formed a square. How?

161. RESCUE THE PLAYER

Remember Problem 54 about a chessboard that had to be put together from 14 parts? The chess player tried to make another problem. He wanted to cut the board into 15 parts like *a* in the diagram and 1 square like *b*. That adds up to 64 squares, but he hasn't been able to do it.

First prove it's impossible. Then show how to cut the chessboard into 10 parts like *c* and 1 like *a*.

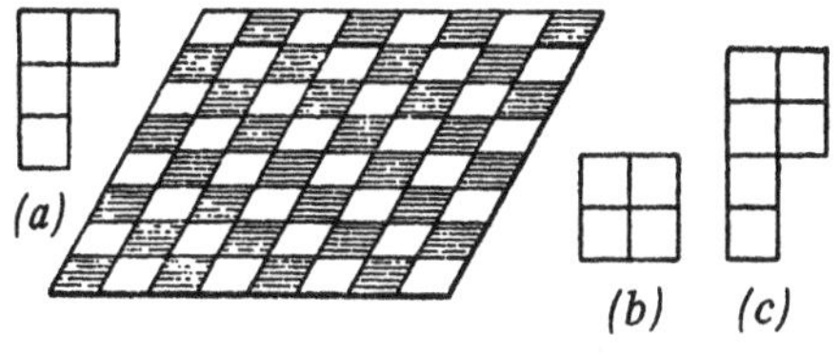

162. A GIFT FOR GRANDMOTHER

A girl has 2 square pieces of checkered cloth: one of 64 squares, the other of 36. She has decided to combine them into a 10-by-10 square babushka for her grandmother.

It is necessary to keep the white and black squares alternate. Also, the edges of the large piece of cloth have fringed borders along two sides and half of another (as shown, left).

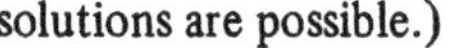

Using straight-line cuts, she cuts each square into 2 parts and makes the babushka out of the 4 parts so that the fringes are on its outside borders. How? (Several solutions are possible.)

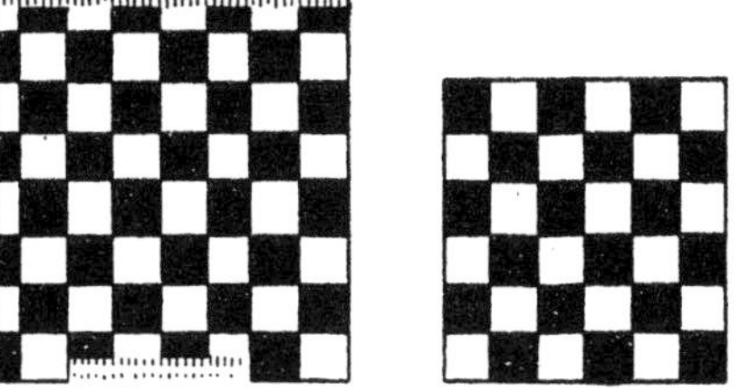

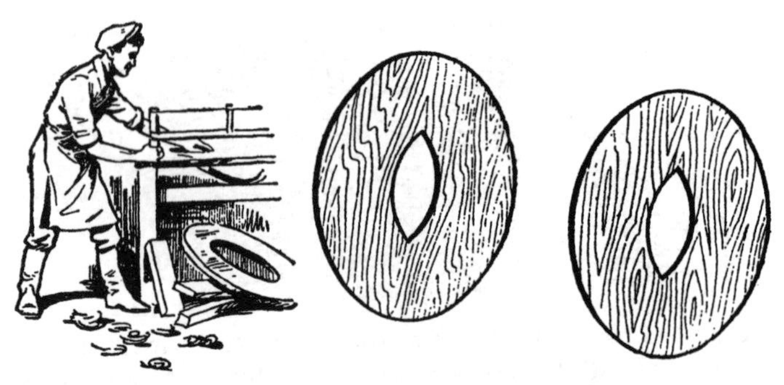

163. THE CABINETMAKER'S PROBLEM

How can a cabinetmaker, using straight-line cuts, saw the two oval frames into parts that will form a circular tabletop from the parts with no waste?

164. A FUR DRESSER NEEDS GEOMETRY!

A fur dresser had to put a patch shaped like a scalene triangle on a piece of fur. Suddenly he realized he had made a terrible mistake. The patch fitted the hole but the fur side faced the wrong way.

The fur dresser, after some thought, cut the triangular patch into 3 parts, each of which would be unchanged when turned over. How?

165. THE FOUR KNIGHTS

Cut the chessboard into 4 congruent parts, each with a knight on it.

166. CUTTING A CIRCLE

Problem: To cut a circle with 6 straight lines into the greatest possible number of parts.

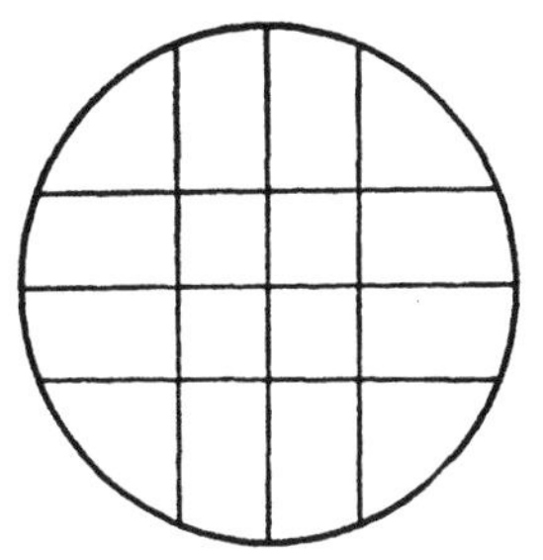

The diagram shows a circle cut into 16 parts, but this is not the maximum. The largest number of parts is $\frac{1}{2}(n^2 + n + 2)$, where n is the number of straight lines.

Cut a circle into 22 parts with 6 straight lines. Try to achieve some symmetry.

167. TRANSFORMING A POLYGON TO A SQUARE

Any 2 squares can be cut into parts that will form a larger square. The well-known Pythagorean theorem will tell us the size of the larger square (see first diagram). But how are the cuts made? In the 2,500 years since Pythagoras, many solutions to this problem have been found. Here is one:

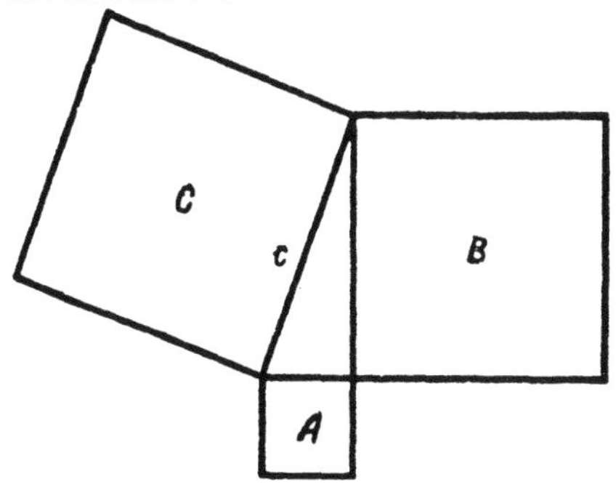

Form *ABCDEF* from the 2 squares (diagram *a* below). Let $FQ = AB$ and cut along *EQ* and *BQ*. Move triangle *BAQ* to *BCP*, and triangle *EFQ* to *EDP*. Square *EQBP* contains all the parts of the 2 given squares.

Can you cut *b* into 3 parts that will form a square?

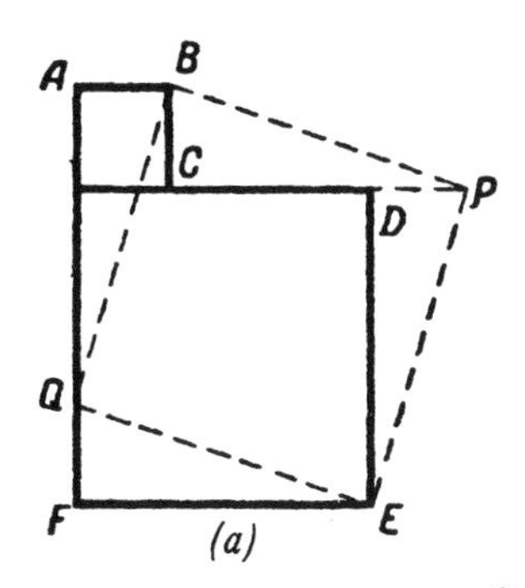

(a)

(b)

168. DISSECTION OF A REGULAR HEXAGON INTO AN EQUILATERAL TRIANGLE

The general methods of cutting a polygon into parts that can be used to form a second polygon are often clumsy and inconvenient. It is more interesting to find how to cut, say, a regular hexagon into the least number of parts that will form an equilateral triangle.

To this day it is not known whether it can be done with 5 parts. Can you find one of the solutions with 6 parts?

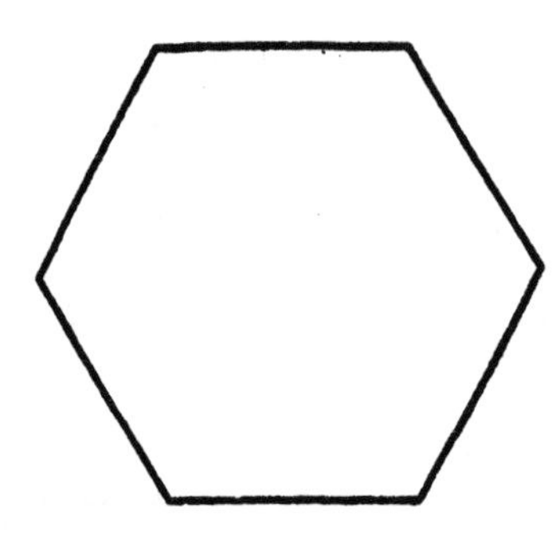

V

Skill Will Find Its Application Everywhere

169. WHERE IS THE TARGET?

The circles show radar screens. A radio wave is sent from the radar station (0 on the screen) and reflected by the target (a ship, for example) to the station, causing a spike to appear on the wavy line. The indicator is marked so that the number under the spike shows the distance in miles from the station to the target. The left screen supplies data from the shore radar station at *A*, the right screen from *B*.

How are the 75 and 90 on the indicators used in locating the target?

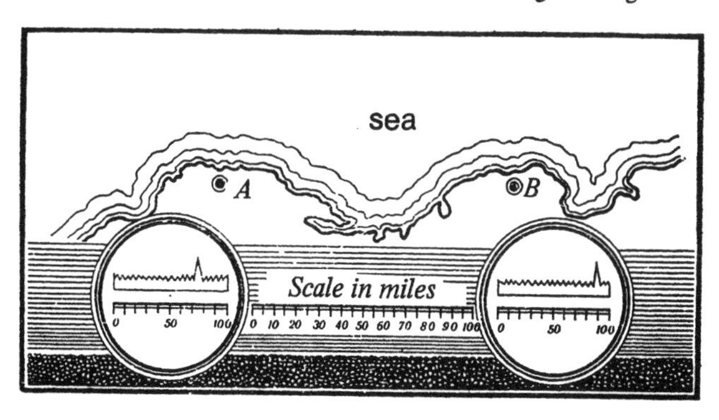

170. THE SLICED CUBE

Imagine a wooden cube 3 inches on a side. Its surface is black but it is not black inside.

How many cuts does it take to divide the cube into cubes with 1-inch sides? How many little cubes are there? How many have 4, 3, 2, 1, and 0 black faces?

171. A TRAIN ENCOUNTER

Two trains, each of 80 cars, must pass on a single track which has a dead-end siding. How can they pass if the siding can only hold a locomotive and 40 cars?

172. A RAILROAD TRIANGLE

(A) Main track *AB* and two small branches *AD* and *BD* form a railroad triangle. When a locomotive goes from *A* to *B*, backs into *BD*, and moves forward out of *AD*, it has reversed its direction on *AB*.

But how does the engineer move the black car to *BD* and the white car to *AD*, and return the locomotive to face right on *AB*, all in 10 moves? The dead end beyond the switch at *C* can only hold the locomotive or 1 car. Each coupling or uncoupling is 1 move.

(B) If the engineer is willing to reverse the final position of his locomotive on *AB* he can solve the problem in 6 moves. How?

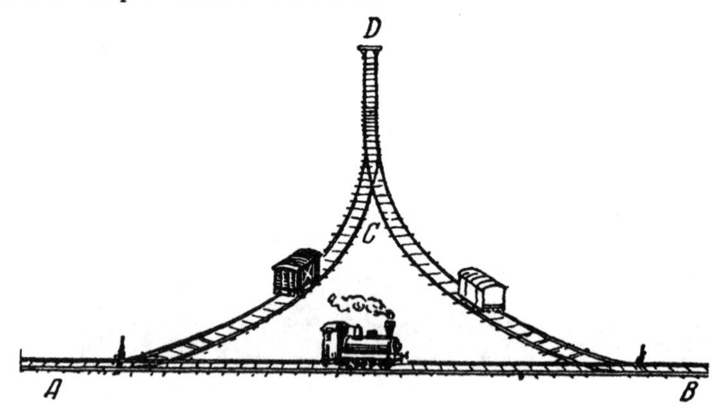

173. A WEIGHING PUZZLE

A balance has only two weights, 1 ounce and 4 ounces. In three weighings, split 180 ounces of grits into two packages of 40 and 140 ounces.

174. BELTS

Wheels *A, B, C,* and *D* are connected with belts as shown. If wheel *A* starts to rotate clockwise as the arrow indicates, can all 4 wheels rotate? If so, which way does each wheel rotate?

Can the wheels turn if all 4 belts are crossed? If 1 or 3 belts are crossed?

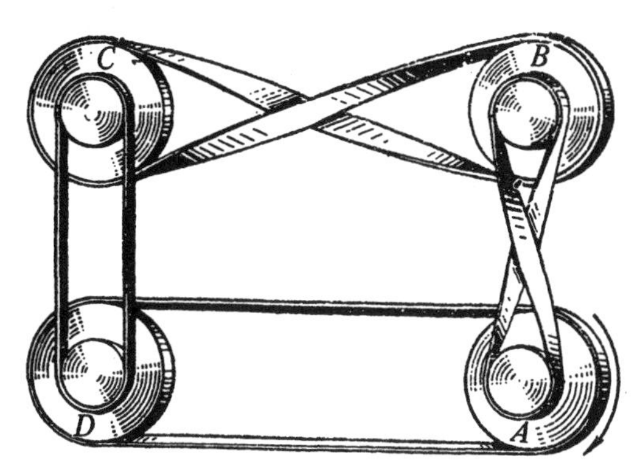

175. SEVEN TRIANGLES

In the diagram, 3 matches are connected with balls of plastic to make an equilateral triangle. Can you form 7 such triangles with 9 matches?

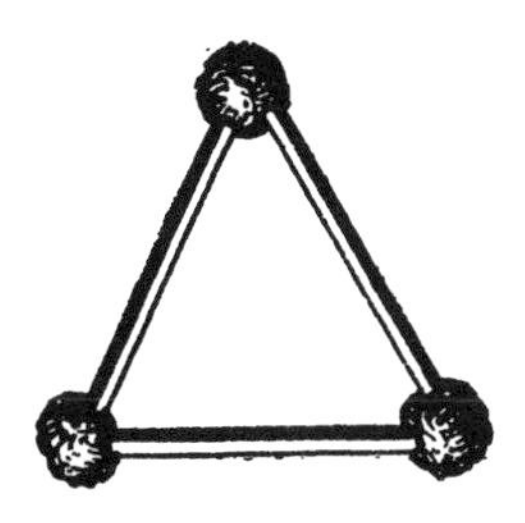

176. THE ARTIST'S CANVASES

An eccentric artist says that the best canvases have the same area as their perimeter. Let us not argue whether such sizes increase the onlooker's appreciation, but only try to find what sides (in integers only) a rectangle must have if its area and perimeter are to be equal.

A school girl from Ordzhonikidze offers an elegant proof that there are only two such rectangles. Can you find the rectangles—and the proof?

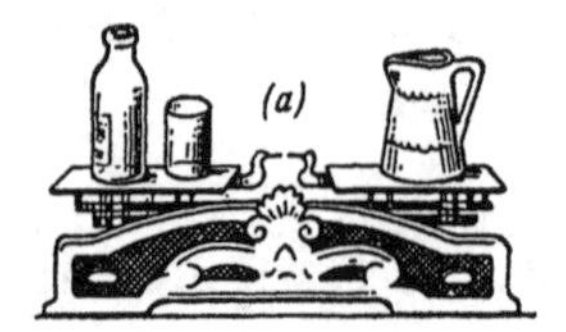

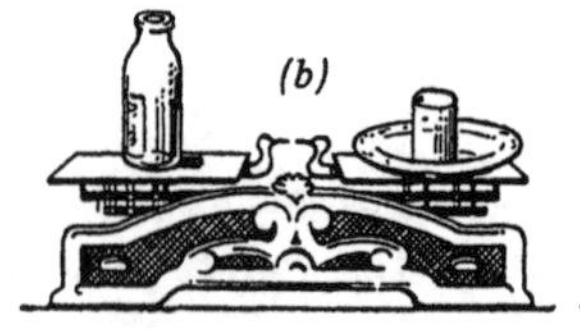

177. HOW MUCH DOES THE BOTTLE WEIGH?

The bottle and glass on the left scale balance the jug on the right scale *(a)*.

The bottle alone balances a glass and a plate *(b)*, and 3 of these plates balance 2 jugs *(c)*.

How many glasses will balance a bottle?

178. SMALL CUBES

A craftsman was making a children's game in which letters and figures are pasted on wooden cubes. But he needed twice the surface area he had available.

How did he get it without adding any cubes?

179. A JAR WITH LEAD SHOT

The builders of an irrigation canal needed a lead plate of a certain size, but had no lead in stock. They decided to melt some lead shot. But how could they find its volume beforehand?

One suggestion was to measure a ball, apply thc formula for the volume of a sphere, and multiply by the number of balls. But this would take too long, and anyway the shot wasn't all the same size.

Another was to weigh all the shot and divide by the specific gravity of lead. Unfortunately, no one could remember this ratio, and there was no manual in the field shop.

Another was to pour the shot into a gallon jug. But the volume of the jug is greater than the volume of the shot by an undetermined amount, since the shot cannot be packed solid and part of the jug contains air.

Do you have a suggestion?

180. WHITHER THE SERGEANT?

A sergeant left point *M* along azimuth 330°. On reaching a small hill, he walked along azimuth 30° until he came to a tree. Here he made a 60° turn to the right. He reached a bridge, then walked beside a river along azimuth 150°. Half an hour later he was at a mill. He changed his direction again, walking along azimuth 210°, his goal being the miller's house. At the house he again turned right and, walking along azimuth 270°, finished his tour.

Using a protractor, draft the sergeant's route neatly and where he got to. He walked 2½ miles along each azimuth.

181. A LOG'S DIAMETER

A plywood sheet is 45 by 45 inches. What is the approximate diameter of the log the sheet was made from?

The diameter d of a circle equals c/π, where c is the circumference, but please do not make a mistake. The diameter of the log is not $45/\pi$.

182. A CALIPER DIFFICULTY

Once Vasili Chapayev, a Red commander in the civil war of 1918, was asked whether his military successes had been lucky. Chapayev replied: "Well, no, one must use one's head . . . have ingenuity . . ."

Indeed, luck is an undependable basis for success. Whether we work or, say, play chess, we may encounter situations that seem hopeless, but determination and ingenuity can save us.

A student had to draw a cylindrical machine part with indentations at its bases. He had no depth gauge, only calipers and a ruler. The problem was, he could find the distance between the indentations with the calipers, but he would have to remove the calipers to measure their spread on the ruler. But to remove the calipers he would have to open the legs, and then there would be nothing to measure.

What did he do?

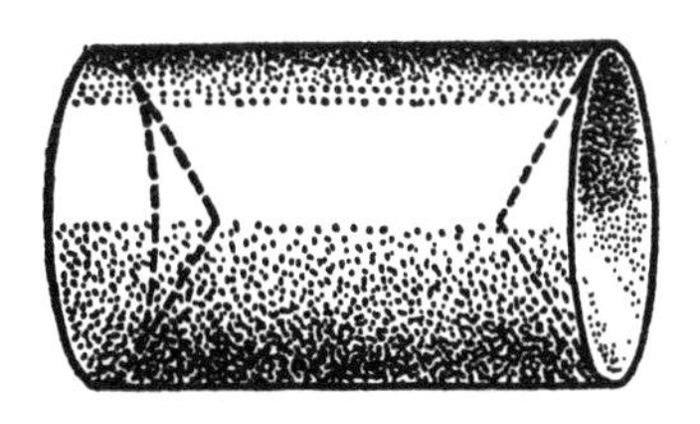

183. WITHOUT A GAUGE

(A) At technical school we study the construction of lathes and machines. We learn how to use instruments and how not to be stumped by difficult situations. Of course the knowledge we brought from high school helps.

My foreman-teacher handed me some wire and asked: "How do you measure the wire's diameter?"

"With a micrometer gauge."

"And if you don't have one?"

On thinking it over, I had an answer. What was it?

(B) Another time I was assigned to make a round hole in a thin sheet of roofing tin.

"I'll go get a drill and chisel," I said to the foreman-teacher.

"I see you have a hammer and a flat file. You can do the job with those."

How?

184. CAN YOU GET 100 PERCENT SAVINGS?

One invention saves 30% on fuel; a second, 45%; and a third, 25%. If you use all three inventions at once can you save 100%? If not, how much?

185. SPRING SCALES

A bar weighs more than 15 pounds and less than 20. Can you find its exact weight with small spring scales, each of which has a maximum load of 5 pounds?

186. INGENUITY IN DESIGN

(A) Link 3 ribbons so that if you cut any link the whole chain comes apart. In an ordinary linkage as shown, the chain only falls apart if the middle link is cut.

(B) Link 5 ribbons so that there is only 1 link whose cutting will make the whole chain come apart.

(C) Link 5 ribbons so that if you cut any link the whole chain comes apart.

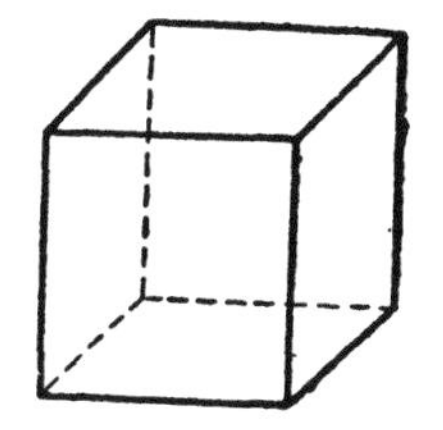

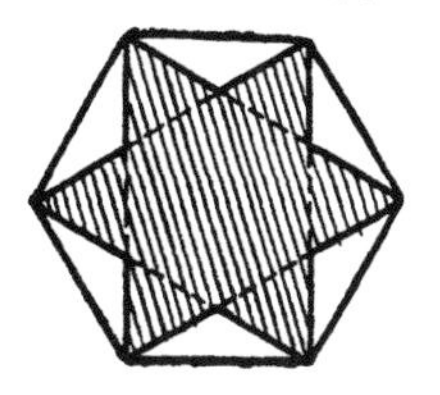

187. CUTTING A CUBE

(A) Can a cube be cut with a plane to produce a regular pentagon?
(B) How about an equilateral triangle? A regular hexagon?
(C) How about a regular polygon with more than 6 sides?

188. TO FIND THE CENTER OF A CIRCLE

Find the center of the circle using only the drafting triangle and pencil as shown.

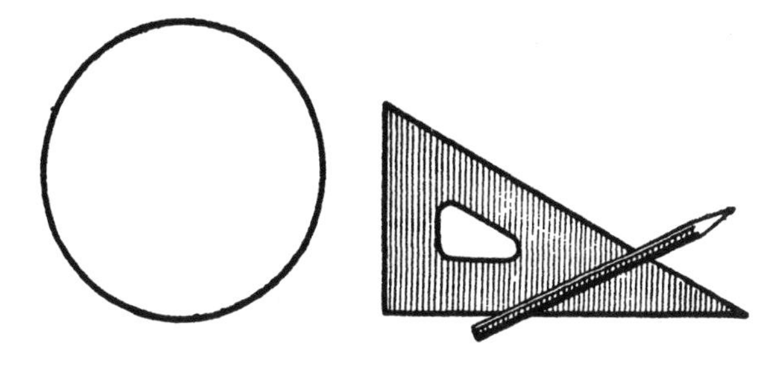

189. WHICH BOX IS HEAVIER?

A cubical box contains 27 congruent large balls; its twin contains 64 congruent smaller balls. All the balls are made of the same material.

Both boxes are filled to the top. In each box, each layer has the same number of balls, and the outside balls of each layer touch the sides.

Which box is heavier?

Try with other numbers that are cubes. Draw a general conclusion.

190. A CABINETMAKER'S ART

At an exhibit of the work of young cabinetmakers, apprentices at a factory school, we were shown a remarkable wooden cube consisting of 2 parts closely fitted

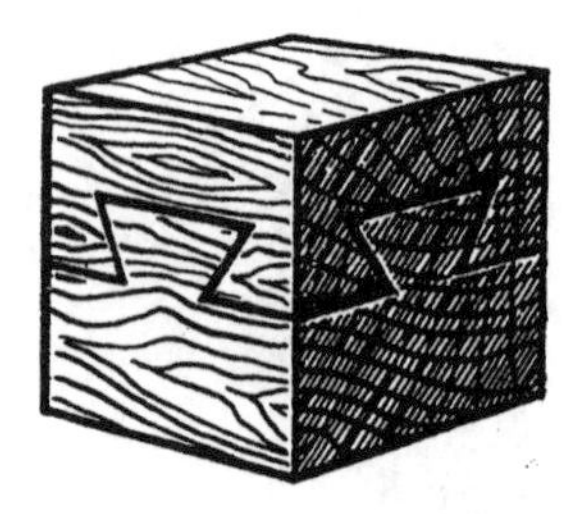

together with dowel pins. The parts are not glued together and apparently can be separated easily. We tried to pull them up and down, to the left and right, and forward and backward—to no avail.

Can you guess how the cube's parts come apart and what shape each has?

191. GEOMETRY ON A BALL

Using a ball—say a croquet ball—a sheet of paper, a pair of compasses, a straightedge without markings, and a pencil, see if you can draw a line segment on the paper equal to the ball's diameter.

192. THE WOODEN BEAM

A wooden beam (which is a rectangular parallelepiped) with edges of 8, 8, and 27 inches is to be sawed into 4 parts out of which a cube can be made.

Naturally, one should draw first and saw later.

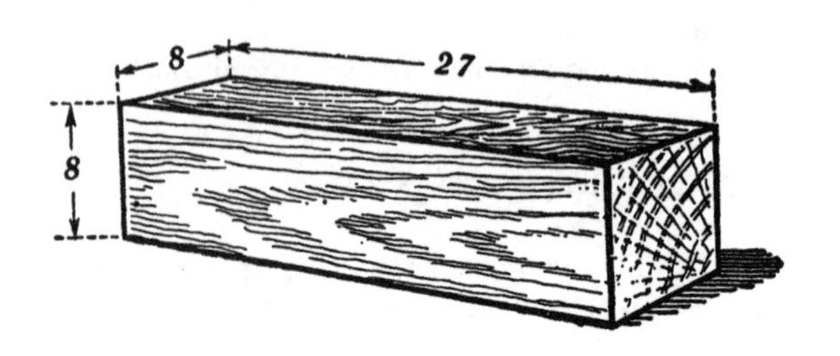

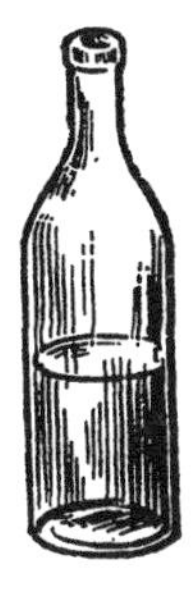

193. A BOTTLE'S VOLUME

If a bottle, partly filled with liquid, has a round, square, or rectangular bottom which is flat, can you find its volume using only a ruler? You may not add or pour out liquid.

194. LARGER POLYGONS

As construction workers can assemble a house from prefabricated parts, we can assemble larger polygons from small ones.

In this problem, we will make larger polygons out of small polygons of the same shape.

It is easy with simple regular polygons, as the squares and triangles in the first diagram show. (Note the turning of the triangles; of course, bending and tearing are not allowed.)

Irregular polygons like *a*, *b*, and *c* in the diagram below can also be used as material for larger similar polygons.

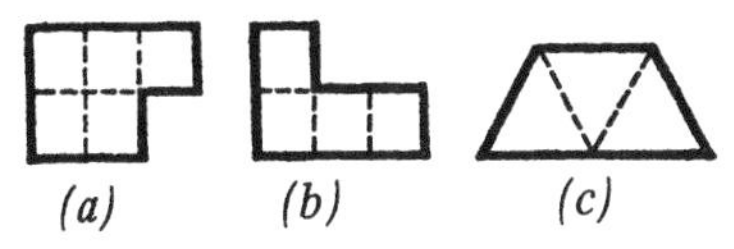

The upper half of the lefthand diagram on page 78 shows how a large *a* and a large *b* can be made out of 4 small *a*'s and 4 small *b*'s, respectively. In the lower half, a large *c* is made out of 16 small *c*'s.

In general, a larger polygon will be 2, 3, 4, 5, . . . times the length of the unit polygon, and so its area will be 4, 9, 16, 25, . . . times that of the unit polygon.

The least number of unit polygons needed to make a similar larger one, thus, is always a square number. The size of this number is unpredictable: sometimes no larger polygon can be formed.

Form larger polygons similar to *a, b,* or *c* in the righthand diagram:

1. from 9 polygons *a;*
2. from 9 polygons *b;*
3. from 4 polygons *c;*
4. from 16 polygons *b;*
5. from 9 polygons *c.*

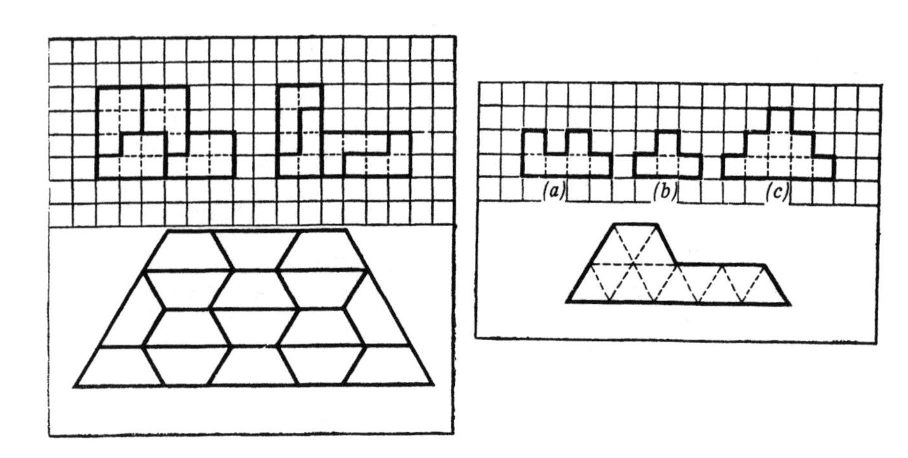

195. LARGER POLYGONS IN TWO STEPS

In this problem, we show a powerful method of making larger polygons from unit polygons—although not necessarily out of the smallest possible number of unit polygons.

We want to make larger polygons from each of the two unit polygons labeled *P* in the upper half of the first diagram. The first step is to make a square. (It takes 4 of each, as shown.) But the unit polygons *P* are themselves made up of squares. The second step, then, would be to assemble the large squares we have just made into larger *P*'s. (It takes 4 of each square, thus 16 of each *P*.) Similarly (lower half of first diagram), larger *P*'s can be made from 36 unit *P*'s in two steps (4 unit *P*'s to a square, 9 squares to a larger *P*).

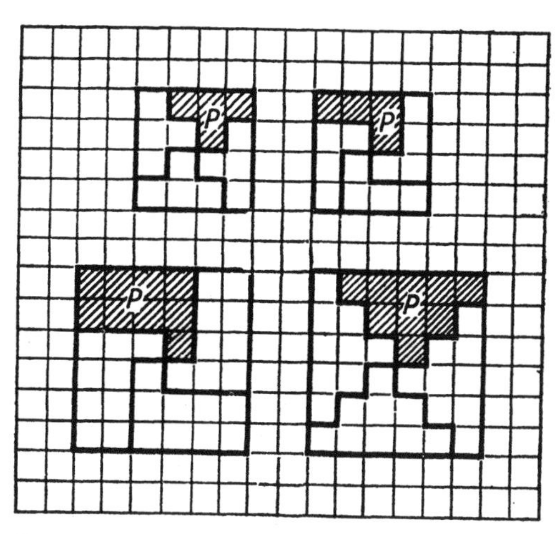

In the second diagram, a larger *P* can be made from 36 unit *P*'s (3 unit *P*'s to an equilateral triangle, 12 equilateral triangles to a larger *P*).

Find other unit polygons that can be formed into larger polygons by this two-step method.

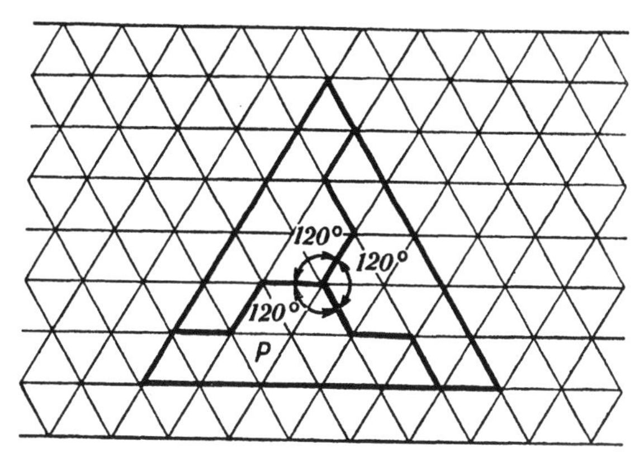

196. A HINGED MECHANISM FOR CONSTRUCTING REGULAR POLYGONS

You can build a simple mechanism with which you can construct any regular n-sided polygon for $n = 5$ through 10.

The mechanism consists of movable rods forming two congruent parallelograms *ABFG* and *BCHK* (first diagram). Rod *DE* is fastened to sliders *D* and *E* which move freely along *AG* and *BK* respectively. $AB = BC = CD = DE$. When *DE* moves the parallelograms are unaffected, and trapezoids *ABCD* and *BCDE* remain congruent. This assures the equality of the 3 angles of an *n*-gon whose four consecutive sides are *AB*, *BC*, *CD*, and *DE*.

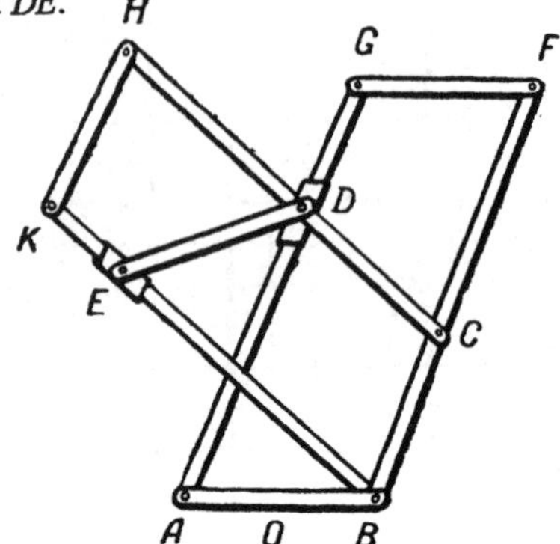

Methods of constructing regular *n*-gons for $n = 5$ through 10 are based on these characteristics (diagrams *a* to *f* below):

a: $\angle DOB = 90°$ in a pentagon.
b: $\angle EAB = 90°$ in a hexagon.
c: $\angle EOB = 90°$ in a heptagon.
d: $\angle EBA = 90°$ in an octagon.
e: $\angle EAB = 60°$ in a nonagon.
f: $\angle DAB = 36°$ in a decagon.

To construct the first four, form right angles Y_1OX, Y_2AX, Y_3OX, and Y_4BX, then place the mechanism with rod *AB* on the straight line *AB*, superposing (in *a* –

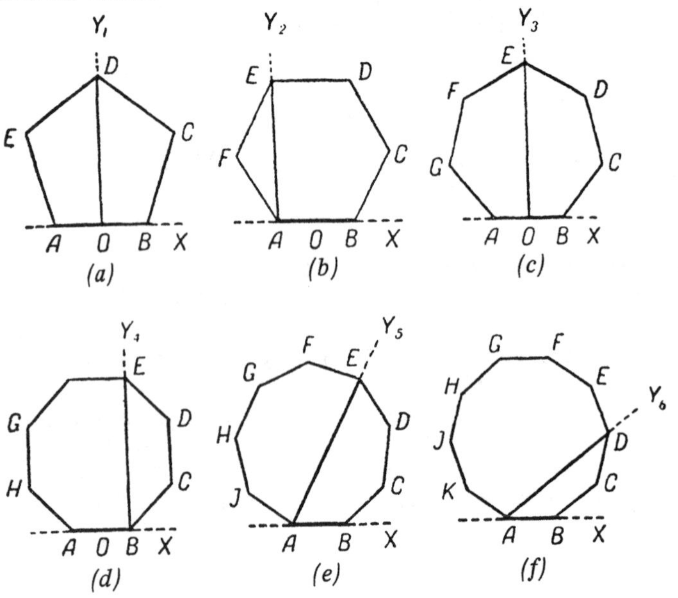

d) O on O, A on A, O on O, and B on B respectively. Keeping rod AB on the paper, turn the remaining rods until D is on line OY_1 (pentagon), or E is on AY_2 (hexagon), or E is on OY_3 (heptagon), or, finally, E is on BY_4 (octagon).

To draw regular n-gons when $n = 9$ or 10 it is first necessary to form rays AY_5 and AY_6 so that $\angle Y_5AX = 60°$ and $\angle Y_6AX = 36°$, then place the mechanism with rod AB on straight line AB, superposing A on A. Keeping rod AB on the paper, turn the remaining rods until E is on AY_5 (nonagon), and D is on AY_6 (decagon). In this way we obtain 4 consecutive sides (and 5 vertices) of the desired n-gon. Having drawn these 4 sides of the n-gon, it is not difficult to complete it simply by turning the pattern.

The length of each n-gon side you construct will equal rod AB. Theoretically such constructions are precise, but in practice the exactness will depend on the precision with which you build your instrument.

Assignment: You can bisect any angle with a straightedge and compasses. With these tools and the hinged mechanism, can you build a $1°$ angle?

VI
Dominoes and Dice

Dominoes

A set of dominoes usually consists of 28 rectangular tiles. Each tile has two squares, each with a 0, 1, 2, 3, 4, 5, or 6. Every combination is represented. The value of a tile is the sum of the values of the two squares. When both squares have the same value the tile is called a doublet.

The basic rule in playing dominoes is that, in adding to a chain, you have to match the value of one square of your tile to the value of a square at one end of the chain.

With a set of dominoes, or a replica cut out of cardboard, you can enjoy solving these curious problems.

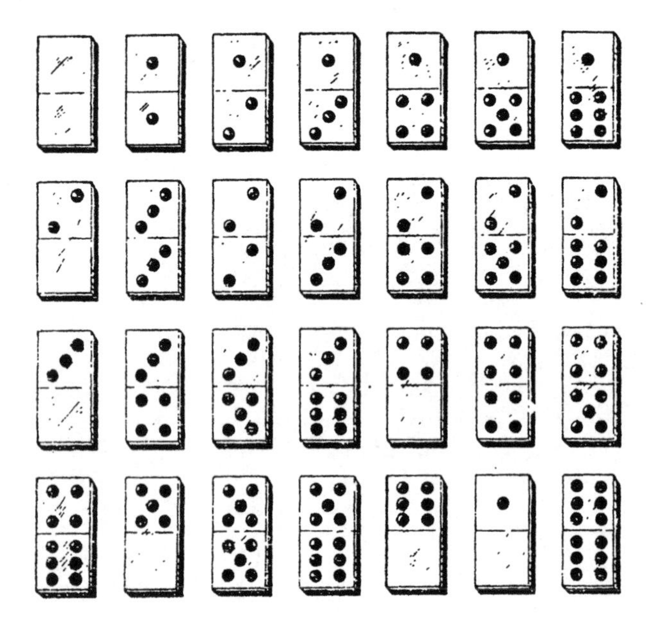

197. HOW MANY DOTS?

If you place all 28 tiles in a continuous chain (adjacent ends of tiles must match) so that 5 dots are at one end, how many dots will be at the other end?

Solve mentally, then check with actual dominoes.

198. A TRICK

Conceal a tile that is not a doublet, and ask a friend to make a chain using all the dominoes. (Of course, you do not tell him that a domino is missing.) You can predict what numbers will be at the ends. Explain.

199. A SECOND TRICK

Show 25 tiles face down in a row and vertical. Leave the room while someone transposes any number of tiles up to 12 from the right end to the left. You return and turn over a tile. Its value is the number of tiles transposed.

The trick is in the arrangement. You placed the 13 tiles in the diagram on the left, and 12 more at random to their right. The tile you turn over when you return is the middle (thirteenth) one. Why?

200. WINNING THE GAME

Sometimes the result of a game of dominoes is foreordained. Suppose *A* and *C* are playing *B* and *D*, each player starting with 7 tiles.

A has: 0-1 0-4 0-5 0-6 1-1 1-2 1-3.

D has: 0-0 0-2 0-3 1-4 1-5 1-6, and one other tile.

A plays 1-1. *B* and *C* pass, since *A* and *D* between them have all the tiles with 1s. *D* can play 1-4, 1-5, or 1-6, to which *A* answers 4-0, 5-0, or 6-0. Again *B* and *C* pass, since they have no tiles with 0s. Indeed, they will have to continue to pass, since *A* and *D* will leave nothing but 0s and 1s to match.

Clearly *A* will win for his team by matching whatever *D* plays. At the end, *D* will be stuck with his seventh tile (which has no 0 or 1). A curious game.

If a game is deadlocked (no one has a playable tile) victory goes to the pair with the smallest total on their remaining tiles.

Suppose *A* and *C* are playing *B* and *D*. Each player has 6 tiles, and 4 tiles remain face down and not to be used.

A has: 2-4 1-4 0-4 2-3 1-3 1-5.

His partner *C* has 5 doublets. *D* has 2 doublets; his tiles total 59.

A plays 2-4; *B* passes; *C* plays a tile; *D* passes; *A* plays a tile; *B* passes; *C* plays a tile and the game is deadlocked. *B* and *D* have lost the game without playing a tile. *A* and *C* have 35 points left and *B* and *D* have 91 points left. The 4 tiles played had a value of 22.

Which 4 tiles were left face down and which 4 tiles were played?

201. HOLLOW SQUARES

(A) Make a hollow square by joining tiles in accordance with the basic rule of the game. Use all 28 tiles. The sum of the points along each side of the square should be 44.

(B) Link all 28 tiles to make a hollow square within a hollow square, as shown. Each of the eight sides should have the same point sum.

Adjacent squares do *not* have to match.

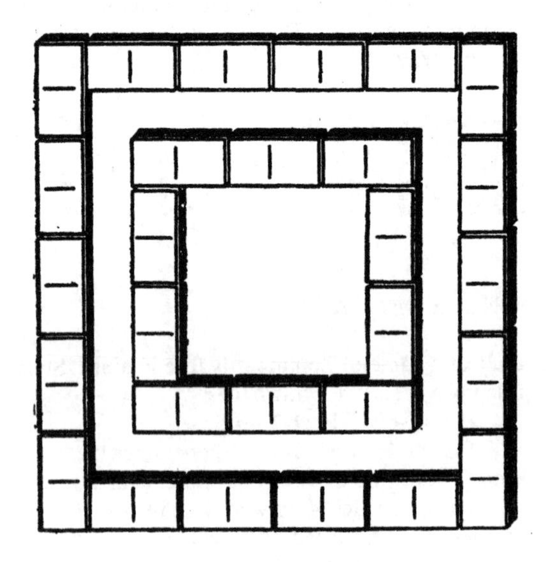

202. WINDOWS

In the diagram, 4 tiles form a "window" with an empty space inside. Each of the four sides contains 11 points.

Make seven other windows out of the 28 tiles. Each window should have the same point sum for each side. (But one window need not equal another.)

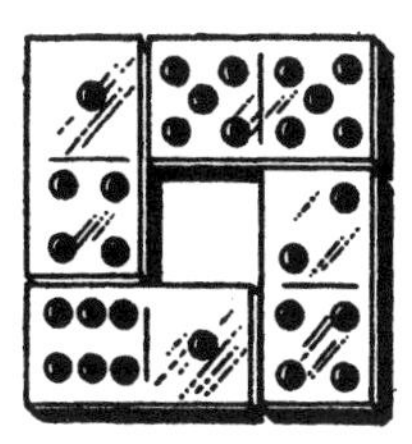

203. MAGIC SQUARES OF DOMINO TILES

You can make not only windows and hollow squares out of dominoes, but solid squares, even magic squares.

The first diagram is a 3-by-3 square. The top row is 7 + 0 + 5 = 12. The second row is 2 + 4 + 6 = 12. The third row adds to 12–as do all three columns and both main diagonals. This magic square of domino tiles contains tiles with values from 0 through 8. Its magic constant is 12.

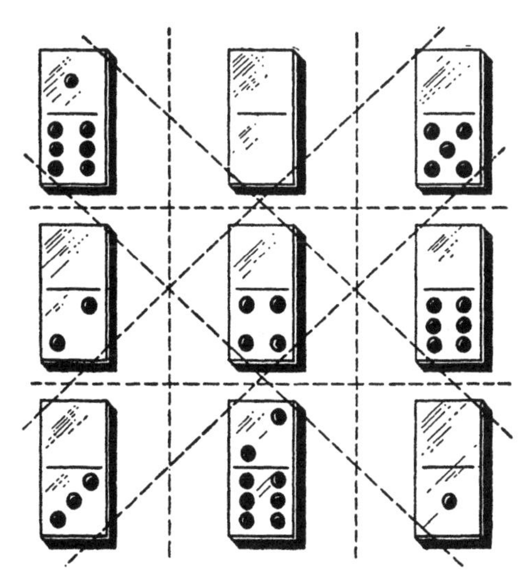

Nine tiles whose values are 1 through 9 form a magic square with the constant 15.

A 4-by-4 magic square can be made out of 16 tiles. (However, values must be repeated, since there are only 13 different domino values.) Here is a 4-by-4 square with a constant of 18:

2-6	1-2	1-3	0-3
1-4	0-2	3-6	1-1
0-5	1-5	0-1	0-6
0-0	2-5	0-4	1-6

(A) Form a magic square with a constant of 21, using the 9 tiles shown below.

(B) Form a magic square with 9 tiles whose values are 4 through 12. What is the constant of the square?

(C) Form a magic square from 16 tiles with the values 1, 2, 3, 3, 4, 4, 5, 5, 6, 6, 7, 7, 8, 8, 9, and 10. (There are several solutions.)

(D) Form a 5-by-5 magic square with a constant of 27 using all the tiles except 5-5, 5-6, and 6-6.

204. A MAGIC SQUARE WITH A HOLE

Using all 28 tiles, form a rectangular array with a central hole, as shown. The sum of each of the eight rows, each of the eight columns, and each of the two diagonals (shown by dotted lines) must be 21. For rows, add whole tiles, but for columns add half tiles (squares), and for each diagonal add the 6 half tiles covered by the dotted line.

The fourth row from the top has been filled in. Its sum is 5 + 6 + 5 + 5 = 21. The 4 corner tiles have been filled in (including the double blank in the lower right corner).

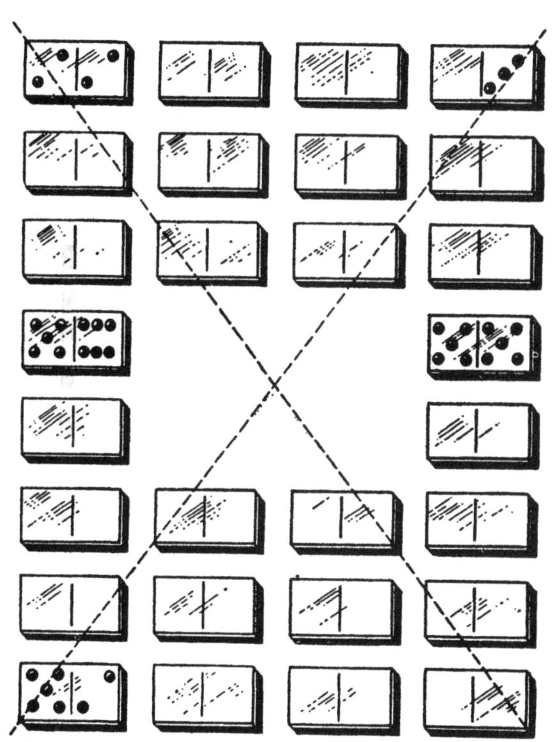

205. MULTIPLYING DOMINOES

The 4 tiles in the picture show the multiplication of a three-digit number, 551, by 4

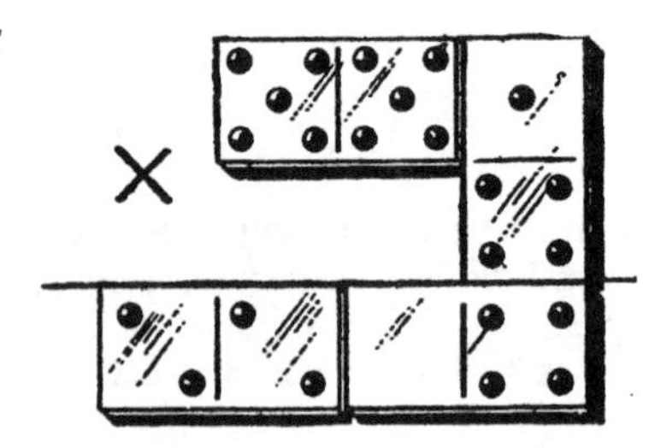

to give the product 2,204. Arrange all 28 tiles to make seven such multiplications.

206. GUESS A TILE

Ask a friend to make a mental note of a tile. Tell him to do these calculations:

1. Multiply the number of points on either half of the tile by 2.
2. Add a number m called out by you.
3. Multiply by 5.
4. Add the number of points on the other half of the tile.

When you get the result, subtract $5m$ from it. The two digits of the resulting number are the numbers on the chosen tile.

Suppose he chose 6-2. He multiplied 6 by 2 and added $m = 3$ to get 15. He multiplied 15 by 5 and added 2 points on the other half of the tile, getting 77. You subtract $5m = 15$. Answer: 62, which means 6-2.

Why does this always work?

Dice

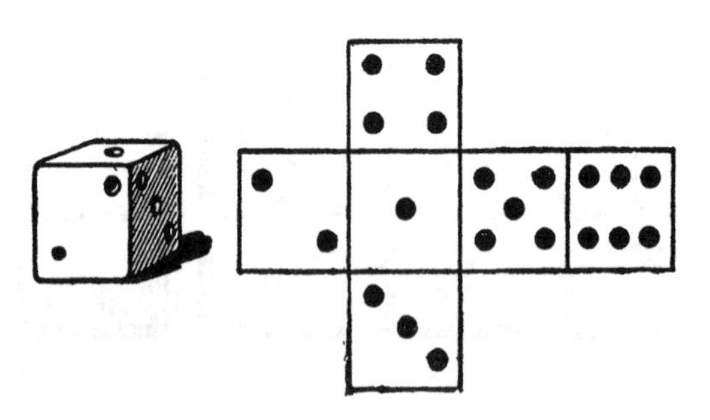

The picture above shows a die and its "unfolded" pattern. The spots are arranged so that the sum of the points on all pairs of opposite faces is 7.

Why is a cube the best shape for a die? First, a die should be a regular solid so that when it is rolled each face has an equal chance to end on top. Of the five regular solids, the cube is the most suitable. It is easy to manufacture and when you throw it, it rolls easily but not *too* easily. The tetrahedron and octahedron (*a* and *b* below) are hard to roll; the dodecahedron and icosahedron (*c* and *d*) are so "round" that they roll almost like balls.

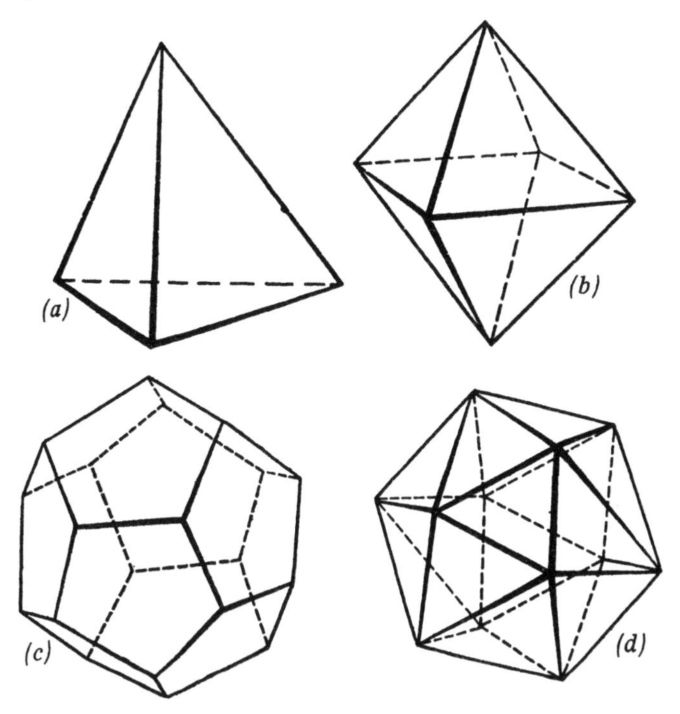

The *principle of* 7 (the sum of opposite faces) is the key to many tricks with dice.

207. A TRICK WITH THREE DICE

The magician turns his back while someone throws 3 dice. He asks a volunteer to add the spots on the top faces of the dice, then lift up any one of the three and add to the previous sum the number on the die's bottom face. He asks the volunteer to roll this same die again, and add to the last total the number of spots on its top face. He turns to the table, reminds the audience he does not know which die was thrown twice, picks up the three dice, shakes them in his hand, and, to the amazement of the audience, guesses the final sum.

The method: Before picking up the dice, add 7 to the spots showing on the top faces of the dice.

Explain.

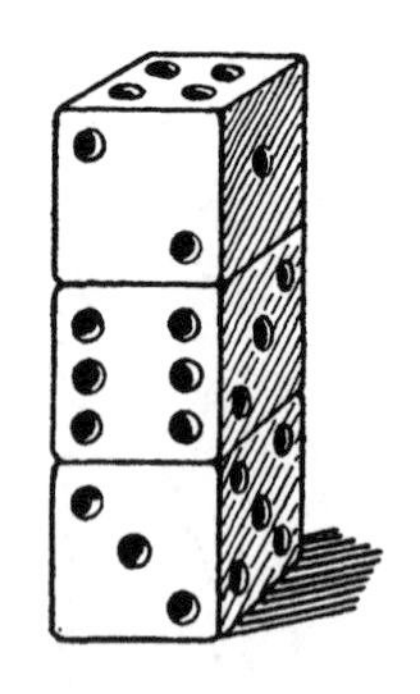

208. GUESSING THE SUM OF SPOTS ON HIDDEN FACES

In the tower of three dice at right, you need only glance at the face on top of the tower and you know the sum of five faces: those on the four faces where two dice touch one another and the face on the bottom of the tower. In the diagram, the sum is 17.

Explain.

209. IN WHAT ORDER ARE THE DICE ARRANGED?

Give a friend three dice, a piece of paper, and a pencil. Turn your back and ask him to roll the dice and then arrange them in a row so their top faces make a three-digit number. For example, the number on the dice shown is 254. Ask him to append the three-digit number given by the bottom faces of the dice. The result is a six-digit number (in our example, 254, 523). He is to divide this number by 111 and tell you the result. In turn, you tell him what is on the top faces of the three dice.

The method: Subtract 7 from the number he tells you, then divide by 9. In our example, 254,523 ÷ 111 = 2,293; 2,293 − 7 = 2,286; 2,286 ÷ 9 = 254.

Explain.

VII
Properties of Nine

Many arithmetical curiosities are connected with the number 9. You may know that a number is divisible by 9 if the sum of its digits is divisible by 9. Example: 354 x 9 = 3,186, and 3 + 1 + 8 + 6 = 18 (divisible by 9). (See Problem 322.)

A boy complained that he could not remember the multiplication table for 9. His father showed him how to help his memory with his fingers, like this:

Lay your hands palms down, extending your fingers. The fingers from left to right are given numbers from 1 through 10, as shown. To find 9 X 7, lift the number 7 finger. There are 6 fingers to its left and 3 to its right. Thus 9 X 7 = 63. This works for 9 X 1, 2, 3, . . . , 10.

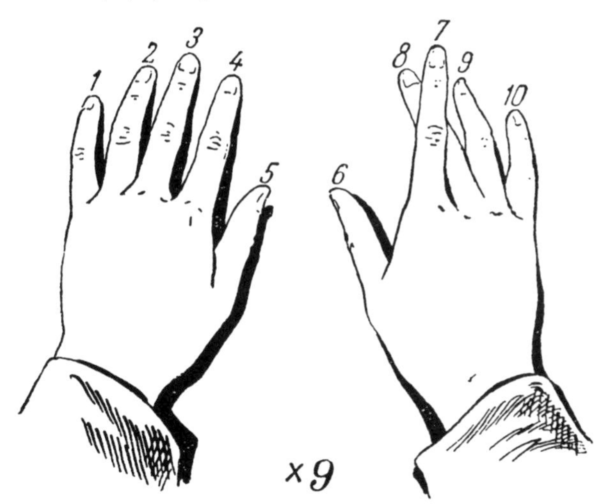

This use of the hands as a calculator is easily explained: the first 10 products of 9 each have digits that add up to 9 (the 9 fingers not raised), and they each have a first digit 1 less than the number multiplied by 9 (fingers to left of raised finger).

More properties: These numbers are always divisible by 9:

1. The difference between a number and the sum of its digits.

2. The difference between two numbers composed of the same digits.
3. The difference between two numbers that have the same digit sums.

The remainders of 7, 8, 9, 10, . . . , when divided by 9, are 7, 8, 0, 1, . . . Let us call these 9-remainders. On your own, express the three statements above in terms of 9-remainders. Some more properties:

4. The 9-remainder of the sum or difference of two numbers equals the 9-remainder of the sum or difference of the 9-remainders of the two numbers.
5. The 9-remainder of the product of two numbers equals the 9-remainder of the product of the 9-remainders of the two numbers.

On your own find an analogous property for division.

210. WHICH DIGIT IS CROSSED OUT?

(A) Ask a friend to write down a number of three or more digits, divide by 9, and tell you the remainder. Then ask him to cross out any digit except 0, divide the remaining number by 9, and tell you the remainder. At once you name the digit crossed out.

The method: If the second remainder is smaller than the first, subtract from the first. If it is larger, subtract it from 9 plus the first. If the remainders are equal, the crossed-out digit is 9. Explain.

(B) Ask a friend to write down a long number, and then write a second number composed of the same digits. He is to subtract the smaller from the larger, cross out any digit (except 0), and tell you the sum of the remaining digits. At once you name the digit crossed out. For example:

$$\begin{array}{r} 72{,}105 \\ \underline{25{,}071} \\ 47{,}034 \end{array}$$

He crosses out 3 and calls out $4 + 7 + 0 + 4 = 15$. The next higher multiple of 9 is 18. You subtract 15 from 18 and announce the crossed-out digit.

How does it work?

(C) Your friend writes down a number (say, 7,146), crosses out a nonzero digit (4, leaving 716), and subtracts the digit sum of the original number ($716 - 18 = 698$). When you hear the result, you tell him the crossed-out digit. How?

211. THE NUMBER 1,313

Ask a friend to write down this easily remembered number, subtract any number from it, and make a five- to seven-digit number with the difference on the left and 100 plus the number subtracted on the right. Now he crosses out any nonzero digit and calls out the resulting number.

You promptly name the crossed-out digit.

What property of 1,313 simplifies your work?

212. GUESS A MISSING NUMBER

(A) I have picked 8 of the numbers from 1 through 9 and hidden them in circles in the diagram below. Each circle has a line passing through it; the tally line *AB* gives sums of the numbers on each line.

Give two methods of finding the number I didn't pick.

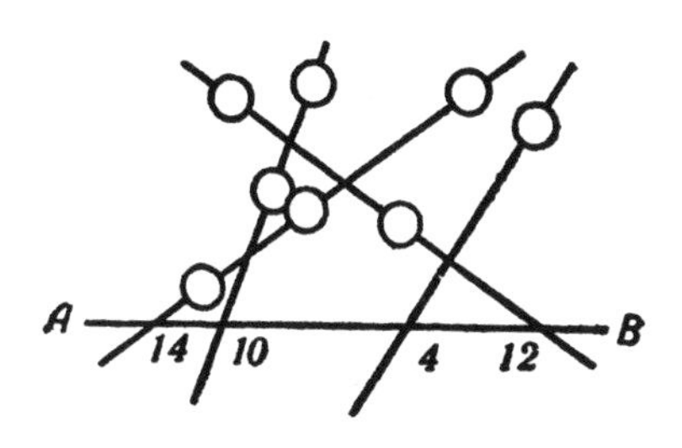

The next two diagrams show how to do the same trick writing the numbers along the sides of a triangle *(a)* or a quadrilateral *(b)*.

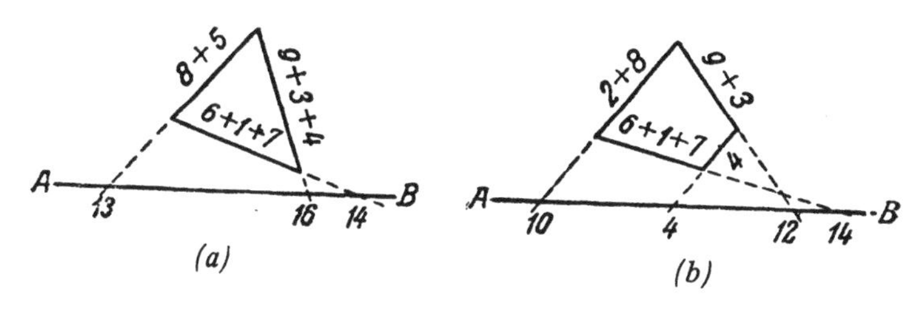

(a) *(b)*

(B) The last diagram shows how to do a similar trick with two-digit numbers. Of the 18 hidden digits (numbers 11, 22, 33, . . . , 99), curved lines connect 17 digits to the tally line *AB*. What is the unconnected digit?

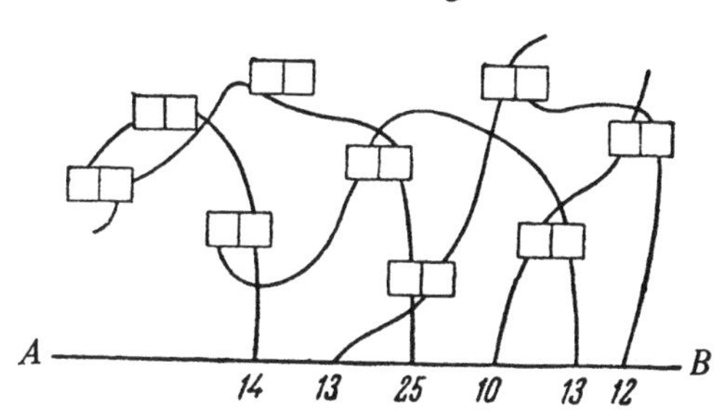

213. FROM ONE DIGIT

A number with two identical digits is multiplied by 99. What is the four-digit product if the third digit is 5?

214. GUESSING THE DIFFERENCE

Find the difference between any asymmetrical three-digit number and the same number reversed (say, 621 − 126 = 495). Tell me the last digit of the difference and I'll tell you the other two.

215. THREE AGES

Transposing the two digits of *A*'s age gives *B*'s age. The difference between their ages is twice *C*'s age and *B* is ten times as old as *C*. What are the three ages?

216. WHAT IS THE SECRET?

While we were exchanging number problems, a guest began writing down an inexhaustible list of numbers, from 435 to 1,207,941,800,554. They all had this property: When we added the digits, then added the digits of the digit sum, and continued until we reached a single digit, this digit was already there at the center of the number. For example, the central digit of the long number given is 1, and the successive digit sums of the number are 46, 10, and 1.

Explain.

VIII
With Algebra and without It

For fifty years, in family circles and schools, people have been scratching their heads over this one:

A lone goose was flying in the opposite direction from a flock of geese. He cried: "Hello, 100 geese!"

The leader of the flock answered: "We aren't 100! If you take twice our number and add half our number, and add a quarter of our number, and finally add you, the result is 100, but . . . well, you figure it out."

The one goose flew on, but could not find the answer. Then he saw a stork on the bank of a pond looking for frogs. Among birds the stork is the best mathematician. He often stands on one leg for hours, solving problems. The goose descended and told his story.

The stork drew a line with his beak to represent the flock. He drew a second line of the same length, a line half as long, a line a fourth as long, and a very small line, rather like a dot, to represent the goose.

"Do you understand?" the stork asked.

"Not yet."

The stork explained the meaning of the lines: two represented the flock, one represented half a flock, one a quarter of a flock, and the dot stood for the goose. He rubbed out the dot, leaving lines that now represented 99 geese. "Since a flock contains four quarters, how many quarters do the four lines represent?"

Slowly, the goose added 4 + 4 + 2 + 1. "Eleven," he replied.

"And if 11 quarters make up 99 geese, how many geese are in a quarter?"

"Nine."

"And how many in the entire flock?"

The goose multiplied 9 by 4 and said:

"Thirty-six."

"Correct! But you couldn't get the right answer yourself, could you? You . . . goose!"

In the following problems, feel free to use any method–algebraic, arithmetic, or graphic.

217. MUTUAL AID

During the rebuilding after World War II, we were short of tractors. The machine and tractor stations would lend each other equipment as needed.

Three machine and tractor stations were neighbors. The first lent the second and third as many tractors as they each already had. A few months later, the second lent the first and third as many as they each had. Still later, the third lent the first and second as many as they each already had. Each station now had 24 tractors.

How many tractors did each station originally have?

218. THE IDLER AND THE DEVIL

An idler sighed: "Everyone says, 'We don't need idlers. You are always in the way. Go to the devil!' But will the devil tell me to get rich?"

No sooner did the idler say this than the devil himself stood in front of him.

"Well," said the devil, "the work I have for you is light, and you will get rich. Do you see the bridge? Just walk across and I will double the money you have now. In fact, each time you cross I will double your money."

"You don't say!"

"But there is one small thing. Since I am so generous you must give me $24 after each crossing."

The idler agreed. He crossed the bridge, stopped to count his money . . . a miracle! It had doubled.

He threw $24 to the devil and crossed again. His money doubled, he paid another $24, crossed a third time. Again his money doubled. But now he had only $24, and he had to give it all to the devil. The devil laughed and vanished.

The moral: When anyone gives you advice you should think before you act.
How much money did the idler start with?

219. A SMART BOY

Three brothers shared 24 apples, each getting a number equal to his age 3 years before. The youngest one proposed a swap:

"I will keep only half the apples I got, and divide the rest between you two equally. But then the middle brother, keeping half his accumulated apples, must divide the rest equally between the oldest brother and me, and then the oldest brother must do the same."

They agreed. The result was that each ended with 8 apples.

How old were the brothers?

220. HUNTERS

Three friends were hunting in the taiga (swampy forest of Siberia). Two of them, while fording a small stream, got their cartridge cases wet. The three friends divided what good cartridges remained, equally.

After each had fired 4 shots, the total cartridges remaining were equal to the number each had after the division.

How many cartridges were divided?

221. TRAINS MEETING

Two freight trains, each 1/6 mile long and traveling 60 miles per hour, meet and pass each other. How many seconds is it between when the locomotives pass each other and the cabooses pass each other?

222. VERA TYPES A MANUSCRIPT

Mother asked Vera to type a manuscript. Vera decided: "I will type an average of 20 pages a day."

She typed the first half of the manuscript rather lazily, at 10 pages a day. To make up for it, she typed the second half at 30 pages a day.

"See, I did average 20 pages a day," Vera concluded. "Half of 10 + 30 is 20."

"No, you didn't," her mother said.

Who was right?

223. A MUSHROOM INCIDENT

Marusya, Kolya, Vanya, Andryusha, and Petya went looking for mushrooms. Only Marusya took her search seriously. The four boys spent most of their time lying in the grass and telling stories. When it was time to return, Marusya had 45 mushrooms and the boys had none.

Marusya was sympathetic. "It won't look good for you boys when we get back to the camp." She gave each boy some mushrooms, leaving none for herself.

On the way back, Kolya found 2 mushrooms and Andryusha doubled the number of the mushrooms he already had. But Vanya and Petya fooled around all the way. As a result, Vanya lost 2 mushrooms and Petya lost half his mushrooms.

At camp they counted up and each boy had the same number of mushrooms.

How many mushrooms did Marusya give each boy?

224. MORE, OR LESS?

Oarsman A rows on a river, x miles with the current and x miles against the current. Oarsman B rows $2x$ miles on a lake where there is no current. Does A take more time than B, or less? (Their rowing strength is the same.)

225. A SWIMMER AND A HAT

A boat is being carried away by a current. A man jumps out and swims against the current for a while, then turns around and catches up with the boat. Did he spend more time swimming against the current or catching up with the boat? (We assume his muscular efforts never change in strength.)

The answer is: Both times were the same. The current carries man and boat downstream at the same speed. It does not affect the distance between the swimmer and the boat.

Now imagine that a sportsman jumps off a bridge and begins to swim against the current. The same moment a hat blows off a man's head on the bridge and begins to float downstream. After 10 minutes the swimmer turns back, reaches the bridge, and is asked to swim on until he catches up with the hat. He does, under a second bridge 1,000 yards from the first.

The swimmer does not vary his effort. What is the speed of the current?

226. TWO DIESEL SHIPS

Two diesel ships leave a pier simultaneously, the *Stepan Razin* downstream and the *Timiryazev* upstream, with the same motive force. As they leave, a life buoy falls off the *Stepan Razin* and floats downstream. An hour later both ships are ordered to reverse course. Will the *Stepan Razin*'s crew be able to pick up the buoy before the ships meet?

227. HOW SHARP-WITTED ARE YOU?

Motorboat M leaves shore A as N leaves B; they move across a lake at constant speed. They meet the first time 500 yards from A. Each returns from the opposite shore without halting, and they meet 300 yards from B.

How long is the lake and what is the relation between the two boats' speeds? Sharp wits can solve the problem with a minimum of calculation.

228. YOUNG PIONEERS

Vitya pledges that his brigade of Young Pioneers will plant half the number of fruit trees the rest of the Pioneers plant. Kiryusha pledges his brigade, the largest in the detachment, will plant as many trees as the rest of the Pioneers (including Vitya's brigade).

Their brigades work the last shift simultaneously. The preceding brigades of the detachment plant 40 trees. Assuming that both pledges are fulfilled exactly, how many trees does the whole detachment plant?

229. HOW MANY TIMES AS LARGE?

Given two numbers, if we subtract half the smaller number from each number, the result with the larger number is three times as large as the result with the smaller number.

How many times is the larger number as large as the smaller number?

230. A DIESEL SHIP AND A SEAPLANE

A diesel ship leaves on a long voyage. When it is 180 miles from shore, a seaplane, whose speed is ten times that of the ship, is sent to deliver mail. How far from shore does the seaplane catch up with the ship?

231. BICYCLE FIGURE RIDERS

Four cyclists do their act on circular paths, each 1/3 mile long. They start simultaneously at the black spots, with speeds of 6, 9, 12, and 15 miles per hour.

By the end of the act (20 minutes), how many times will they have simultaneously returned to the spots where they started?

232. SPEED OF WORK BY TURNER BYKOV

State prize-winning lathe turner P. Bykov reduced the time to process a metal part from 35 minutes to 2½. He increased his cutting speed by 1,690 inches per minute—to how much?

233. JACK LONDON'S JOURNEY

Jack London tells how he raced from Skagway in a sled pulled by 5 huskies to reach the camp where a comrade was dying.

For 24 hours the huskies pulled the sled at full speed. Then 2 dogs ran off with a pack of wolves. London, left with 3 dogs, was slowed down proportionally. He reached camp 48 hours later than he had planned. If the runaway huskies had stayed in harness for 50 more miles, London writes, he would have been only 24 hours late.

How far is the camp from Skagway?

234. FALSE ANALOGY

Scientific discoveries are sometimes made by using analogy. If certain features of two objects are similar, perhaps other features are also similar. Analogy, however, is only a tool for good guesses. The guesses have to be tested.

Analogy has its place in mathematics also, but so, alas, does false analogy:

"By how much is 40 larger than 32?"

"By 8."

"By how much is 32 smaller than 40?"

"By 8."

"By what percentage is 40 larger than 32?"

"By 25 percent."

"By what percentage is 32 smaller than 40?"

"By 25 percent."

But it is only 20 percent smaller.

(A) Suppose your monthly income increases 30 percent. By what percentage does your purchasing power increase?

(B) Suppose that your monthly income does not change, but prices go down 30 percent. By what percentage does your purchasing power increase?

(C) When a secondhand bookstore holds a sale with a 10 percent reduction in prices, it makes an 8 percent profit on each book sold. What was the profit before the sale?

(D) If a metal worker reduces the time per part by p percent, how much does he increase his productivity?

235. A LEGAL TANGLE

Ancient Roman mathematical works were utilitarian. Here is a Roman inheritance problem:

A dying Roman, knowing his wife was pregnant, left a will saying that if she had a son, he would inherit two-thirds of the estate and the widow one-third, but if she had a daughter, the daughter would get one-third and the widow two-thirds.

Soon after his death his widow had twins—a boy and a girl. This is a possibility the will maker had not foreseen. What division of the estate keeps as close as possible to the terms of the will?

236. TWO CHILDREN

(A) I have 2 children. They aren't both boys. What is the probability that both children are girls?

(B) An artist has 2 children. The older is a boy. What is the probability that both children are boys?

237. WHO RODE THE HORSE?

One day a young man and an older man left the village for the city, one on horse, one in a car. Soon it was apparent that if the older man had ridden three times as far as he had, he would have half as far to ride as he had, and if the young man had ridden half as far as he had, he would have three times as far to ride as he had.

Who rode the horse?

238. TWO MOTORCYCLISTS

Two motorcyclists started at the same time, covered the same distance, and returned home at the same time. But one rode twice as long as the other rested on his trip, and the other rode three times as long as the first one rested on his trip.

Who rode faster?

239. IN WHICH PLANE DID VOLODYA'S FATHER FLY?

Volodya asked, "What plane did you fly during the air parade?"

His father sketched a formation of 9 planes.

"The number of planes to the right of me multiplied by the number of planes to the left of me is 3 less than it would have been if my plane had been 3 places to the right of me."

How did Volodya solve the problem?

240. EQUATIONS TO SOLVE IN YOUR HEAD

$$6{,}751x + 3{,}249y = 26{,}751;$$
$$3{,}249x + 6{,}751y = 23{,}249.$$

Is this a joke? Not if you can multiply the first equation by 6,751 and the second by 3,249 in your head, and not if you use a second, simpler method.

241. TWO CANDLES

Two candles have different lengths and thicknesses. The long one can burn 3½ hours; the short one, 5 hours.

After 2 hours burning, the candles are equal in length. Two hours ago, what fraction of the long candle's height gave the short candle's height?

242. WONDERFUL SAGACITY

Nikanorov the bookkeeper asked each of four children to think of a four-digit number. "Now please transfer the first digit to the end and add the new number to the old one. For example, 1,234 + 2,341 = 3,575. Tell me your results."

Kolya: "8,612."

Polya: "4,322."

Tolya: "9,867."

Olya: "13,859."

"Everyone except Tolya is wrong," said the bookkeeper.

How did he know?

243. CORRECT TIMES

A wall clock loses 2 minutes in an hour. A table clock gets 2 minutes ahead of the wall clock in an hour. An alarm clock falls 2 minutes behind the table clock in an hour. A wristwatch gets 2 minutes ahead of the alarm clock in an hour.

At noon all four timepieces were set correctly. To the nearest minute, what time will the wristwatch show when the correct time is 7:00 p.m.?

244. WATCHES

My watch is 1 second fast per hour and Vasya's is 1½ seconds slow per hour. Right now they show the same time. When will they show the same time again? When will they show the same *correct* time again?

245. WHEN?

(A) A craftsman goes to lunch not long after noon. As he leaves he observes the exact placement of a clock's hands. On his return he notices that the minute and hour hands have exchanged places.

When does he return?

(B) I go for a walk, for more than 2 hours but less than 3. When I return the minute and hour hands have exchanged places.

How much longer than 2 hours did the walk last?

(C) A boy begins to solve a problem at the time between 4:00 and 5:00 p.m. when the clock's hands are together. He finishes when the minute hand is exactly opposite the hour hand.

How many minutes does it take him to solve the problem, and when does he finish it?

246. WHEN DOES THE CONFERENCE START AND END?

A conference begins between 6:00 and 7:00 p.m., and ends between 9:00 and 10:00 p.m. The minute and hour hands have exchanged places.

When does the conference start and end?

247. A SERGEANT TEACHES HIS SCOUTS

Sergeant Semochkin uses every opportunity to teach his soldier-scouts observation and sharpness. He will suddenly ask them: "How many supports did the bridge have—the one we crossed today?"

Or he will set them a puzzle:

"Say two of you have to cover the same distance. The first scout runs half the time and walks the second half. The second scout runs halfway there and walks the rest. Neither one runs or walks faster than the other.

"Who gets there first? If they walked and then ran, who would get there first?"

248. TWO DISPATCHES

The first dispatch said:

"Train N passes me in t_1 seconds."

The second dispatch said:

"Train N crossed a bridge a yards long in t_2 seconds."

If train N's speed is constant, what is it, and how long is the train?

249. NEW STATIONS

Every station on the N railroad sells tickets to every other station. When it added some new stations, 46 additional sets of tickets had to be printed.

How many is "some"? How many stations were there before?

250. SELECT FOUR WORDS

B E
O A K
R O O M
I D E A L
S C H O O L
K I T C H E N
O V E R C O A T
R E V O L V I N G
D E M O C R A T I C
E N T E R T A I N E R
M A T H E M A T I C A L
S P O R T S M A N S H I P
K I N D E R G A R T E N E R
I N T E R N A T I O N A L L Y

The words run from 2 through 15 letters. Choose four different words with a, b, c, and d letters so that $a^2 = bd$ and $ad = b^2c$.

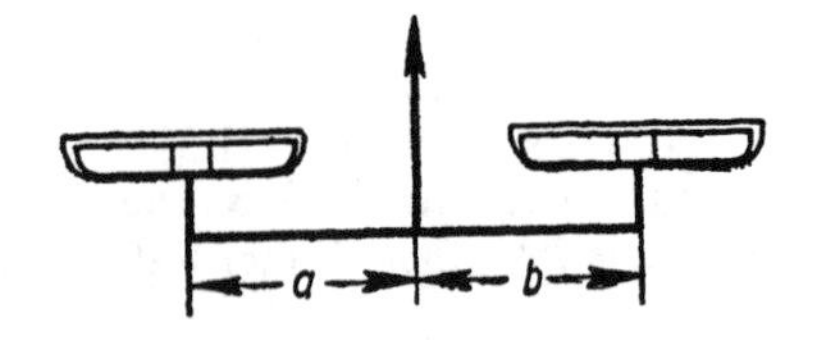

251. FAULTY SCALES

Good scales should have equal arms ($a = b$ in the picture), but in one grocery stall they didn't. Pending replacement, the manager wondered if he could give correct weight this way:

"I'll balance a 1-pound weight on the left with sugar on the right, and then I'll balance the 1-pound weight on the right with some more sugar on the left, and the sugar will add up to exactly 2 pounds."

Will it? Is there another way?

252. AN ELEPHANT AND A MOSQUITO

Does the weight of an elephant equal the weight of a mosquito?

Let x be the weight of an elephant, and y that of a mosquito. Call the sum of the two weights $2v$, then $x + y = 2v$.

From this equation we can obtain two more:

$$x - 2v = -y; x = -y + 2v.$$

Multiply:

$$x^2 - 2vx = y^2 - 2vy.$$

Add v^2:

$$x^2 - 2vx + v^2 = y^2 - 2vy + v^2, \text{ or } (x - v)^2 = (y - v)^2.$$

Take square roots:

$$x - v = y - v;$$

$$x = y.$$

That is, the elephant's weight (x) equals the mosquito's weight (y).
What is wrong here?

253. A FIVE-DIGIT NUMBER

There is an interesting five-digit number A. With a 1 after it, it is three times as large as with a 1 before it. What is it?

254. LIVE TO 100 WITHOUT AGING

My age, which plus yours is 86, is 15/16 the age you will be when I am 9/16 the age you would be if you were twice as old as I will be when I am twice your age.
How old am I and how old are you?
Solution: Work the problem to untangle the curlicues running through it.
1. One day I will be $2x$ and you will be x. (See first diagram.)

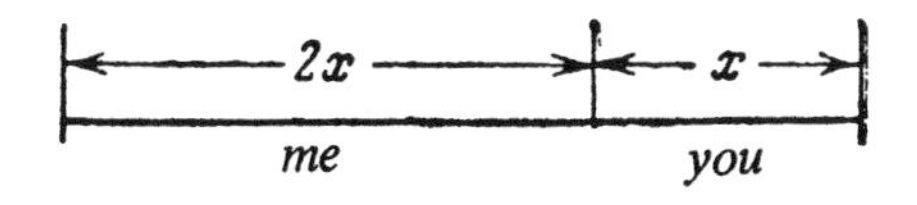

It follows that I am always older than you by x years.
2. When you are twice $2x$ I am $5x$, and when I am 9/16 of your $4x$, I am $2\frac{1}{4}x$ and you are $1\frac{1}{4}x$ (second diagram).

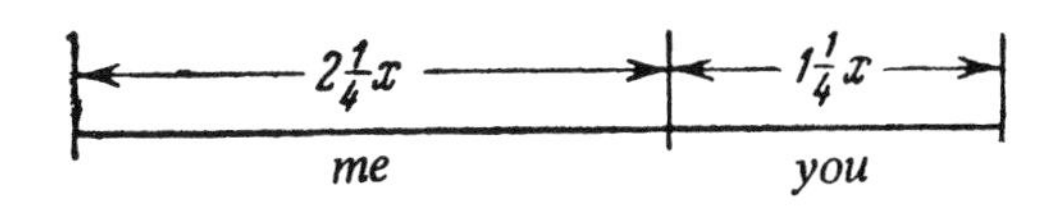

3. When I am 15/16 of your $1\frac{1}{4}x$, I am $75x/64$ and you are $11x/64$ (third diagram).

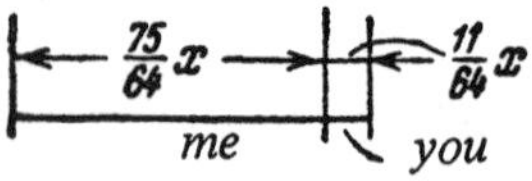

Our ages add to 86, so right now I am 75 and you are 11, according to the problem. Actually, I am a long way from 75, and you are probably older than 11. But if I am twice as old as you were when I was your age, and when you are my age the sum of our ages will be 63, how old am I now and how old are you?

255. THE LUCAS PROBLEM

This problem was invented by Edouard Lucas, a French nineteenth-century mathematician.

"Every day at noon," Lucas said, "a ship leaves Le Havre for New York and another ship leaves New York for Le Havre. The trip lasts 7 days and 7 nights. How many New York-Le Havre ships will the ship leaving Le Havre today meet during its journey to New York?"

Can you answer graphically?

256. A SINGULAR TRIP

Two boys go on a bicycle trip. En route one of the bicycles breaks down and has to be left behind for repairs. They decide to share the remaining bicycle: They start simultaneously, one on bicycle, one on foot. At a certain point the cyclist dismounts, leaves the bicycle behind, and continues on foot. His friend, when he reaches the waiting bicycle, mounts it and rides until he catches up with his friend, who takes the bicycle, and so on.

How far from their destination should the bicycle be left behind the last time so they reach the destination simultaneously? The distance from breakdown to destination is 60 miles, and they each walk 5 miles per hour and bicycle 15 miles per hour.

257. A CHARACTERISTIC OF SIMPLE FRACTIONS

Write down some simple positive fractions. Make a new fraction whose numerator equals the sum of the numerators written down and whose denominator equals the sum of the denominators written down. Is the new fraction larger than the smallest one written down, and smaller than the largest? Must it always be?

IX
Mathematics with Almost No Calculations

All problems are solved by reasoning, and problems stressing deduction rather than calculation have a special appeal and value. They teach you to analyze, and to seek unorthodox ways of solving a problem.

258. SHOES AND SOCKS

I went to the closet while my sister was asleep, so I left the light off.

I found my shoes and socks, but I must confess they were in no kind of order—just a jumbled pile of 6 shoes of three brands, and a heap of 24 socks, black and brown.

How many shoes and socks did I have to take with me to be sure I had a pair of matching shoes and a pair of matching socks?

259. APPLES

Three kinds of apples are mixed in a box. How many apples must you take to be sure of at least 2 apples of one kind? At least 3 apples of one kind?

260. A WEATHER FORECAST

It is raining at midnight—will we have sunny weather in 72 hours?

261. ARBOR DAY

On Arbor Day the Young Pioneers of the fourth grade started early and planted 5 trees before the sixth-graders came. But they planted them on the side assigned to the sixth grade.

The fourth-graders had to cross the street and start over, so the sixth-graders finished first. To pay their debt, they crossed the street and planted 5 trees. They planted 5 more trees, and all the work was finished.

Were the sixth-graders ahead by 5 or 10 trees?

262. MATCH NAMES AND AGE

[In Russia the wife of a Mr. Serov is Mrs. Serova.]

Three Young Pioneers were talking. The leader said: "Burov, Gridnev, and Klimenko arrive tomorrow. Their first names are Kolya, Petya, and Grisha, but not necessarily in that order."

"I think Kolya's last name is Burov."

"You are wrong," said the leader. "I'll give you some hints.

"The father of Nadya Serova, whom you know well, is a brother of Burov's mother.

"Petya started grade school when he was 7. He wrote me recently, 'Finally, this year I am beginning sixth-grade algebra.'

"Our beehive keeper, Semyon Zakharovich Mokrousov, is Petya's grandfather.

"Gridnev is 1 year older than Petya. And Grisha is 1 year older than Petya."

Give both names of the three boys—and their ages.

263. A SHOOTING MATCH

Andryusha, Borya, and Volodya each fired 6 shots, and each got 71 points.

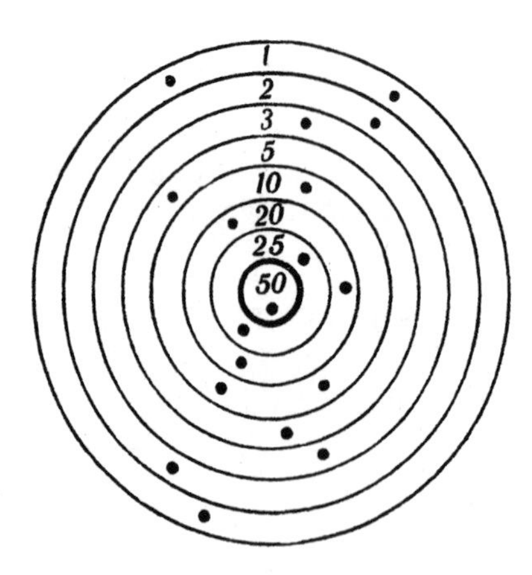

Andryusha's first 2 shots got 22 points and Volodya's first shot got only 3 points. Who hit the bull's-eye?

264. A PURCHASE

"Your pencils, notebooks, and colored paper cost $1.70."

"I bought 2 pencils at 2 cents each and 5 pencils at 4 cents each—and 8 notebooks and 12 sheets of colored paper, I don't remember the prices. But the bill can't be $1.70."

Why not?

265. PASSENGERS IN A RAILROAD COMPARTMENT

Six passengers sharing a compartment are from Moscow, Leningrad, Tula, Kiev, Kharkov, and Odessa.

1. *A* and the man from Moscow are physicians.
2. *E* and the Leningrader are teachers.
3. The man from Tula and *C* are engineers.
4. *B* and *F* are World War II veterans, but the man from Tula has never served in the armed forces.
5. The man from Kharkov is older than *A*.
6. The man from Odessa is older than *C*.
7. At Kiev, *B* and the man from Moscow got off.
8. At Vinnitsa, *C* and the man from Kharkov got off.

Match initials, professions, and cities. Also, are these facts both sufficient and necessary to solve the problem?

266. A CHESS TOURNAMENT

An infantryman, flyer, tankman, artilleryman, cavalryman, mortarman, sapper, and communications man, playing in the Soviet Army chess tournament, had the ranks (not necessarily in order) of colonel, major, captain, lieutenant, senior sergeant, junior sergeant, corporal, and private. Can you match them up?

1. In round 1 the colonel played the cavalryman.
2. The flyer arrived just in time for round 2.
3. In round 2 the infantryman played the corporal.
4. In round 2 the major played the senior sergeant.
5. After round 2 the captain quit the tournament. He was the only player to leave the tournament.
6. The junior sergeant missed round 3 because of illness.
7. The tankman missed round 4 because of illness.
8. The major missed round 5 because of illness.
9. In round 3 the lieutenant beat the infantryman.
10. In round 3 the artilleryman drew the colonel.
11. In round 4 the sapper beat the lieutenant.
12. In round 4 the sergeant beat the colonel.

13. Before the last round the cavalryman and the mortarman finished an adjourned game from round 6.

267. VOLUNTEERS

Six Komsomol members volunteered to saw large logs into ½-yard logs to heat the school. The three leaders of pairs were Volodya, Petya, and Vasya.

Volodya and Misha would saw up 2-yard logs, Petya and Kostya, 1½-yard logs, and Vasya and Fedya, 1-yard logs. (These are all first names.)

The next day the school bulletin praised the good work of teams led by Lavrov, Galkin, and Medvedev. Lavrov and Kotov sawed logs into 26 small logs, Galkin and Pastukhov, 27, and Medvedev and Yevdokimov, 28. (These are all last names.)

What is Pastukhov's first name?

268. WHAT IS THE ENGINEER'S LAST NAME?

On the Moscow-Leningrad train are three passengers named Ivanov, Petrov, and Sidorov. By coincidence the engineer, the fireman, and a conductor have the same last names.

1. Passenger Ivanov lives in Moscow.
2. The conductor lives halfway between Moscow and Leningrad.
3. The passenger with the same last name as the conductor lives in Leningrad.
4. The passenger who lives nearest to the conductor earns exactly three times as much a month as the conductor.
5. Passenger Petrov earns 200 rubles a month.
6. Sidorov (a member of the crew) recently beat the fireman at billiards.

What is the engineer's last name?

269. A CRIME STORY

(From the American journal *Scripta Mathematica*)

An elementary school teacher in New York State had her purse stolen. The thief had to be Lillian, Judy, David, Theo, or Margaret.

When questioned, each child made three statements:

Lillian: (1) I didn't take the purse. (2) I have never in my life stolen anything. (3) Theo did it.

Judy: (4) I didn't take the purse. (5) My daddy is rich enough, and I have a purse of my own. (6) Margaret knows who did it.

David: (7) I didn't take the purse. (8) I didn't know Margaret before I enrolled in this school. (9) Theo did it.

Theo: (10) I am not guilty. (11) Margaret did it. (12) Lillian is lying when she says I stole the purse.

Margaret: (13) I didn't take the teacher's purse. (14) Judy is guilty. (15) David can vouch for me because he knows me since I was born.

Later, each child admitted that two of his statements were true and one was false. Assuming this is true, who stole the purse?

270. HERB GATHERERS

Two Pioneer groups collected and sold some valuable medicinal herbs to the local Soviet medical agency. The agency added a small sum as a bonus. The first group got most of the bonus because it collected more herbs.

For fun, a Pioneer coded the calculations splitting the bonus, replacing each digit except one with a star.

Can you decode them?

(A) Add the packages of herbs collected by the first group and the second group:

```
   *
+  *
----
 * *
```

(B) Divide the bonus (in cents) by the total number of packages:

```
          * *
     ________
* 7 /  * * *
       * *
       ---
         * *
         * *
         ---
```

(C) How much bonus does the first group get?

```
  * *
X   *
-----
  * *
```

(D) How much bonus does the second group get?

```
  * *
X   *
-----
  * *
```

271. A HIDDEN DIVISION

Olya, editor of the mathematical bulletin *Think!*, writes down the division of a seven-digit number by a two-digit one, and lays the sheet aside. Two chess players start putting captured men on the digits of her calculation. By the time the game is over, they cover every digit except the remainder.

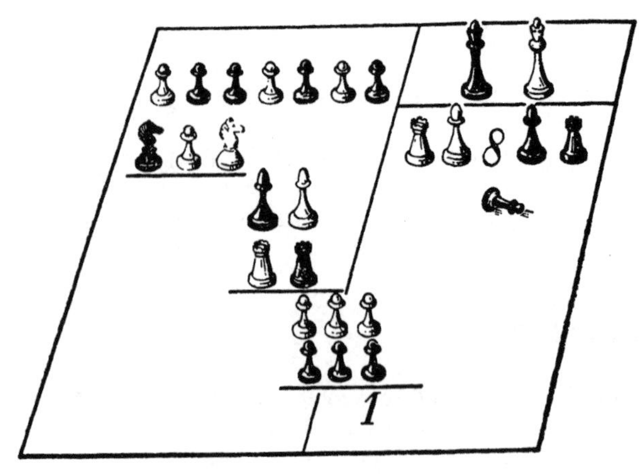

Olya decided to leave it as a decoding problem, after uncovering an 8 in the quotient so there would be no duplicate solutions.

This one is easier than it looks.

272. CODED OPERATIONS

In these seven cryptarithms, digits are represented by letters and asterisks. Identical letters stand for identical digits, different letters stand for different digits. An asterisk stands for any digit.

(A)

```
      A B C
      B A C
    * * * *
  * * A
* * * B
* * * * * *
```

(B)

```
      * * *
      * 2 *
      * * *
  * * * *
  * 8 *
* * 9 * 2 *
```

(C)

```
                          * * 7 * *
* * * * * 7 * | * * 7 * * * * * * *
                * * * * * *
                * * * * 7 7 *
                * * * * * * *
                    * 7 * * * *
                    * 7 * * * *
                    * * * * * * *
                    * * * * 7 * *
                        * * * * * *
                        * * * * * *
```

(D) (There are four solutions.)

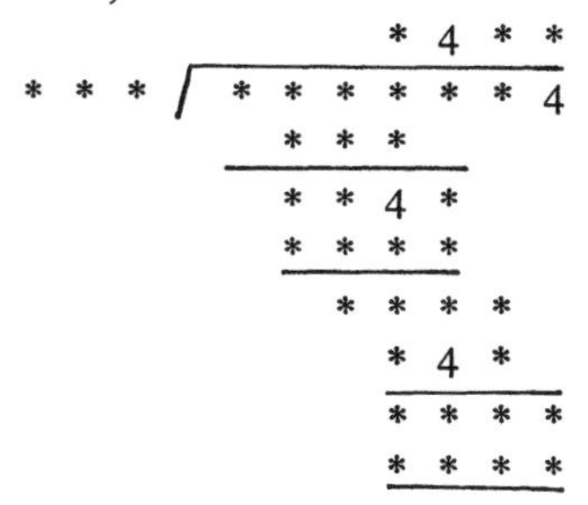

(E) (There are two solutions.)

D O + R E = M I;
F A + S I = L A;
R E + S I + L A = S O L.

(F) (Although even the number of digits in the divisor is not given, there is only one solution to this fairly easy division.)

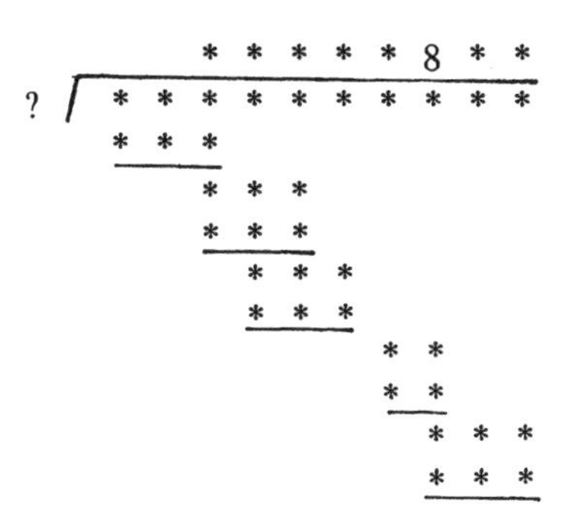

(G)

```
          A T O M
          A T O M
        -----------
        * * * * *
      * * * * *
    * * * * *
  * * * * *
  -------------------
  * * * * A T O M
```

273. A PRIME CRYPTARITHM

In this remarkable cryptarithm, each digit is a prime (2, 3, 5, or 7). No letters or digits are provided as clues; but there is only one solution.

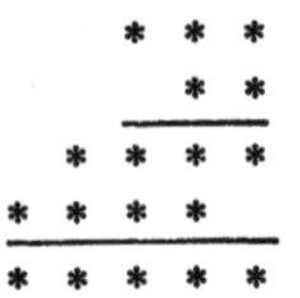

274. THE MOTORCYCLIST AND THE HORSEMAN

A motorcyclist was sent by the post office to meet a plane at the airport.

The plane landed ahead of schedule, and its mail was taken toward the post office by horse. After half an hour the horseman met the motorcyclist on the road and gave him the mail.

The motorcyclist returned to the post office 20 minutes before he was expected. How many minutes early did the plane land?

275. ON FOOT AND BY CAR

Here is a variation of the preceding problem:

An engineer goes every day by train to the city where he works. At 8:30 a.m., as soon as he gets off the train, a car picks him up and takes him to the plant.

One day the engineer takes a train arriving at 7:00 a.m., and starts walking toward the plant. On the way, the car picks him up and he arrives at the plant 10 minutes early.

When does he meet the car?

276. PROOF BY CONTRADICTION

When two statements A and B are mutually exclusive, only one is true. Proving A true by proving B false is called *proof by contradiction.*

Example: The sum of two numbers is 75. The first is larger than the second by 15. Prove by contradiction that the second number is 30.

Solution: Suppose the second number is not 30. Then it is either greater than 30 or less than 30. If it is greater than 30, the first number is greater than 45 and their sum is greater than 75, which is impossible. If it is less than 30, the first number is less than 45 and their sum is less than 75, which is also impossible.

Therefore, the second number is 30.

Use proof by contradiction!

(A) The product of two integers is greater than 75. Prove that at least one integer is greater than 8.

(B) The product of a certain two-digit number by 5 is a two-digit number. Prove that the first digit of the multiplicand is 1.

277. TO FIND A FALSE COIN

(A) (An easy question.) Of 9 coins of the same denomination, 8 weigh the same, and 1—a counterfeit—is lighter than the others. Find the counterfeit in two weighings on a balance, without using weights.

(B) (A bit more difficult.) The same, but with 8 coins in all.

(C) (Difficult.) Of 12 coins, 11 are the same weight and the counterfeit is either lighter or heavier. In three weighings, find which coin is counterfeit and whether it is lighter or heavier.

Problems to solve on your own:

(D) Tell in three weighings which one, *if any,* of 13 machine parts is nonstandard in weight, and whether it is lighter or heavier. A fourteenth machine part of standard weight is supplied.

(E) Find a general solution to (D) with n weighings of $\frac{1}{2}(3^n - 1)$ machine parts (and an added machine part of standard weight).

278. A LOGICAL DRAW

Three puzzle competitors are blindfolded. A white piece of paper is glued to each one's forehead and they are told that not all the pieces of paper are black. The blindfolds are removed and the prize goes to the first man to deduce whether the paper on his forehead is white or black.

All three announce white at the same time. Why?

279. THREE SAGES

Three ancient Greek philosophers took a nap under a tree. While they were asleep, a prankster smeared their faces with charcoal. On awakening, they began to laugh, each thinking the other two were laughing at each other.

Suddenly one man stopped laughing. How did he realize his own face was also smeared?

280. FIVE QUESTIONS

A mathematical statement should be complete, but it should not include unnecessary words. Brevity and precision are distinctive and pleasing features of mathematical language.

(A) Can you find the unnecessary words in these assertions?

1. The sum of the two acute angles of a right triangle is 90°.
2. If a leg of a right triangle is half the hypotenuse, the acute angle opposite is 30°.

(B) Give one- or two-word equivalents:

1. The part of a secant which is not outside a circle.
2. The polygon with the least number of sides.
3. A chord that passes through the center of a circle.
4. An isosceles triangle whose base is equal to the other sides.
5. Circles with a common center.

(C) In triangle *ABC, AB* = *BC* and *AD* = *DC.* Give at least three names for line *BD.*

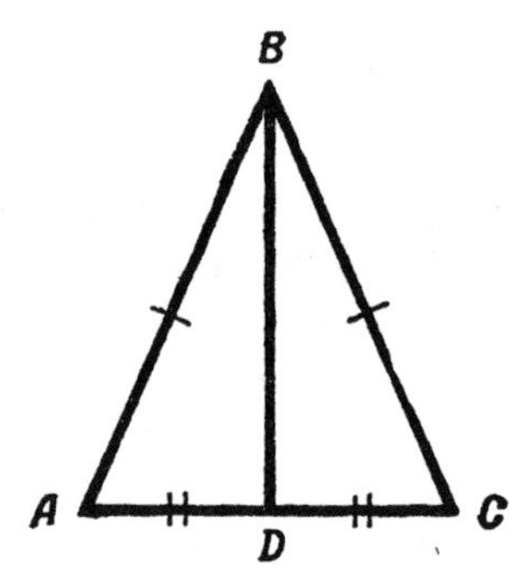

(D) Here are seven related terms: parallelogram, geometric figure, square, polygon, plane figure, rhombus, convex quadralateral. Arrange them so that the concept expressed by each term includes the concept expressed by the following term.

(E) The sum of the exterior angles of a convex polygon is 4 right angles. What is the largest number of acute interior angles a convex polygon can have?

281. REASONING WITHOUT EQUATIONS

Some problems, algebraic in appearance, can be solved by logic.

(A) A two-digit number, read from right to left, is 4½ times as large as from left to right. What is it?

1. It is greater than 9 because it has two digits.
2. It is less than 23 because 23 × 4½ is greater than 100.
3. It is an even number because it is an integer when it is multiplied by 4½.
4. Nine times half of it is its reverse, so its reverse is divisible by 9.

5. It has the same digits as its reverse, so it too is divisible by 9. (See Chapter VII.)

Complete the solution.

(B) The product of four consecutive integers is 3,024. What are the integers?

1. None of the four integers is 10. (Or the product would end in 0.)
2. At least one integer is less than 10. (Or the product would have at least five digits.)
3. Then all four integers are less than 10.
4. none of the four integers is 5. (Or the product would end in 5 or 0.)

Complete the solution.

282. A CHILD'S AGE

A child's age increased by 3 years gives a number which has an integral square root. Decreased by 3 years, the child's age gives the square root.

How old is the child? Find a similar relationship with a difference other than 3.

283. YES OR NO?

Ask a friend to pick a number from 1 through 1,000. After asking him ten questions that can be answered yes or no, you tell him the number.

What kind of questions?

X
Mathematical Games and Tricks

Games

284. ELEVEN MATCHES

On the table are 11 matches (or other objects). The first player picks up 1, 2, or 3 matches. The second player picks up 1, 2, or 3, and so on. The player who picks up the last match loses.

(A) Can the first player always win?

(B) Can he if there are 30 matches instead of 11?

(C) Can he in general, with n matches to be picked up 1 through p at a time (p not greater than n)?

285. WINNER PICKS UP THE LAST MATCH

You pick up 1 through 6 of 30 matches. The second player picks up 1 through 6 matches, and so on. The player who picks up the last match wins. How do you get to pick up the last match?

286. AN EVEN NUMBER WINS

Two players pick up 1 through 4 of 27 matches until they are all picked up. You are the first player. To win, you must have an even number of matches at the end.

How do you win this game?

287. WYTHOFF'S GAME

[This game was invented in 1907 by W. A. Wythoff, who apparently did not know that it had earlier been played in China as the game of *tsyanshidzi* (picking stones).–*M.G.*]

There are two piles of pebbles (or other objects). Players take turns picking up pebbles.

1. You can pick up any or all the pebbles in one pile, or
2. You can pick up pebbles from both piles, but only if you take the same number from each pile.

The player who takes the last pebble wins.

Winning positions include (1, 0) (Meaning 1 pebble in the first pile, none in the second) and (n, n). You pick up, respectively, 1 pebble (rule 1) and $2n$ pebbles (rule 2).

(1, 2) is a losing position: The table shows the position after each of *A*'s four possible moves. In each case, *B* can pick up all the remaining pebbles:

A	*B*	*A*	*B*	*A*	*B*	*A*	*B*
0	0	1	0	1	0	0	0
2	0	0	0	1	0	1	0

For example, in the third variant *A* picks up a pebble from the second pile (rule 1). *B* picks up both remaining pebbles (rule 2).

But though (1, 2) loses, not only (1, 0) and (1, 1) but any other $(1, n)$ wins. *A* simply picks up $(n - 2)$ pebbles from the second pile, leaving (1, 2).

Are there other losing positions besides (1, 2)?

288. HOW TO WIN?

Checkers are placed in squares *d, f,* and *h* of the 8-square board shown. A move for *A* or *B* is to move a checker to the left to any square, occupied or not, over another checker or checkers if desired. The player who puts the last checker on square *a* wins.

A can *always* win. See if you can show how.

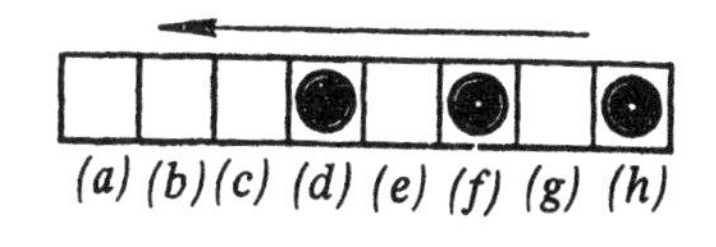

(No answer to this problem is given in the back of the book.)

289. MAKE SQUARES

Each player has an assortment of 18 cardboard pieces (*a* in the diagram shows the shapes and the number of each).

The board is a 6-by-6 square. There are many ways to fit pieces together to form a 2-by-2 square (see *b*). The heavy lines form 9 sections, and a player can only use 4 of these.

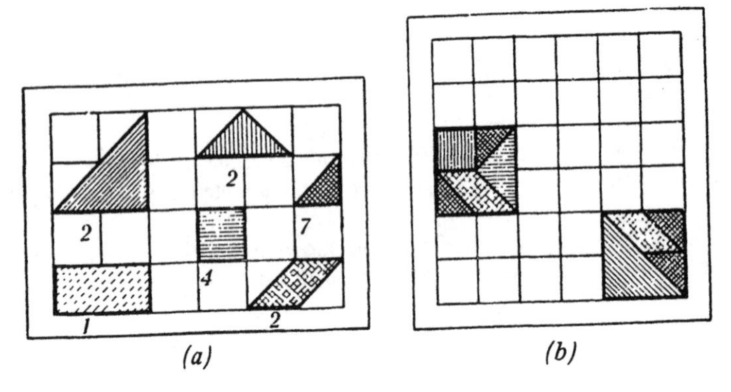

(a) (b)

Each move is placing a piece on the board, with no overlapping or moving of pieces. The player who makes the most 2-by-2 squares wins. (The maximum is 4.)

(No answer to this problem is given in the back of the book.)

290. WHO WILL BE FIRST TO REACH 100?

A calls 7, *B* 12, *A* 22, *B* 23, and so on. Each call is higher by any number from 1 through 10.

Whoever calls 100 wins. How does *A* win?

291. THE GAME OF SQUARES

A figure made up of preferably an odd number of small squares, drawn on checkered graph paper, is the field of play.

The border of the figure is considered already drawn. The two players take turns marking interior sides of squares. The player who draws the fourth side of a square initials it. He gets an extra move: he can and must draw one more line. He can take any number of squares in one turn.

The game ends when all the squares are complete. The player who has initialed the most squares wins.

The theory of the game is complex. No general strategy is known, but knowledge of some simple positions will give you an advantage.

1. The player who starts a 2-by-2 square loses all 4 squares. No matter which line *x* he draws, his opponent draws the appropriate *y* winning a square, then the next 2 lines (going counterclockwise in *a* of the diagram).

2. The figure in *b* is almost as bad. If *A* marks any line but *x* he loses all 5

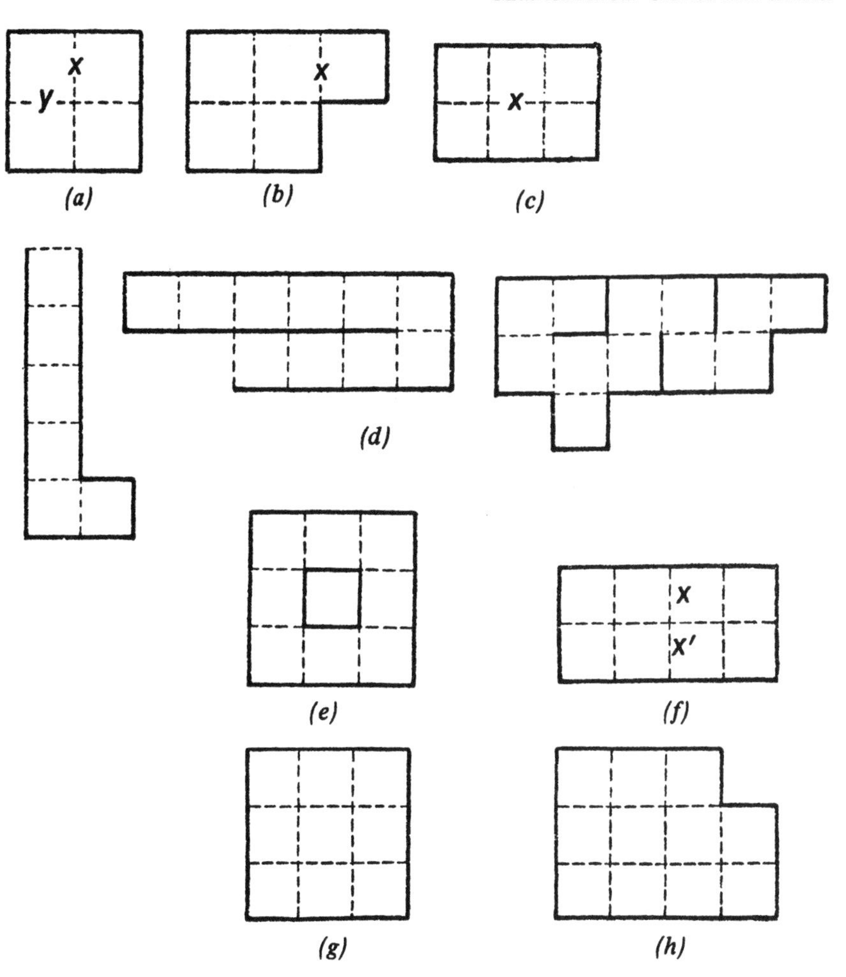

(a) (b) (c) (d) (e) (f) (g) (h)

squares; by marking x he at least wins 1 square, though, being forced to draw another line, he must start the remaining 2-by-2 square.

3. *A* can win all 6 squares in *c*, but only if he starts with x.

4. A channel 1 square wide *(d)*–no matter how it twists–is won completely by *A*. If the channel surrounds a completed square *(e)*, *B* takes all the squares.

5. *A* can win 4 squares in *(f)* if he first draws x of x^1. Any other opening move loses.

The art of the game is in breaking the field of play into simpler figures, then choosing which ones to start, and which ones to stay out of.

In *g*, *A* can win at least 8 squares by what first move?

On your own, how does *A* open *h*? How many squares can he win?

292. MANCALA

Mancala is an old folk game played throughout Africa. In one variant of mancala, the field of play has 12 hollows (or "holes"), each with 4 balls in it. [For a complete description of all forms of mancala, see H. J. R. Murray, *A History of Board-Games* (New York: Oxford University Press, 1952), Chapters 7 and 8.–*M.G.*]

One player sits on the *AF* side, the other on the *af* side.

A move consists of taking all the balls out of 1 hole on your side and distributing them 1 each into the subsequent holes. (The order of the holes is:

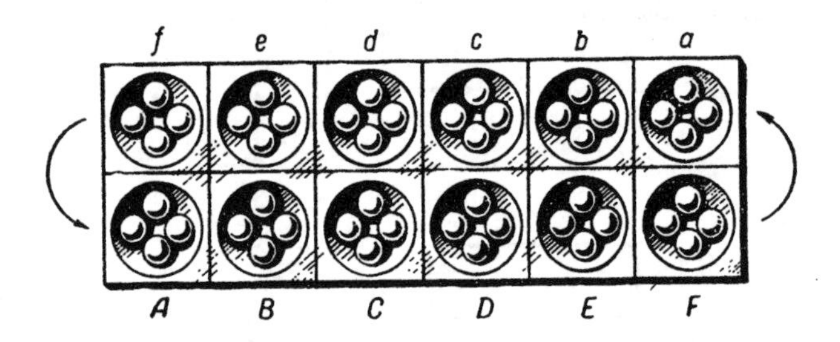

ABCDEFabcdef–counterclockwise. Say *P* empties hole *D*, placing 1 ball each in *E*, *F*, *a*, and *b*. Then *p* might empty *a* (which now holds 5 balls), distributing them to *b*, *c*, *d*, *e*, and *f*. The position is:

f	*e*	*d*	*c*	*b*	*a*
4	4	4	0	5	5
5	5	5	5	6	0
A	*B*	*C*	*D*	*E*	*F*

If 12 or more balls are taken out of a hole, the hole is skipped when it is reached–the twelfth ball goes in the next hole.

Each player tries to place the last ball of a set in his opponent's last hole *(f* of *F)* so that 2 or 3 balls are now in this hole. He then takes as loot all balls in consecutive preceding holes on his opponent's side that contain 2 or 3 balls. For example, consider this position:

f	*e*	*d*	*c*	*b*	*a*
2	1	2	3	1	2
0	0	0	0	0	6
A	*B*	*C*	*D*	*E*	*F*

1. *P* moves from *F* (his only move), giving:

f	*e*	*d*	*c*	*b*	*a*
3	2	3	4	2	3
0	0	0	0	0	0
A	*B*	*C*	*D*	*E*	*F*

P's last ball entered *f*, which contains 3 balls as loot for *P*. He also takes 2 and 3 balls from holes *e* and *d*. (He cannot skip over *c* to *b* and *a*.) He wins 8 balls.

2. In this position:

f	*e*	*d*	*c*	*b*	*a*
0	1	2	0	1	2
1	0	0	0	7	7
A	*B*	*C*	*D*	*E*	*F*

P, moving from *F*, wins nothing, since his last ball goes in *A*, on his own side of the board. Moving from *E*, he also wins nothing. His last ball enters *f* but does not result in 2 or 3 balls in that hole.

3. An empty hole does not guarantee safety:

f	*e*	*d*	*c*	*b*	*a*
0	0	0	0	0	0
1	0	0	0	0	17
A	*B*	*C*	*D*	*E*	*F*

All the holes on *P*'s side are empty, but *P* wins 12 balls. His move from *F* results in:

f	*e*	*d*	*c*	*b*	*a*
2	1	1	1	1	0
2	2	2	2	2	2
A	*B*	*C*	*D*	*E*	*F*

The last ball falls in *f*, and *P* takes all the balls from *P*'s side.

The game ends when the players agree there are not enough balls left to form loot, or when a player cannot make a move.

(No answer is given in the back of the book to this or the next two problems.)

293. AN ITALIAN GAME

This game, which combines some elements of poker and bingo, is played with a standard deck of cards.

Each player has a 5-by-5 square. As 25 cards are called, the player chooses a cell to enter each corresponding number in. (Enter kings as 13, queens as 12, jacks as 11, aces as 1.) When the square is filled, he scores each of twelve rows, columns, and main diagonals according to the table. The highest score wins.

A sample square is shown in the diagram. Player's pair of 5s is worth 10 points in the third row, but a pair of kings (13s) is worth 20 points on a diagonal.

	1	1	7	1	7	(80)
	2	10	2	13	2	(40)
	5	12	13	5	7	(10)
	3	3	3	11	3	(160)
	4	12	4	13	12	(20)
(20)	(50)	(10)		(10)	(10)	(160)

294. GAMES WITH ALMOST-MAGIC SQUARES

An almost-magic square is one whose rows and columns—but not necessarily diagonals—add up to the square's constant. (See magic squares, Chapter XII.) Games can be set up and played solitaire or by several people. The diagram shows three examples.

6	8	7	0	9
6	5	8	9	2
9	2	7	9	3
5	7	7	4	7
4	8	1	8	9

(a)

7	3	4	2
3	4	4	5
5	4	4	3
1	5	4	6

(b)

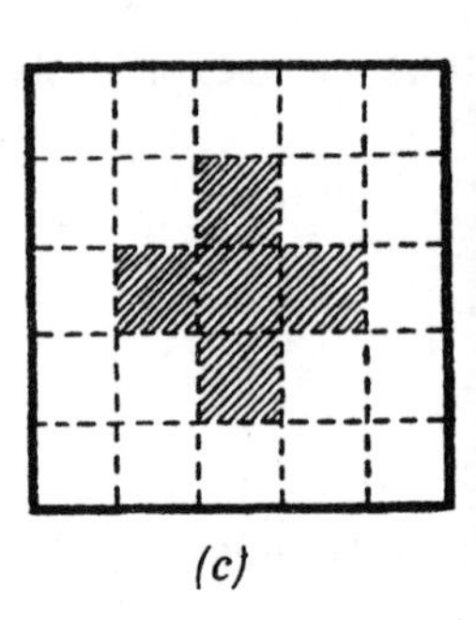

(c)

In *a,* players drew 5-by-5 squares. They were asked to use the numbers from 0 through 9 at least once each to fill the 25 cells. The almost-magic square shown has a constant of 30.

(Hint: It is easier to shift numbers on cardboard around than to keep erasing numbers.)

In *b,* players drew 16-cell squares. Numbers 1 through 7 were to be used at least once each so that, in addition to rows and columns, the shaded central cells would add to the constant (16, as filled in).

(Hint: The winner of a game can be the one with the highest constant.)

On your own, fill *c* with numbers 0 through 8 used at least once each (but use 0 and 6 only once). The shaded cells should add up to the same constant as the rows and columns.

295. NUMBER CROSSWORD PUZZLES

These puzzles are done just like crosswords, except that the Across and Down clues define numbers instead of words. Heavy lines (rather than black squares) indicate the ends of numbers.

(A) See the first two diagrams.

[1]		[2]	[3]
[4]	[5]		
[6]			[7]
[8]		[9]	

Across

1. Difference between a number of four consecutive increasing digits, and the same number reversed
4. Number with consecutive increasing digits
6. 3 Down times 8 Across
8. Prime number
9. Multiple of 13

Down

1. Cube of a digit in 1 Across
2. Two cubed, then 1 Across times 7 Down (last three digits)
3. 6 Across divided by 8 Across
5. Three consecutive digits
7. A factor of 3 Down times a factor of 1 Across

Let's get started. You may be surprised to learn that there is only one answer to 1 Across. But whether you subtract 1,234 from 4,321 or 6,789 from 9,876 the answer is 3,087. We show it filled in on the first row of the second diagram.

[1] 3	0	[2] 8	[3] 7
[4] 4	[5] 5	6	7
[6] 3			[7]
[8]		[9]	

(B) See the third diagram.

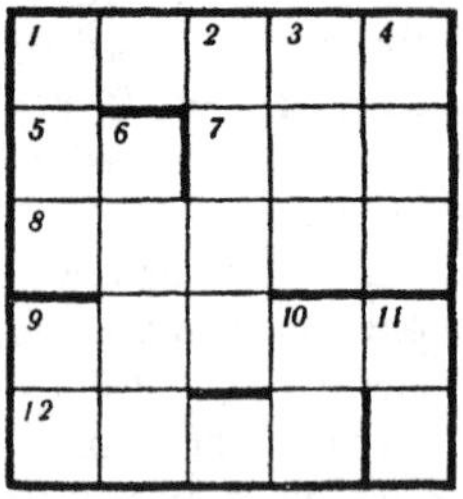

Across

1. Number with five different digits, none in common with 8 Across, which in turn consists of five different digits
5. Highest two-digit factor of 3 Down
7. 3 Down reversed
8. See 1 Across
9. One-ninth the sum of 1 Across and 8 Across
12. Product of three two-digit prime numbers, two of them factors of 6 Down

Down

1. First digit equals sum of next two
2. Year in second half of eighteenth century
3. Difference between 1 Across and 8 Across
4. Last digit is the product of first two digits
6. Reverse is multiple of 3 Down, and a product of three two-digit primes
9. Factor of 6 Down but not of 3 Down
10. Same as 5 Across
11. Smallest two-digit factor of 3 Down

(C) See the last diagram.

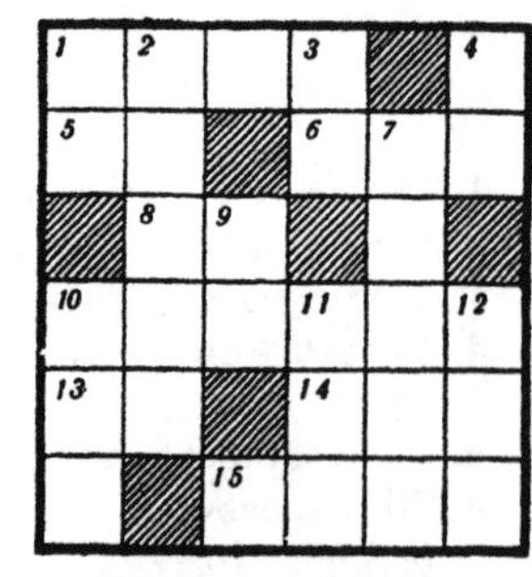

Across

1. Square of a prime number
5. Half the highest common factor of 10 Down and 11 Down
6. Cube of a square
8. Square root of 1 Across
10. A symmetrical square (same left to right as right to left)
13. Larger by 1 than 9 Down
14. Five times as large as 8 Across
15. Square of a number larger by 1 than 13 Across

Down

1. Smaller by 8 than the smallest integer which, when divided by 2, 3, 4, 5, and 6 has remainders 1, 2, 3, 4, and 5
2. Number whose digit sum is 29
3. Prime number
4. Prime factor of 11 Down
7. Quadrupled product of one-tenth 15 Across and 13 Across
9. Twice 4 Down
10. 11 Down reversed
11. Square root of 10 Across
12. Multiple of the highest prime factor of 13 Across

Tricks

296. GUESSING "THOUGHT" NUMBERS

We show seven ways to do the trick.

(A) Think of a number. Subtract 1. Double the difference and add the "thought" number. Tell me the result and I will guess the original number.

Method: Add 2 to the result and divide by 3. The quotient is the "thought" number.

Example: The "thought" number is 18: $18 - 1 = 17$; $17 \times 2 = 34$; $34 + 18 = 52$. I think: $52 + 2 = 54$; $54 \div 3 = 18$.

Proof: Call the "thought" number x: we get $x - 1$, $2(x - 1) + x = 3x - 2$.

Adding 2 gives $3x$, and dividing by 3 gives x.

(B) Ask a friend to think of a number. Tell him to multiply or divide by several numbers which you call out at random, without telling you the result.

Then ask him to divide by the "thought" number and then add the "thought" number. From his result, you immediately name the "thought" number.

The trick is simple. As he multiplies and divides, you do the same—only you start with 1. No matter how many multiplications and divisions are made, his result is yours multiplied by the "thought" number; when he divides by the "thought" number his result is the same as yours.

Then when he adds the "thought" number and calls out the result, he is calling out your result plus the "thought" number. Subtract your result and you have the number he thought of.

(C) For any odd number n, we will call $(\frac{1}{2}n + 1)$ the larger part of n. The larger part of 13 is 7, the larger part of 21 is 11, and so on.

Think of a number. Add half of it—or if it is an odd number add its larger part. Then add half the sum—or if the sum is odd add its larger part. Divide by 9 and announce the quotient. Announce if there is a remainder and whether it is less than, equal to, or greater than 5.

The "thought" number equals 4 times the quotient, plus:

0, if no remainder;

1, if remainder is less than 5;

2, if remainder is 5;

3, if remainder is greater than 5.

Example: The "thought" number is 15, then $15 + 8 = 23$, $23 + 12 = 35$, $35 \div 9 = 3$ (remainder 8). What is announced: "The quotient is 3, the remainder is more than 5."

You think: $(3 \times 4) + 3 = 15$, the "thought" number.

Prove by using algebra. Hint: Every "thought" number can be represented as $4n$, $4n + 1$, $4n + 2$, or $4n + 3$, where n is 0 or a positive integer.

(D) Again, think of a number, add half (or the larger part); to the sum add half (or the larger part). Do not divide by 9 as in (C), but announce all the digits in the answer, and their place in the answer, except one. (You are not allowed to conceal a 0.) Announce, also, which steps if any a larger part was used in.

Example: $28 + 14 = 42$, $42 + 21 = 63$. Conceal the 3. "The first digit is 6, and no larger parts were used."

To find the "thought" number, add the digits announced, and add:

0, if no larger part used;
6, if larger part used in first step only;
4, if larger part used in second step only;
1, if larger part used twice.

Subtract the sum from the next higher mutiple of 9. The result is the digit he concealed.

Now that you have his answer, divide by 9. Discarding any remainder, multiply by 4, and, as in (C), add these numbers:

0, if no remainder when you divided by 9;
1, if remainder was less than 5;
2, if remainder was 5;
3, if reaminder was greater than 5.

Example: "The first digit is 6, and no larger parts were used." Subtract 6 from 9: he concealed 3, and his answer was 63. Divide by 9 and get 7 (no remainder). Multiply by 4: the "thought" number was 28.

Another example: The "thought" number is 125, and $125 + 63 = 188$, $188 + 94 = 282$. Conceal the first 2. "The second and third digits are 8 and 2. A larger part was used in the first addition."

To find the "thought" number add 8 and 2, and 6 for the larger part used. Sixteen is 2 short of 18, so the answer is 282.

Divide by 9 and get 31 (remainder of 3). Multiply by 4 and add 1, getting 125.

Why are 6, 4, or 1 added when larger parts are used?

(E) Think of a number from 1 through 99. Square it. Add a number (announce it) to the "thought" number. Square the sum, and announce the difference between the squares.

To find the "thought" number, divide half the answer by the added number, and subtract half the added number.

Example: $53^2 = 2{,}809$. "I add 6." $59^2 = 3{,}481$; $3{,}481 - 2{,}809 = 672$. "The answer is 672."

To find the "thought" number, $336 \div 6 = 56$, $56 - \frac{1}{2}(6) = 53$.

Prove that it works.

(F) Think of a number from 6 through 60. Announce the remainders when the "thought" number is divided by 3; when it is divided by 4; when it is divided by 5.

To find the "thought" number, divide $(40r_3 + 45r_4 + 36r_5)$ by 60. (The r's are the announced numbers.) If the remainder is 0, the "thought" number is 60; otherwise, the remainder is the "thought" number.

Example: The "thought" number is 14. The remainders are $r_3 = 2, r_4 = 2, r_5 = 4$.

You calculate:

$S = (40 \times 2) + (45 \times 2) + (36 \times 4) = 314$; $314 \div 60 = 5$, with remainder 14. The "thought" number is 14.

Prove the trick by algebra.

(G) Once you understand the mathematical bases of the above tricks you can change them to suit your taste. For example, (F), instead of 3, 4, and 5 you can use 3, 5, and 7 with "thought" numbers from 8 through 105. How does the formula change?

Answer: $70r_3 + 21r_5 + 15r_7$, where r_3, r_5, and r_7 are the remainders after dividing the "thought" number by 3, 5, and 7. The "thought" number equals the remainder after dividing S by 105 (or it is 105 if the remainder is 0).

297. WITHOUT ASKING QUESTIONS

There are mathematical laws which enable you to determine the results of calculations using "thought" numbers without asking a single question.

(A) Say your friend's "thought" number is 6. Tell him to multiply it by 4 and add 5, but don't ask the result. Tell him to divide by 3. (He has 13 now, but doesn't tell you.)

In your mind you divide the first number you supplied by the third: $4 \div 3 = 1\frac{1}{3}$. Tell him to subtract $1\frac{1}{3}$ times his "thought" number. What is his answer?

In your mind you divide the second number you supplied by the third: $15 \div 3 = 5$. You know that his answer is $[13 - (6 \times 1\frac{1}{3}) = 5]$, although you haven't asked a single question.

How does it work?

(B) Each spectator picks a "thought" number from 51 through 100. In turn, you, as the "magician," write a number from 1 through 50 and seal it in an envelope.

In your mind, subtract your number from 99. Announce the result and tell the spectators to add it to their number, cross out the first digit of the sum, and add that same digit to the result. The answer is to be subtracted from their "thought" number.

Although they don't know it, their final answers are all the same and as, in turn, they look in the envelope, they see that you wrote down their final answer in advance.

How?

298. I FOUND OUT WHO TOOK HOW MANY

While you back is turned, A thinks of a number n and takes $4n$ coins (or other items) from a pile. B takes $7n$ and C takes $13n$. C gives A and B as many coins as each already has. B does the same to A and C, and A does the same to B and C.

Ask one of them how many coins he has. Divide by 2 and announce how many A took. Divide how many A took by 4, multiply by 7, and announce how many B took. Divide how many B took by 7, multiply by 13, and announce how many C took.

Explain.

299. ONE, TWO, THREE ATTEMPTS

Think of two positive integers, add their sum to their product, and tell me the result. As a jumper often succeeds only after several attempts, I, too, may not guess your "thought" numbers on the first try, but I will guess them.

My method is simple. I add 1 to your result. Then I factor the sum into all possible pairs (except 1 and the sum), and subtract 1 from each factor. This gives me a set of number pairs one of which is your "thought" number.

If I have only one pair, I can guess your numbers at once. For example, you think of 4 and 6. Adding their sum (10) to their product (24) gives 34. I add 1 to make 35. Since 35 has only the factor pair (5 x 7), I know your original numbers were 5 – 1 = 4 and 7 – 1 = 6.

Prove the correctness of this procedure.

300. WHO TOOK THE PENCIL, WHO TOOK THE ERASER?

I turned my back on two boys, Zhenya and Shura, and asked that one take a pencil and the other an eraser. I said: "Whoever is holding the pencil–your number is 7. Whoever is holding the eraser–your number is 9."

(One number should be prime, the other composite but not divisible by the first.)

"Zhenya, multiply your number by 2. Shura, multiply your number by 3."

(One number should be a divisor of the composite number you called out, as 3 is a divisor of 9. The other number should not share any factor but 1 with the first.)

"Add your products and tell me the sum."

If the sum is divisible by 3, Shura took the pencil. If not, Shura took the eraser. Why?

301. GUESSING THREE CONSECUTIVE NUMBERS

A friend picks three consecutive numbers (say, 31, 32, 33), none greater than 60. Ask him to announce a multiple of 3 that is less than 100 (say, 27). He is to add the four numbers and multiply by 67 (123 × 67 = 8,241). He announces the last two digits and you tell him the three "thought" numbers and the missing digits.

Method: Divide his multiple of 3 by 3 and add 1. Subtract the result from his announced two-digit number to get his first "thought" number: 41 – (9 + 1) = 31.

As for the missing digits, simply double the two-digit number he announced: 41 × 2 = 82.

Explain.

302. GUESSING SEVERAL "THOUGHT" NUMBERS

Think of several one-digit numbers. Multiply the first by 2, add 5, multiply by 5, and add 10. Add the second and multiply by 10; add the third and multiply by 10, and so on until you finally add the last "thought" number without multiplying by 10. Announce the answer and how many "thought" numbers there are.

To find the "thought" numbers, subtract 35 (if there are two "thought" numbers), or 350 (three numbers), or 3,500 (four numbers), and so on, from the answer. The digits of the result are the "thought" numbers.

Example: The "thought" numbers are 3, 5, 8, and 2. (2 × 3) + 5 = 11; (11 × 5) + 10 = 65; 10(65 + 5) = 700; 10(700 + 8) = 7,080; 7,080 + 2 = 7,082; 7,082 − 3,500 = 3,582.

Show how it works.

303. HOW OLD ARE YOU?

"You don't want to tell me? All right, just tell me the result if, from the number ten times as large as your age you subtract the product of any one-digit number and 9. Thank you, now I know your age."

Example: Age 17, product 3 × 9 = 27. Announce 170 − 27 = 143.

Method of guessing: Remove the last digit of the number called out and add this digit to what remains. In this case, 143 without 3 is 14, and 14 + 3 = 17.

(Use this trick only with those older than 9.)

Easy—and mystifying! But, to avoid embarrassment, learn why the trick works before you use it.

304. GUESS HIS AGE

To vary the last trick, ask him to multiply his age by 2, add 5, multiply by 5, and announce the answer.

Remove the last digit (which is always a 5), and subtract 2 from the remaining number to find his age.

Example: 21 × 2 + 5 = 47; 47 × 5 = 235. Then 235 becomes 23, and 23 − 2 = 21.

305. A GEOMETRICAL "VANISH"

Here is an amusing paradox that most people find hard to explain. If you move part of a diagram, a line segment will disappear before your very eyes.

Draw the 13 lines shown in the figure at left. Cut along *MN*. Slide the top piece to the left 1 line (figure at right).

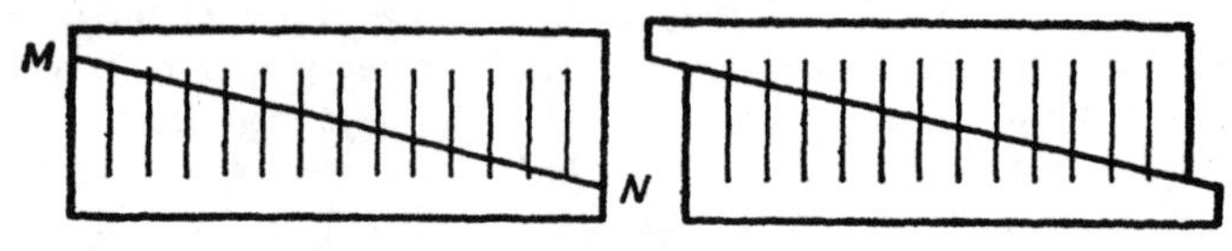

But wait! Where did the thirteenth line go?

XI
Divisibility

Among arithmetical operations the oddest is division. Consider division by 0. For other operations 0 is a number with the same rights as any number. Zero can be added, subtracted, and multiplied, but no number or algebraic expression can be divided by 0. Inattentiveness to this rule leads to "proofs" of absurd theorems.

Theorem: Any quality equals its half.

Proof: Let $a = b$. Multiply by a:

$$a^2 = ab.$$

Subtract b^2:

$$a^2 - b^2 = ab - b^2.$$

Factor:

$$(a + b)(a - b) = b(a - b).$$

Divide by $(a - b)$:

$$a + b = b.$$

Since $b = a$, we can substitute a for b: $2a = a$. Dividing by 2 means $a = \frac{1}{2} a$. A whole equals its half.

The error should be obvious.

Another oddity is that the sum, difference, or product of two integers is always an integer, but the quotient may not be.

The development of the theory of divisibility of integers led to a great expansion of the theory of numbers. Work on the problems of this chapter may stimulate you to study number theory.

306. THE NUMBER ON THE TOMB

Scholars discovered 2,520 in hieroglyphs engraved on the stone lid of a tomb in an Egyptian pyramid. Why was such an honor paid this number?

Perhaps because it is divisible by every integer from 1 through 10. It is the lowest number so divisible. Demonstrate this.

307. NEW YEAR'S GIFTS

Our union's executive committee was arranging a New Year's tree for the children. [There are no Christmas trees in the USSR officially, only "New Year's" trees.—*A.P.*] After distributing candies and cookies in the gift packages, we began the oranges. But we calculated that if we put 10 oranges to a package, one package would only get 9; if we put 9 oranges to a package, one package would only get 8; if 8, 7; if 7, 6; and so on down to 2 oranges per package with one package getting only 1 orange.

How many oranges did we have?

308. IS THERE SUCH A NUMBER?

Is there a number which when divided by 3 gives a remainder of 1; when divided by 4, gives a remainder of 2; when divided by 5, gives a remainder of 3; and when divided by 6 gives a remainder of 4?

309. A BASKET OF EGGS

A woman was carrying a basket of eggs to market when a passerby bumped her. She dropped the basket and all the eggs broke. The passerby, wishing to pay for her loss, asked:

"How many eggs were in your basket?"

"I don't remember exactly," the woman replied, "but I do recall that whether I divided the eggs by 2, 3, 4, 5, or 6, there was always 1 egg left over. When I took the eggs out in groups of 7, I emptied the basket."

What is the least number of eggs that broke?

310. A THREE-DIGIT NUMBER

I am thinking of a three-digit number. If you subtract 7 from it, the result is divisible by 7; if 8, divisible by 8; and if 9, divisible by 9. What is the number?

311. FOUR DIESEL SHIPS

Four diesel ships left a port at noon, January 2, 1953.

The first ship returns to this port every 4 weeks, the second every 8, the third every 12, and the fourth every 16.

When did all four ships meet again in the port?

312. THE CASHIER'S ERROR

The customer said to the cashier: "I have 2 packages of lard at 9 cents; 2 cakes of soap at 27 cents; and 3 packages of sugar and 6 pastries, but I don't remember the prices of the sugar and pastries."

"That will be $2.92."

The customer said: "You have made a mistake."

The cashier checked again and agreed.

How did the customer spot the error?

313. A NUMBER PUZZLE

Find the number t and the digit symbolized by a in:

$$[3(230 + t)]^2 = 492{,}a04.$$

314. A TEST OF DIVISIBILITY BY 11

It is not always necessary to test divisibility by actual division. We already know (Chapter VII) that if the digit sum of a number is divisible by 9, the number is divisible by 9. You also know that a number ending in 0 is divisible by 10; a number ending in 5 or 0 is divisible by 5; an "even number" (ending in 2, 4, 6, 8, or 0) is divisible by 2; but do you know this simple test of divisibility by 11?

Add the digits in even positions (second, fourth, and so on), and add the digits in odd positions (first, third, and so on). If the difference between sums is 0 or a multiple of 11, the number is divisible by 11. Otherwise it is not.

Let us introduce the term modulus, which is similar to "remainder." Thus 18 has a remainder of 7 when divided by 11, and 18 = 7 mod 11 ("seven modulo eleven") = −4 mod 11. The digits 0, 1, 2, . . . 9 are, of course, 0, 1, 2, . . . , 9 mod 11; but 0, 10, 20, . . . , 90 are, by actual division, 0, 10, 9, . . . , 2 mod 11 = 0, −1, −2, . . . , −9 mod 11; and 0, 100, 200, . . . , 900 are again 0, 1, 2, . . . , 9 mod 11, and so on.

The sum of two numbers has the same modulus as the sum of the moduli of the numbers, so, given a number:

$$N = a + 10b + 100c + 1{,}000d = \ldots;$$

$$\begin{aligned} N \bmod 11 &= a \bmod 11 + 10b \bmod 11 + 100c \bmod 11 + 1{,}000d \bmod 11 + \ldots \\ &= a \bmod 11 - b \bmod 11 + c \bmod 11 - d \bmod 11 + \ldots \\ &= a \bmod 11 + c \bmod 11 + \ldots - (b \bmod 11 + d \bmod 11 + \ldots. \end{aligned}$$

Since $a, b, c, d, \ldots$ are the digits of N, starting from the right end, our test of divisibility by 11 is correct.

Now tell me:

(A) If $37{,}a10{,}201$ is divisible by 11, what is the digit a?

(B) If $[11(492 + x)]^2 = 37{,}b10{,}201$, what are digit b and number x?

315. A TEST OF DIVISIBILITY BY 7, 11, AND 13

Seven, 11, and 13 are three consecutive prime numbers. Their product is 1,001; and when a product is that close to a power of 10, a test of divisibility is not far away.

The test is: Separate the number from right to left in groups of 3 digits. (The commas conventionally printed in a number do this for you.) Add the groups in even positions and in odd positions. If the difference between sums is divisible by 7, 11, or 13, the entire number is divisible by 7, 11, or 13 respectively. (If the difference is 0, the number is divisible by 7, 11, and 13.)

Example: Separate 42,623,295 into 42, 623, and 295:

$$623 - (42 + 295) = 286.$$

Since 286 is divisible by 11 and 13 but not by 7, then 42,623,295 is divisible by 11 and 13 but not by 7.

Now 1,000 = $10^3 + 1$, and this is a factor of $10^6 - 1$ and of $10^9 + 1$. Using these facts, can you demonstrate that the test works on numbers which separate into four groups?

Can you demonstrate the test in general, on your own, in two ways? One can be based on the solution to the problem just given. One can be based on the similarity to Problem 314. (Try using modulo 1,001 instead of modulo 11.)

316. SHORTENING THE SHORTCUT FOR DIVISIBILITY BY 8

Since 10 is divisible by 2; 100 by 4; 1,000 by 8; 10,000 by 16; and so on, we have tests like these:

If the last digit of a number is divisible by 2, the number is divisible by 2. (The rest of the number is divisible by 10 and so is divisible by 2.)

If the number formed by the last two digits of a number is divisible by 4, the entire number is divisible by 4.

If the number formed by the last three digits of a number is divisible by 8, the entire number is divisible by 8.

But it is easier to divide a two-digit number by 4 than to divide a three-digit number by 8, so we supply a shortcut of the shortcut:

Add the two-digit number which begins a three-digit number to half the last digit. If the sum is divisible by 4, the three-digit number is divisible by 8. Example (with 592):

$$59 + 1 = 60; 60 \div 4 = 15;$$
$$592 \div 8 = 74.$$

Prove the test is correct.

(We must admit that for ten even numbers of 968 or more, you will have to test a three-digit number for divisibility by 4, but this three-digit number is never greater than 103.)

317. A REMARKABLE MEMORY

When a friend writes a three-digit number, you quickly add three or even six more digits so that the resulting six- or nine-digit number will be divisible by 37.

Suppose he writes 412. Add 143 on the left or right to form 143,412 or 412,143. Both numbers are divisible by 37.

The explanation does not lie in your remarkable phenomenal memory for numbers divisible by 37. You may have a most ordinary memory, but you do know a simple test of divisibility by 37:

Separate a number from right to left into groups of three digits (the last group on the left may be incomplete). Consider each group an independent number. Add the numbers. If the sum is divisible by 37, the entire number is divisible by 37. For example, 153,217 is divisible by 37 because $153 + 217 = 370$, and 370 is divisible by 37.

On your own, prove this. (Hint: 37 is a factor of $999 = 10^3 - 1$.)

To do the trick rapidly, note that 111, 222, 333, . . . 999 are all divisible by 37. You tacked 143 onto your friend's 412 because they add to 555. If he chose 341, you would add 103, or 214, or 325, and so on.

To make a nine-digit number, pretend you are making a six-digit number and split the three-digit "tack" in two. Instead of adding 325 onto 341, add 203 and 122 (sum 325): 203,341,122, which is divisible by 37.

Prove that a nine-digit number whose three triplets of digits have a sum of the form *AAA* (three identical digits) is divisible by 37.

318. A TEST OF DIVISIBILITY BY 3, 7, AND 19

The product of the prime numbers 3, 7, and 19 is 399. If a number $100a + b$ (where b is a two-digit number and a is any positive integer) is divisible by 399 or by any of its divisors, then $a + 4b$ is divisible by the same number.

On your own, can you prove this? (Hint: Use $400a + 4b$ as a link.) Can you formulate and prove its converse?

Devise a simple test of divisibility by 3, 7, and 19.

319. THINGS OLD AND NEW ABOUT DIVISIBILITY BY 7

The Russian people like the number 7. You will find it in our folk songs and proverbs:

Measure cloth seven times before you cut it once.

Seven misfortunes, one reckoning.

One plows, seven come along with spoons (about idlers who like to eat at the expense of those who work).

A baby had seven nurses, yet lost his eye.

You already know two tests of divisibility by 7 (in conjunction with other numbers). There are many other such tests. Here is one:

Multiply the first digit on the left by 3 and add the second digit. Multiply by 3 and add the third digit, and so on until you add the last digit.

To simplify your calculations, whenever a result is 7 or more subtract the highest multiple of 7 that gives 0 or a positive integer. If and only if the final result is divisible by 7, the given number is divisible by 7.

Example: Test 48,916:

$$\begin{array}{ll} 4 \times 3 = 12, & 12 - 7 = 5; \\ 5 + 8 = 13, & 13 - 7 = 6; \\ 6 \times 3 = 18, & 18 - 14 = 4; \\ 4 + 9 = 13, & 13 - 7 = 6; \\ 6 \times 3 = 18, & 18 - 14 = 4; \\ 4 + 1 = 5; & \\ 5 \times 3 = 15, & 15 - 14 = 1; \\ 1 + 6 = 7. & \end{array}$$

Therefore, 48,916 is divisible by 7.

On your own, prove the validity of the test. [Hint: Is $a + 10b + 10^2c + \ldots - (a + 3b + 3^2c + \ldots)$ divisible by 7?]

(No answer to this problem is given in the back of the book.)

320. A SECOND TEST OF DIVISIBILITY BY 7

Do the same as in Problem 319, but work from right to left with 5 as multiplier.

Example: Test 37,184:

$$\begin{array}{ll} 4 \times 5 = 20, & 20 - 14 = 6; \\ 6 + 8 = 14, & 14 - 14 = 0; \\ 0 \times 5 = 0; & \\ 0 + 1 = 1; & \\ 1 \times 5 = 5; & \\ 5 + 7 = 12, & 12 - 7 = 5; \\ 5 \times 5 = 25, & 25 - 21 = 4; \\ 4 + 3 = 7. & \end{array}$$

Therefore, 37,184 is divisible by 7. Can you prove that the test works? (No answer to this problem is given in the back of the book.)

321. TWO UNUSUAL SEVEN-DIVISION THEOREMS

Theorem 1: If a two-digit number represented as AB is divisible by 7, then $BA + A$ is divisible by 7.

For example 14 is divisible by 7, and 41 + 1 is divisible by 7.

(Note: In comparing $10a + b$ with $10b + 2a$, try multiplying the first by 2 and the second by 3.)

Theorem 2: If a three-digit number represented as ABC is divisible by 7, then $CBA - (C - A)$ is divisible by 7.

For example, 126 is divisible by 7, and 621 − (6 − 1) = 616 is divisible by 7.

Or 693 is divisible by 7, and 396 − (3 − 6) = 399 is divisible by 7.

(No answer to this problem is given in the back of the book.)

322. GENERAL TESTS OF DIVISIBILITY

When we tested for divisibility by 11 = 10 + 1 (problem 314), we added and subtracted alternate digits (each of which could be called a group of one digit).

When we tested for divisibility by 1,001 = 10^3 + 1, as well as by its prime factors 7, 11, and 13 (Problem 315), we added and subtracted alternate groups of three digits.

Similarly, to test for divisibility by 101 = 10^2 + 1, we add and subtract alternate groups of two digits. To test for divisibility by 10,001 = 10^4 + 1, and by its prime factors 73 and 137, we add and subtract alternate groups of four digits.

For example, test 837,362,172,504,831. We make four-digit groups 837,3621,7250,4831. The odd-numbered groups make 837 + 7,250 = 8,087; the even-numbered, 3,621 + 4,831 = 8,452. The difference of the sums is 365 = 73 × 5. So the fifteen-digit number is divisible by 73 but not by 137.

In general, to test for divisibility by 10^n + 1, and by its smaller prime factors if any, add and subtract alternate groups of n digits, working from right to left.

The test for divisibility by 9 + 10 − 1 was basically similar (Chapter VII). We added all the digits (each of which could be called a group of one digit). It will not surprise you that if the sum is divisible by 3 (a prime factor of 9), the number is divisible by 3.

We can make similar tests for 99 = 10^2 − 1, 999 = 10^3 − 1, and so on. To simplify, we divide each of these numbers by 9 (which we already have a test for). Divisibility by other prime factors is unaffected.

Thus we have a second test for divisibility by 11: Add the groups of two digits (right to left).

The test for divisibility for 111 (and its relevant prime factor 37) is to add the groups of three digits, as we did in Problem 317.

The number 1,111 = 101 × 11 gives us nothing new, but 11,111 = 271 × 41 gives us a test for these two primes.

In general, to test for divisibility by $\frac{1}{9}(10^n - 1)$, and by its smaller prime factors if any, add the groups of n digits, working from right to left.
(No answer to this problem is given in the back of the book.)

323. A DIVISION CURIOSITY

To conclude this chapter, I present four arrangements of the ten digits:

2,438,195,760; 4,753,869,120; 3,785,942,160; 4,876,391,520.

Each is divisible by 2, 3, 4, 5, 6, 7, 8, 9, 10, 11, 12, 13, 14, 15, 16, 17, and 18.
(No answer to this problem is given in the back of the book.)

XII
Cross Sums and Magic Squares

Cross Sums

Let us arrange the digits from 1 through 9 in two rows with equal sums. They cannot be parallel rows, because the sums of the nine digits (45) is not divisible by 2. But they can be intersecting rows:

```
    5
    9
3 7 1 8 4
    6
    2
```

Each row adds to 23.

We will call such intersecting rows with identical sums "cross sums," by analogy with crosswords. After solving the cross-sum problems offered, see if you can invent some of your own, symmetrical if possible. Most cross-sum problems have more than one solution.

324. A STAR

Can you place the integers from 1 through 12 in the circles of the 6-pointed star so that the sum of the numbers in each of the six rows is 26?

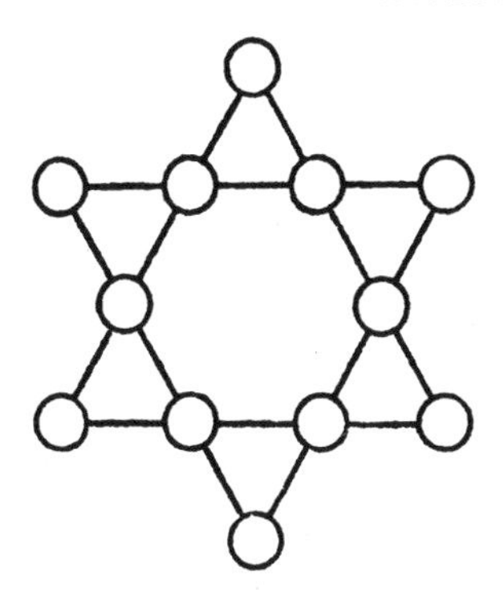

325. A CRYSTAL

We have a crystal lattice whose "atoms" are joined in ten rows of 3 atoms each. Select thirteen integers, of which twelve are different, and place them in the "atoms" so that each row totals 20. (The smallest number needed is 1, the largest 15.)

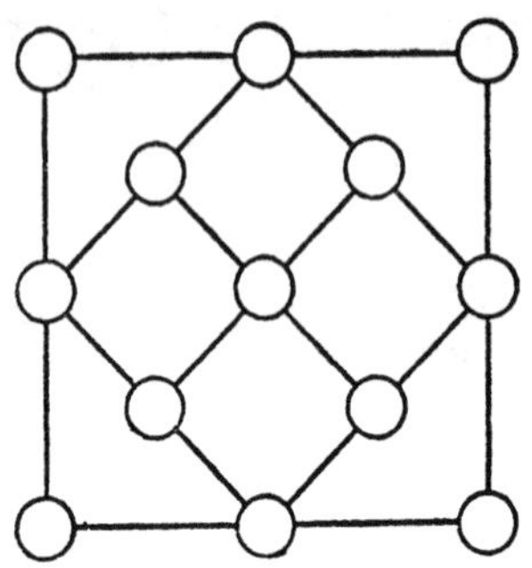

326. AN ORNAMENT FOR A WINDOW

A store selling semiprecious stones used a 5-pointed star made of circular spots held together by wire.

The fifteen spots hold from 1 through 15 stones (each number used once). Each of the five circles holds 40 stones, and at the five ends of the star there are 40 stones.

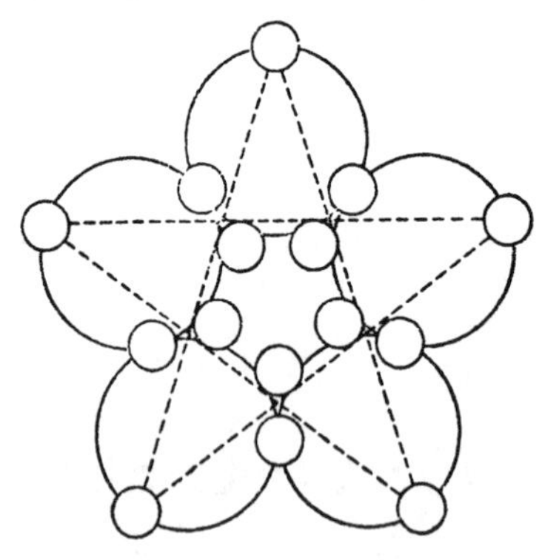

327. THE HEXAGON

Enter integers from 1 through 19 in the spots of the hexagon so that each row of three (on the rim, and outward from the center) adds to 22.

Rearrange to add to 23.

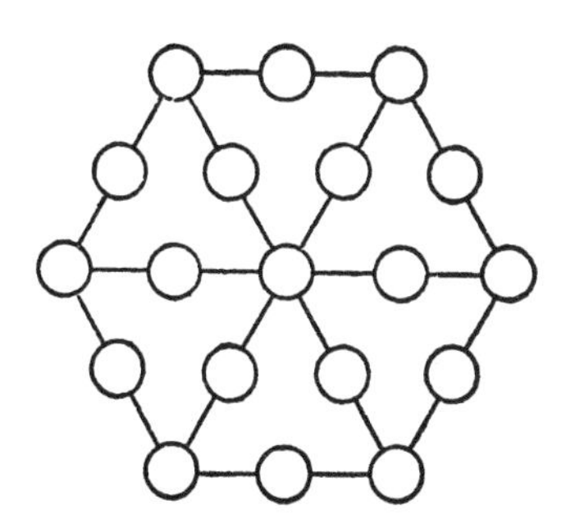

328. A "PLANETARIUM"

In a small "planetarium" there are 4 "planets" in each orbit and 4 along each radius (diagram, left). The planetarium has 5 planets in each orbit and 5 along each radius

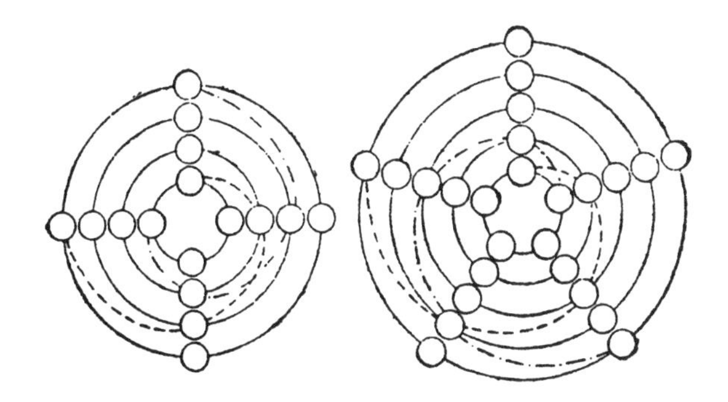

(diagram, right). Weights of planets in the small planetarium are expressed by integers from 1 through 16; in the large planetarium, from 1 through 25.

Arrange the weights so that each planetary system is in equilibrium, with sums of 34 for the small planetarium and 65 for the large, for:

1. Weights along each radius.
2. Weights around each orbit.
3. Weights spiraling in from the outside orbit to the inside orbit in both directions (see dotted lines for examples).

And (small planetarium only) in each adjacent pair of orbits of each adjacent pair of arms. The four weights add to 34.

There are 28 identical sums to be formed in the small planetarium and 20 in the large.

Surprisingly, there are many solutions.

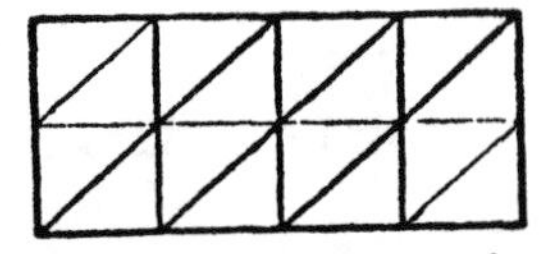

329. OVERLAPPING TRIANGLES

The rectangular ornament consists of 16 small triangles, which are to contain the integers from 1 through 16. Can you see the 6 larger, overlapping right triangles they form? The sum of integers in each of these triangles is to be 34.

330. INTERESTING GROUPINGS

In the large triangle, find the overlapping 3 triangles with 4 cells and 3 trapezoids with 5 cells. The cells contain 1 through 9 so that each triangle adds to 17 and each trapezoid to 28.

Find four arrangements with triangles of 20 and trapezoids of 25.

Find one arrangement with triangles of 23 and trapezoids of 22.

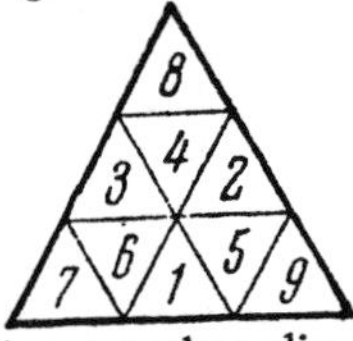

Hint: To avoid constant erasing, number slips of paper and slide them about from cell to cell until you get the desired result.

Magic Squares

331. WAYFARERS FROM CHINA AND INDIA

The magic square is an ancient and beautiful kind of cross sum.

The Chinese apparently invented magic squares: they are mentioned in a Chinese book written four to five thousand years before our era.

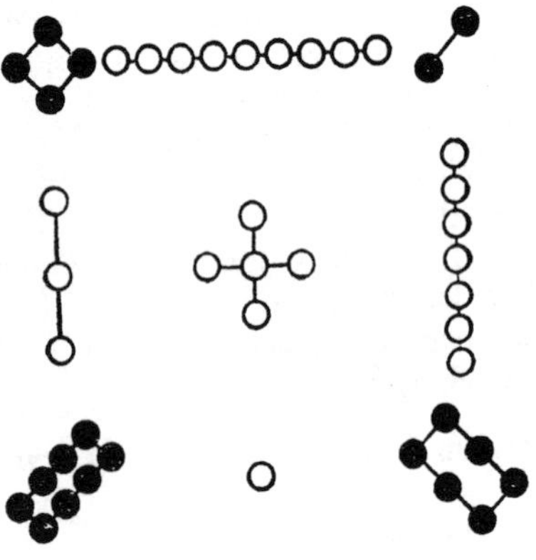

The world's oldest magic square, a Chinese creation, is shown in the first diagram. The black circles depict even (feminine) numbers or yin; the white circles, odd (masculine) numbers or yang. The numerals of the square are also shown. The first nine numbers are arranged to add to 15 across each row, down each column, and along each main diagonal. Thus 15 is the magic constant of the square.

The 4-by-4 magic square shown is two thousand years old and comes from India. The first sixteen numbers are used, with a magic constant of 34.

4	9	2
3	5	7
8	1	6

1	14	15	4
12	7	6	9
8	11	10	5
13	2	3	16

Magic squares reached Europe at the beginning of the fifteenth century. A magic square appears in one of Albrecht Dürer's best engravings, "Melencolia I" (1514). His square is the one from India, slightly altered.

Let us look at six more properties of our square:

1. The corners add to 34.
2. The five 2-by-2 squares in the corners and the center add to 34.
3. In each row one pair of adjacent numbers adds to 15, and the other to 19.
4. Add the squares of the numbers in each row:

$1^2 + 14^2 + 15^2 + 4^2 = 438$ and $13^2 + 2^2 + 3^2 + 16^2 = 438$;
$12^2 + 7^2 + 6^2 + 9^2 = 310$ and $8^2 + 11^2 + 10^2 + 5^2 = 310$.

Outer and inner pairs of equal sums resulted.

5. The same is true of the columns. Each outer colunn's squared numbers add to 378; each inner column's, to 370.

6. Draw on the square (*a* in the diagram) a smaller square, as shown by the broken lines. Take pairs of opposite sides in the new square. The sums are 34:

$12 + 14 + 3 + 5 = 15 + 9 + 8 + 2.$

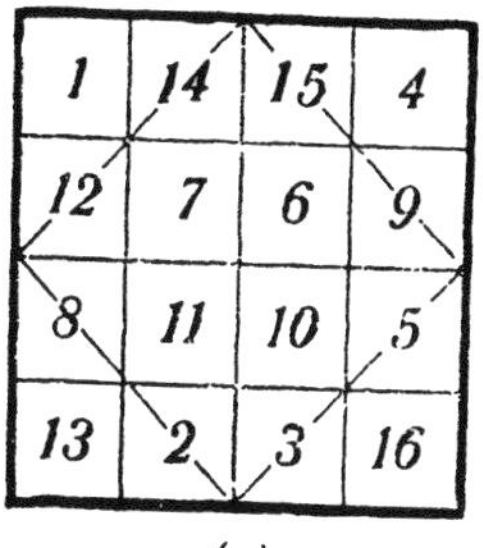

1	14	15	4
12	7	6	9
8	11	10	5
13	2	3	16

(a)

12	7	6	9
1	14	15	4
8	11	10	5
13	2	3	16

(b)

And the sums of the squares and of the cubes of these numbers are equal:

$$12^2 + 14^2 + 3^2 + 5^2 = 15^2 + 9^2 + 8^2 + 2^2.$$
$$12^3 + 14^3 + 3^3 + 5^3 = 15^3 + 9^3 + 8^3 + 2^3.$$

When we exchange two rows (*b* in diagram), the rows and columns still, of course, add to 34, but the main diagonals do not. It is now a semimagic square.

Problem: Interchange rows and columns of the magic square from India to produce a magic square with these properties:

1. The sums of the squares of the numbers on the main diagonals are equal.
2. The sums of the cubes of the numbers on the main diagonals are equal.

332. HOW TO MAKE A MAGIC SQUARE

A magic square of the third order has 3 cells in each row, one of the fourth order has 4 cells in each row, and so on. Hundreds of clever methods of forming magic squares of orders above 3 have been devised.

Squares of odd order: Let us use one of the known methods for forming a square of order 5. The same method will produce squares of the third, seventh, or any other odd order.

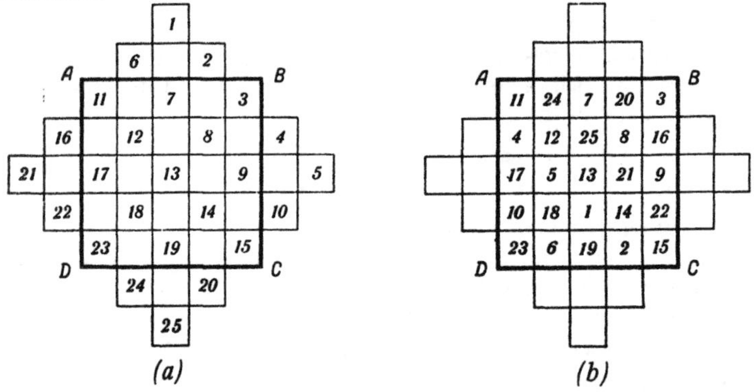

(a) (b)

Draw the order-5 matrix *ABCD* (*a* in diagram above) with 25 cells, and on each side add 4 cells as shown. Enter the integers from 1 through 25 on five slanting lines. Each integer outside *ABCD* is moved 5 (*n* for order *n*) cells along its row or column into the square. For example, 6 is moved under 18, 24 over 12, 16 to the right of 8, and 4 to the left of 12.

The result is the magic square in *b*, with the magic constant 65. Each number plus its "opposite" (on the other side of the central cell in the same relation to the central cell) is 26:

$$1 + 25 = 19 + 7 = 18 + 8 = 23 + 3 = 6 + 20 = 2 + 24 = 4 + 22,$$

and so on. It is a symmetric square.

Form magic squares of the third and (on your own) seventh orders, using the method just described.

Squares of orders that are multiples of 4: Here is one rather easy method:

1. Number the cells consecutively, as shown in the 4-by-4 square *(a below)* and the 8-by-8 square *(c)*.
2. Divide the square with two vertical and two horizontal lines so that in each corner there is a square of order $n/4$ and in the center a square of order $n/2$.

1	2	3	4
5	6	7	8
9	10	11	12
13	14	15	16

(a)

16	2	3	13
5	11	10	8
9	7	6	12
4	14	15	1

(b)

1	2	3	4	5	6	7	8
9	10	11	12	13	14	15	16
17	18	19	20	21	22	23	24
25	26	27	28	29	30	31	32
33	34	35	36	37	38	39	40
41	42	43	44	45	46	47	48
49	50	51	52	53	54	55	56
57	58	59	60	61	62	63	64

(c)

64	63	3	4	5	6	58	57
56	55	11	12	13	14	50	49
17	18	46	45	44	43	23	24
25	26	38	37	36	35	31	32
33	34	30	29	28	27	39	40
41	42	22	21	20	19	47	48
16	15	51	52	53	54	10	9
8	7	59	60	61	62	2	1

(d)

3. Within these 5 squares, interchange all pairs of numbers symmetrically opposite the square's center. Outside the 5 squares, leave the numbers as they are.

The results are shown in *b* of the 4-by-4 and 8-by-8 diagrams. Magic squares so formed are symmetric.

Two problems to be done on your own:

In forming a magic square of order $4ka$, we reverse step 3. We leave the numbers in the 5 squares as is. In the remaining four rectangles we interchange all pairs of numbers symmetrically opposite the square's center. Result: a magic square.

Form a magic square of order 12.

Squares of even orders that are not multiples of 4: To make magic squares of orders 6, 10, 14, 18, . . . , one of the best methods is to put a frame around a magic square of order $4n$, as shown here. Within the original square (here the order-4

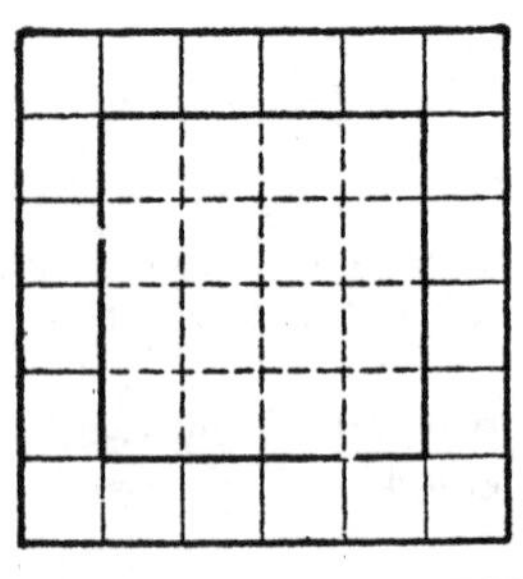

(a)

1	9				2
6					
10					
					7
					8
		3	4	5	

(b)

1	9	34	33	32	2
6					31
10					27
30					7
29					8
35	28	3	4	5	36

(c)

26	12	13	23
15	21	20	18
19	17	16	22
14	24	25	11

(d)

1	9	34	33	32	2
6	26	12	13	23	31
10	15	21	20	18	27
30	19	17	16	22	7
29	14	24	25	11	8
35	28	3	4	5	36

square we made before) each number is raised by $(2n - 2)$, where n is the order of the square we wish to form (here, 6). In this case, 1 becomes 1 + 10 = 11, 2 becomes 12, 3 becomes 13, and so on. The new order-4 square is *a* in the last diagram. It is always possible to place 1 through 10 and 27 through 36 (see *b*) so that the result is a magic square with magic constant $(n^3 + n)/2$. Here $n = 6$, so the constant is 111.

On your own, make another magic square of order 6 and a magic square of order 10.

333. TESTING YOUR WITS

In the white cells of the order-7 square arrange the integers from 30 through 54 so that each row and column adds to 150, and the main diagonals add to 300. Instead of placing numbers at random, try to work according to a system.

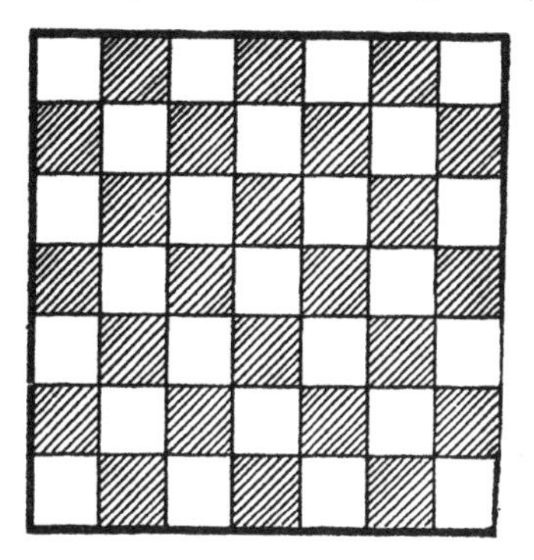

334. THE MAGIC GAME OF "15"

A square box contains blocks numbered from 1 through 15. There is an empty space. The usual way to play is to place the blocks in the box at random, then try to slide them around until the numbers are in consecutive order (first figure).

This game is not very interesting. Its mathematical value can be enriched by introducing an additional requirement: Move the blocks until you form a magic square (the empty space is 0).

Arrange the blocks with 14 and 15 interchanged, as shown in the second figure. Try to obtain a magic square with constant 30 in no more than 50 moves.

1	2	3	4
5	6	7	8
9	10	11	12
13	15	14	

A magic square can be formed from the layout of the first figure but it will not be the same as a magic square formed from the second. In fact, the second figure cannot be reached from the first. Half the possible positions are based on the first, half on the second, according to research done in the last half of the nineteenth century, when the "15" game was a craze throughout Europe.

There is a curious way to determine whether a given position belongs to the family of the first figure or of the second. Lift pairs of blocks out of the box and exchange their positions before putting them back. It is not difficult to choose blocks so that you can establish consecutive order (the first figure) in rather few moves. If the number of moves required is even, the position was of the first family; if odd, of the second.

335. UNORTHODOX MAGIC SQUARES

The usual magic square of order n contains the integers from 1 through n^2, each number appearing once. In this problem the cells may be filled with any numbers.

1	2	3	4
5	6	7	8
8	7	6	5
4	3	2	1

(A) Let a 4-by-4 square be filled, as shown in the figure below, with integers from 1 through 8, each occurring twice. Rearrange the numbers to form a magic square with constant 18. These should also add to 18:

1. The four corners.
2. Each of the nine 2-by-2 squares. No such square may contain the same number twice.
3. The four corners of each of the four 3-by-3 squares. No set of four corners may contain the same number twice.

(B) Using the odd numbers from 1 through 31, an order-4 magic square with constant 64 and additional properties:

1. The four corners of the 4-by-4 square, the four 3-by-3 squares, the nine 2-by-2 squares, and the six 2-by-4 rectangles must add to 64.
2. Draw a tilted square whose corners are the centers of the sides of the given square. Each pair of opposite sides must add to 64.
3. The sums of the squares of the numbers in two rows must be equal, and the sums of the squares of the numbers in the other two rows must be equal.
4. The sums of the squares of the numbers in two columns must be equal, and the sums of the squares of the numbers in the other two columns must be equal.

96	11	89	68
88	69	91	16
61	86	18	99
19	98	66	81

(C) A prankster devised a magic square with the constant 265 (see figure above). I call it a topsy-turvy square. Why?

336. THE CENTRAL CELL

Form a magic square of order 3, using 1 through 9. It will either be the square shown in the first diagram or one of its three rotations or a mirror image of one of the four rotational positions.

4	9	2
3	5	7
8	1	6

$S = 15$

Prove that the central cell is one-third the magic constant, and that in an orthodox magic square of order 3 the central cell is always 5. (Use the notation of the diagram below.)

a_1	a_2	a_3
a_4	a_5	a_6
a_7	a_8	a_9

337. AN ARITHMETICAL CURIOSITY

Many curious relationships hold among integers. Consider this set: 1, 2, 3, 6, 7, 11, 13, 17, 18, 21, 22, 23. At first glance, they do not seem remarkable. Separate them into two groups:
1, 6, 7, 17, 18, 23; and 2, 3, 11, 13, 21, 22.

Compare the sums of the numbers:
$1 + 6 + 7 + 17 + 18 + 23 = 72; 2 + 3 + 11 + 13 + 21 + 22 = 72.$

Compare the sums of the squares of the numbers:
$1^2 + 6^2 + 7^2 + 17^2 + 18^2 + 23^2 = 1{,}228;$
$2^2 + 3^2 + 11^2 + 13^2 + 21^2 + 22^2 = 1{,}228.$

The sums of the cubes, the fourth powers, and the fifth powers are also equal.

If you increase or diminish all twelve numbers by the same integer, the properties do not change. For example, subtract 12 from each number leaving:

−11, −6, −5, 5, 6, 11; and −10, −9, −1, 1, 9, 10.

Since in each group the negative numbers exactly match the positive ones, not only sums of the numbers are equal, but the sums of the cubes and fifth powers. Checking the squares and even the fourth powers is not difficult.

This formula gives you as many sets of twelve such numbers as you wish:

$$(m-11)^n + (m-6)^n + (m-5)^n + (m+5)^n + (m+6)^n + (m+11)^n = (m-10)^n + (m-9)^n + (m-1)^n + (m+1)^n + (m+9)^n + (m+10)^n,$$

where m is any integer and $n = 1, 2, 3, 4$, or 5.

(No answer to this problem is given in the back of the book.)

338. REGULAR MAGIC SQUARES OF ORDER 4

Numbers 1 through 15 can be represented as the sum of one or more digits from the set 1, 2, 4, 8, without repetitions: $1 = 1, 2 = 2, 3 = 1 + 2, 4 = 4, 5 = 1 + 4$, and so on through $15 = 1 + 2 + 4 + 8$.

If we take an order-4 magic square and reduce each number by 1, its cell will contain numbers 0 through 15 (first diagram). Draw the matrices of four order-4

9	14	2	5
15	4	8	3
0	11	7	12
6	1	13	10

1			1
1			1
	1	1	
	1	1	

	2	2	
2			2
	2	2	
2			2

	4		4
4	4		
		4	4
4		4	

8	8		
8		8	
	8		8
		8	8

squares. In the first matrix, put a 1 in each cell corresponding to a number on the magic square that is expressed by a sum (as explained) that includes digit 1. In the second matrix, do the same with 2s; the third, with 4s; and in the fourth, with 8s. The results are shown in the four squares.

0	4	15	11
9	13	2	6
14	10	5	1
7	3	8	12

		1	1
1	1		
		1	1
1	1		

		2	2
		2	2
2	2		
2	2		

	4	4	
	4		4
4		4	
4			4

		8	8
8	8		
8	8		
		8	8

An order-4 magic square is called regular if each of these four matrices is itself a magic square. Thus the square in the first diagram above is regular because the four matrices are all magic. The square in the diagram above is irregular because its second and third matrices (of the four matrices in the last diagram) are not magic along their main diagonals.

It has been shown that there are 528 regular order-4 magic squares, if we do not count rotations and reflections as different squares.

(No answer to this problem is given in the back of the book.)

339. DIABOLIC MAGIC SQUARES

A magic square is called diabolic or pandiagonal if it has a constant sum not only along its rows, columns, and main diagonals, but also its so-called broken diagonals. The six broken diagonals of an order-4 square are as shown in the first diagram:

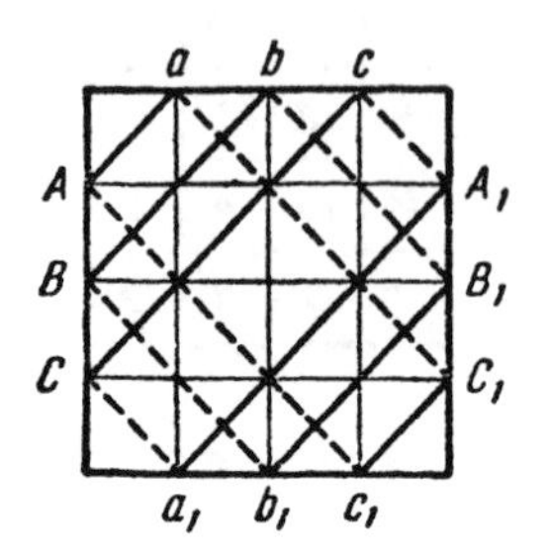

Aa and a_1A_1, Bb and b_1B_1, Cc and c_1C_1;
cA_1 and Ac_1, bB_1 and Bb_1, aC_1 and Ca_1.

A square of order 5 has eight broken diagonals (second diagram), and the number increases by 2 with each increase in order.

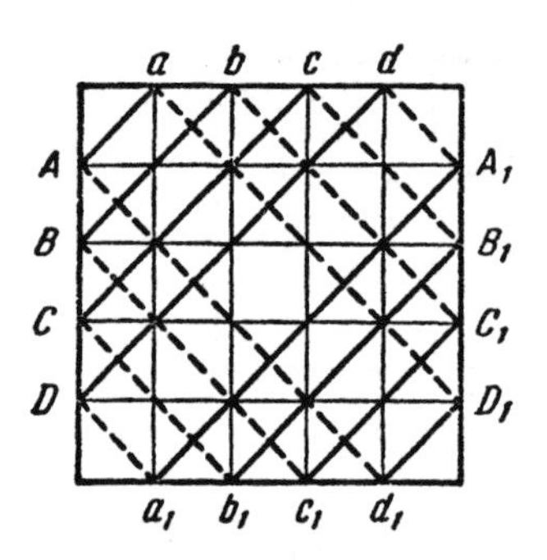

	0	1	2	3	4
0	1	8	15	17	24
1	20	22	4	6	13
2	9	11	18	25	2
3	23	5	7	14	16
4	12	19	21	3	10

There is only one basic magic square of order 3, and it is not diabolic. It has been proved that there are no diabolic squares of order $(4k + 2)$, with k an integer (for example, orders 6 and 10).

Diabolic squares exist for all other orders. The last diagram shows a diabolic square or order 5. Its rows, columns, main diagonals, and broken diagonals (twenty sets in all) have the constant sum of 65.

This diabolic square is none other than the large planetarium of Problem 328, in another form.

(No answer to this problem is given in the back of the book.)

XIII
Numbers Curious and Serious

This chapter presents many entertaining number oddities—and some of the toughest problems in the book.

340. TEN DIGITS

(A) Almost all of mankind uses the decimal system of counting with digits 1 through 9 and 0. How many different ten-digit integers can be written, using each digit only once? One million? Or less? How can we answer without writing down all the possible numbers?

(B) Let us look at six of these ten-digit numbers:

1,037,246,958; 1,286,375,904; 1,370,258,694;

1,046,389,752; 1,307,624,958; 1,462,938,570.

Each number has ten different digits, and divided by 2 produces a number of nine different digits which, when divided by 9, produces a number of eight different digits.

There is a number with ten different digits which when divided by 9, produces a palindromic quotient (same from right to left as left to right). Can you find it?

(C) Consider the numbers:

$a = 123{,}456{,}789$;

$b = 987{,}654{,}321$.

They are the smallest and largest nine-digit numbers consisting of nonrepeating digits without 0.

The difference $(b - a)$ contains the same nine digits:

987,654,321 − 123,456,789 = 864,197,532.

Multiply a and b in turn by all the one-digit numbers except 0 and 1. What characteristic of the products enables you to separate the multipliers into two groups: 2, 4, 5, 7, 8; and 3, 6, 9?

By the way, $b = 8a + 9$.

(D) Multiply 12, 345,679 (whose digits are in increasing sequence with 8 missing) by any one-digit number, then by 9. Each digit of the final result is the first multiplier. For example:

$$\begin{array}{r} 12{,}345{,}679 \\ 7 \\ \hline 86{,}419{,}753 \\ 9 \\ \hline 777{,}777{,}777 \end{array} \qquad \begin{array}{r} 12{,}345{,}679 \\ 8 \\ \hline 98{,}765{,}432 \\ 9 \\ \hline 888{,}888{,}888 \end{array}$$

Why?

341. MORE NUMBER ODDITIES

(A) A telegraph tape broke in the middle of the number 9,801. To amuse myself, I added 98 and 01, squared the result–and got the initial number: $(98 + 01)^2 =$ 9,801.

It works for 3, 025 and one other four-digit number. What is the best way to find the third number and prove there is no fourth?

(B) Consider this array:

A

1	3	5	7	9	11	13	. . .
1	4	7	10	13	16	19	. . .
1	5	9	13	17	21	25	. . .
1	6	11	16	21	26	31	. . .
1	7	13	19	25	31	37	. . .
1	8	15	22	29	36	43	. . .
1	9	17	25	33	41	49	. . .
.	.	.	.	.	.	.	*C*

The first number in each row is 1, and each row is an arithmetic progression. In the first row the difference between successive numbers is 2; in the second, 3; in the third, 4; and so on. The array extends to the right and downward to infinity.

If we add the numbers in each right-angled corridor (once called a gnomon), the sum is n^3, where n is the number of its row. In the second corridor, $1 + 4 + 3 = 2^3$; the third, $1 + 5 + 9 + 7 + 5 = 3^3$, and so on.

Any number on diagonal *AC* is the square of the row number. The sum of the numbers in any square array whose diagonal is a segment on *AC* is a square. For example, the sum of the numbers in the square with the diagonal 25, 36, 49 is $25 + 31 + 37 + 29 + 36 + 43 + 33 + 41 + 49 = 324 = 18^2$.

Check some other square arrays on diagonal *AC*.

(C) Some curious properties of 37:

1. $37 \times 3, 6, 9, \ldots, 27 = 111, 222, 333, \ldots, 999$.

2. The sum of its digits × 37 = the sum of the cubes of its digits:

$$(3+7) \times 37 = 3^3 + 7^3.$$

3. The sum of the squares of its digits less the product of its digits is 37:

$$(3^2 + 7^2) - (3 \times 7) = 37.$$

4. Take a three-digit multiple of 37, for example 37 × 7 = 259. An end-around carry produces 925, and another 592. Both are divisible by 37. Another example: 185, 518, 851.

On your own, show why this is true. (Hint: If $100a + 10b + c$ is divisible by 37, is $1{,}000a + 100b + 10c$? Is $a + 100b + 10c$?)

Five-digit multiples of 41 have the same property: 15,498, 81,549, 98,154, 49,815, and 54,981 are all divisible by 41.

342. REPEAT THE OPERATION

(A) Write four random positive integers in a row (say, 8, 17, 3, 107). Take the positive differences between the first and second numbers, the second and third, the third and fourth, and the fourth and first:

$$17 - 8 = 9,\quad 17 - 3 = 14,\quad 107 - 3 = 104,\quad 107 - 8 = 99.$$

Call these the first differences (9, 14, 104, 99).

The second differences are (5, 90, 5, 90), the third differences are (85, 85, 85, 85), and the fourth differences (0, 0, 0, 0).

Call the chosen row A_0 and the differences $A_1, A_2, A_3, \ldots$ For (93, 5, 21, 50):

A_0 = (93, 5, 21, 50); A_4 = (58, 30, 4, 32);
A_1 = (88, 16, 29, 43); A_5 = (28, 26, 28, 26);
A_2 = (72, 13, 14, 45); A_6 = (2, 2, 2, 2);
A_3 = (59, 1, 31, 27); A_7 = (0, 0, 0, 0).

Seven steps. (1, 11, 130, 1,760) takes six steps. Usually eight steps is enough. Are there any rows that never produce a row of 0s?

Note that such rows do exist when a row is not composed of 2^n integers:

A_0 = (2, 5, 9); A_5 = (1, 1, 0);
A_1 = (3, 4, 7); A_6 = (0, 1, 1);
A_2 = (1, 3, 4); A_7 = (1, 0, 1);
A_3 = (2, 1, 3); A_8 = (1, 1, 0).
A_4 = (1, 2, 1);

$A_8 = A_5$; the differences A_5, A_6, and A_7 will repeat endlessly.

(B) Take any integer and add up the squares of its digits. Keep doing it and, eventually, you will reach 1 or 89. For 31:

$3^2 + 1^2 = 10;$
$1^2 + 0^2 = 1.$

You get 1 from powers of 10 and, in general, numbers made up of 1, 3, 6, and 8 (each used no more than once), and such numbers with any number of 0s: 13, 103, 3,001, 68, 608, 8,006, and so on.

All other numbers produce 89. Try 48:

$$4^2 + 8^2 = 80; \qquad 5^2 + 2^2 = 29;$$
$$8^2 + 0^2 = 64; \qquad 2^2 + 9^2 = 85;$$
$$6^2 + 4^2 = 52; \qquad 8^2 + 5^2 = 89.$$

Continuing, we get:

$$8^2 + 9^2 = 145; \qquad 4^2 = 16;$$
$$1^2 + 4^2 + 5^2 = 42; \qquad 1^2 + 6^2 = 37;$$
$$4^2 + 2^2 = 20; \qquad 3^2 + 7^2 = 58;$$
$$2^2 + 0^2 = 4; \qquad 5^2 + 8^2 = 89.$$

The intermediate numbers are 145, 42, 20, 4, 16, 37, and 58. You could say that one of these is the final number rather than 89: It makes no difference.

Can you prove that, starting with a number of three or more digits, you will eventually reach a one- or two-digit number? The Moscow mathematician I. Y. Tanatar points out that once this is done, you can check the method number by number.

On your own, investigate repeated summing of cubes and fourth powers of integers.

343. A NUMBER CAROUSEL

From the bottomless treasure chest of numbers I take the six-digit number shown on the clock face. Multiply it by 1, 2, 3, 4, 5, and 6:

$$142{,}857 \times \begin{cases} 1 = 142{,}857; \\ 2 = 285{,}714; \\ 3 = 428{,}571; \\ 4 = 571{,}428; \\ 5 = 714{,}285; \\ 6 = 857{,}142. \end{cases}$$

Each product can be read clockwise from the figure. Also, the first three plus the last three digits of each product always equals 999,999.

Look at the seven-digit products:

$$142{,}857 \times \begin{cases} 8 = 1{,}142{,}856 & (142{,}856 + 1 = 142{,}857); \\ 9 = 1{,}285{,}713 & (285{,}713 + 1 = 285{,}714); \\ 10 = 1{,}428{,}570 & \ldots\ldots \\ 11 = 1{,}571{,}427 & \ldots\ldots \\ \ldots\ldots & \ldots\ldots \\ 69 = 9{,}857{,}133 & (857{,}133 + 9 = 857{,}142). \end{cases}$$

The operations in parentheses, which always produce a cyclic permutation of 142,857, are sums of the last six digits and the first.

Consider $142{,}857 \times 7 = 999{,}999$.

Such a product indicates that 142,857 is the period of the fraction 1/7 in decimals. Dividing, 1/7 = 0.142857142857142857 . . . .

Note that when a fraction *a*/*b* is changed to a repeating decimal, its period cannot have more than (*b* – 1) digits, although its number of digits can be a factor of (*b* – 1). If it does have *b* – 1 digits, it is a complete period. Whenever 1/*n* has a complete decimal period, the period is a cyclic number with the same properties as 142,857, the complete period of 1/7. For example, 1/17 has the complete decimal period 0.588,235,294,117,647, which multiplied by any number from 1 through 16, gives a cyclic permutation of itself. Multiplied by 17 it gives a number consisting of sixteen 9s.

(No answer to this problem is given in the back of the book.)

344. INSTANT MULTIPLICATION DISKS

The disk shown will give products of the number on the inner wheel (052,631,578,947,368,421) and the numbers 1 to 18 on the outer wheel. Turn the inner wheel so its 0 faces 2, and the product can be read on the inner wheel: 105,263,157,894,736,842.

This works because the number on the inner wheel is the complete decimal period of 1/19, and its products with 1 through 18 are cyclic permutations of itself.

The first nine primes with complete decimal periods are 7, 17, 19, 23, 29, 47, 59, 61, and 97. Disks can be made for each.

(No answer to this problem is given in the back of the book.)

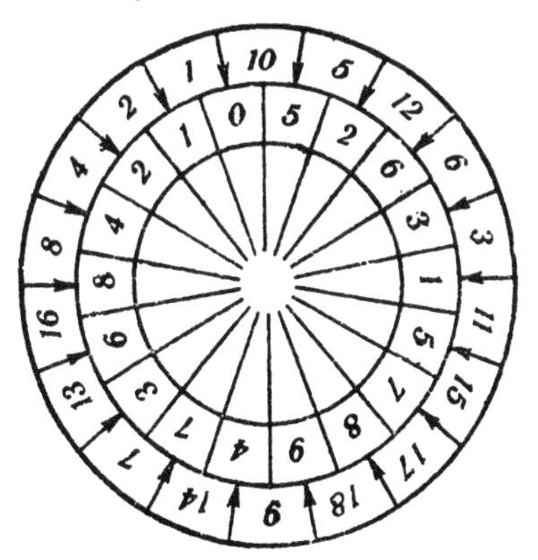

345. MENTAL GYMNASTICS

You don't have to be a lightning calculator to multiply 142,857 mentally by any number through 7,000. In Problem 343, we showed that the seven-digit product of

142,857 X 11 = 1,571,427 can be changed to a cyclic permutation of 142,857 by transposing the first digit to the last slot: 1 + 571,427 = 571,428.

Similarly, two digits can be transposed on eight-digit products, and three digits in the nine-digit products.

142,857 X 111 = 15,857,127; 15 + 857,127 = 857,142;
142,857 X 1,111 = 158,714,127; 158 + 714,127 = 714,285.

This makes mental multiplication feasible. Suppose you are asked to multiply 142,857 X 493. Divide 493 by 7, getting 70 3/7. Your first two digits are 70 and the last six are 70 less than a cyclic permutation of 142,857. For 3/7, the permutation is 428,571. Then the answer is:

142,857 X 493 = 70,428,501.

If you are asked to multiply by 378, when you divide by 7 you get 54 with no remainder. Change it to 53 7/7. Then the first two digits are 53 and the last six, 53 less than (as you know) 999,999. The product is 53,999,946.

(No answer to this problem is given in the back of the book.)

346. PATTERNS OF DIGITS

Digits can form combinations that call to mind the intricate patterns of snowflakes.

(A) Below are some plain multiplications with remarkable results.

```
    77             77
  X 77           X 77
  ----           ----
    49              7
  4949    or      777
    49           ----
  ----            847 X 7 = 5929
  5929
```

```
   666            666
 X 666          X 666
 -----          -----
    36              6
  3636            666
363636    or    66666
  3636          -----
    36          73926 X 6 = 443556
------
443556
```

```
            777777777777
          X 777777777777
          --------------
                 49
                4949
               494949
              49494949
             4949494949
            494949494949
           49494949494949
          4949494949494949
         494949494949494949
        49494949494949494949
       4949494949494949494949
      494949494949494949494949
       4949494949494949494949
        49494949494949494949
         494949494949494949
          4949494949494949
           49494949494949
            494949494949
             4949494949
              49494949
               494949
                4949
                 49
      ------------------------
      604938271603728395061729
```

or

```
         777777777777
       X 777777777777
       --------------
                7
               777
              77777
             7777777
            777777777
           77777777777
          7777777777777
         777777777777777
        77777777777777777
       7777777777777777777
      777777777777777777777
     77777777777777777777777
     ----------------------
     8641975308624691358024
                          X
    -----------------------
    60493827160372839506172
```

(B) Each digit from 1 through 9 appears once and only once in each of these equations:

$1{,}738 \times 4 = 6{,}952$; $483 \times 12 = 5{,}796$;
$1{,}963 \times 4 = 7{,}852$; $297 \times 18 = 5{,}346$;
$198 \times 27 = 5{,}346$; $157 \times 28 = 4{,}396$;
$138 \times 42 = 5{,}796$; $186 \times 39 = 7{,}254$.

(C) These equations have the same digits on both sides:

$42 \div 3 = 4 \times 3 + 2$;
$63 \div 3 = 6 \times 3 + 3$;
$95 \div 5 = 9 + 5 + 5$;
$(2 + 7) \times 2 \times 16 = 272 + 16$;
$5^{6-2} = 625$;
$(8 + 9)^2 = 289$;
$2^{10} - 2 = 1{,}022$;
$2^{8-1} = 128$;
$4 \times 2^3 = 4^3 \div 2 = 34 - 2$;

$\sqrt{121} = 12 - 1$;
$\sqrt{64} = 6 + \sqrt{4}$;
$\sqrt{49} = 4 + \sqrt{9} = 9 - \sqrt{4}$;
$\sqrt{169} = 16 - \sqrt{9} = \sqrt{16} + 9$;
$\sqrt{256} = 2 \times 5 + 6$;
$\sqrt{324} = 3 \times (2 + 4)$;
$\sqrt{11{,}881} = 118 - 8 - 1$;
$\sqrt{1{,}936} = -1 + 9 + 36$;
$\sqrt[3]{1{,}331} = 1 + 3 + 3 + 1 + 3$.

(D) In each of these equations a number is multiplied by the sum of its two parts, giving the sum of the cubes of the two parts:

$37 \times (3 + 7) = 3^3 + 7^3$;
$48 \times (4 + 8) = 4^3 + 8^3$;
$111 \times (11 + 1) = 11^3 + 1^3$;
$147 \times (14 + 7) = 14^3 + 7^3$;
$148 \times (14 + 8) = 14^3 + 8^3$.

(E) Numbers can grow like crystals:

$16 = 4^2$;
$1{,}156 = 34^2$;
$111{,}156 = 334^2$;
$11{,}115{,}556 = 3{,}334^2$;
$1{,}111{,}155{,}556 = 33{,}334^2$;
$111{,}111{,}555{,}556 = 333{,}334^2$.

There is only one other two-digit square $(10a + b)$ like 16. No matter how many times we insert $(10a + b - 1)$ in the center, the new number is a square.

Can you find the number?

(F) Nine is a different kind of crystal square. Write it as 09. Keep adding 1 on the left and 8 in the second position on the right:

$09 = 3^2$;
$1{,}089 = 33^2$;
$110{,}889 = 333^2$;
$11{,}108{,}889 = 3{,}333^2$.

The digits added are 1 more than 0 and 1 less than 9; similarly, with 36 we add 1 more than 3 and 1 less than 6:

$$36 = 6^2;$$
$$4{,}356 = 66^2;$$
$$443{,}556 = 666^2.$$

Find another square of the second type.

347. ONE FOR ALL AND ALL FOR ONE

(A) Instead of writing the numbers from 1 through 10 with the usual ten digits, they can be written with just one:

$$1 = 2 + 2 - 2 - \tfrac{2}{2}; \qquad 6 = 2 + 2 + 2 + 2 - 2;$$
$$2 = 2 + 2 + 2 - 2 - 2; \qquad 7 = 22 \div 2 - 2 - 2;$$
$$3 = 2 + 2 - 2 + \tfrac{2}{2}; \qquad 8 = 2 \times 2 \times 2 + 2 - 2;$$
$$4 = 2 \times 2 \times 2 - 2 - 2; \qquad 9 = 2 \times 2 \times 2 + \tfrac{2}{2};$$
$$5 = 2 + 2 + 2 - \tfrac{2}{2}; \qquad 10 = 2 + 2 + 2 + 2 + 2.$$

Write the numbers 11 through 26, using five 2s for each. Besides the signs used above, you can use exponents and parentheses.

(B) Write the numbers from 1 through 10, using four 4s for each.

(C) Numbers 2 through 9 can be expressed by fractions in which every digit except 0 appears once and only once. For example:

$$2 = \frac{13{,}458}{6{,}729} \qquad 4 = \frac{15{,}768}{3{,}942}.$$

Form similar fractions that equal 3, 5, 6, 7, 8, and 9.

(D) Using all ten digits, 9 can be expressed by a fraction in six different ways. Here are three:

$$9 = \frac{97{,}524}{10{,}836} = \frac{57{,}429}{06{,}381} = \frac{95{,}823}{10{,}647}.$$

Can you find others? (Hint: Try shuffling the digits around in the given fractions without transferring any between numerator and denominator.)

348. EVEN NUMBERS CAN BE ODD

(A) Products can be the reverse of sums:

$$9 + 9 = 18; \qquad 9 \times 9 = 81;$$
$$24 + 3 = 27; \qquad 24 \times 3 = 72;$$
$$47 + 2 = 49; \qquad 47 \times 2 = 94;$$
$$497 + 2 = 499; \qquad 497 \times 2 = 994.$$

(B) Pairs of two-digit numbers can have the same product when both numbers are reversed:

$$12 \times 42 = 21 \times 24; \quad 24 \times 63 = 42 \times 36;$$
$$12 \times 63 = 21 \times 36; \quad 24 \times 84 = 42 \times 48;$$
$$12 \times 84 = 21 \times 48; \quad 26 \times 93 = 62 \times 39;$$
$$13 \times 62 = 31 \times 26; \quad 36 \times 84 = 63 \times 48;$$
$$23 \times 96 = 32 \times 69; \quad 46 \times 96 = 64 \times 69.$$

Can you find the other four pairs?

(C) Squares of consecutive numbers can use the same digits:

$$13^2 = 169; \quad 157^2 = 24{,}649; \quad 913^2 = 833{,}569;$$
$$14^2 = 196. \quad 158^2 = 24{,}964. \quad 914^2 = 835{,}396.$$

(D) Is there an integer with these properties?

1. It is the fourth power of the sum of its digits.
2. If we separate it into three groups of two digits each, the sum of the two-digit numbers is a square.
3. If we reverse its digits and again separate it into three groups of two digits each, the sum of the two-digit numbers is the same square.

Yes: 234,256.

(E) A constellation of six numbers 2, 3, 7, 1, 5, 6, has this interesting property:

$$2 + 3 + 7 = 1 + 5 + 6;$$
$$2^2 + 3^2 + 7^2 = 1^2 + 5^2 + 6^2.$$

Countless constellations have the property:

$$x_1 + x_2 + x_3 = y_1 + y_2 + y_3,$$
$$x_1^2 + x_2^2 + x_3^2 = y_1^2 + y_2^2 + y_3^2.$$

On your own, find one.

Such constellations can contain eight or ten numbers, extending the property through cubes:

$$0 + 5 + 5 + 10 = 1 + 2 + 8 + 9,$$
$$0^2 + 5^2 + 5^2 + 10^2 = 1^2 + 2^2 + 8^2 + 9^2,$$
$$0^3 + 5^3 + 5^3 + 10^3 = 1^3 + 2^3 + 8^3 + 9^3,$$
$$1 + 4 + 12 + 13 + 20 = 2 + 3 + 10 + 16 + 19,$$
$$1^2 + 4^2 + 12^2 + 13^2 + 20^2 = 2^2 + 3^2 + 10^2 + 16^2 + 19^2,$$
$$1^3 + 4^3 + 12^3 + 13^3 + 20^3 = 2^3 + 3^3 + 10^3 + 16^3 + 19^3.$$

More than two hundred years ago two academicians of St. Petersburg, Christian Goldbach and the Swiss genius Leonard Euler, developed many equations that would generate such constellations.

For sets of six numbers:

$$x_1 = a + c;\ x_2 = b + c;\ x_3 = 2a + 2b + c;$$
$$y_1 = c;\ y_2 = 2a + b + c;\ y_3 = a + 2b + c.$$

(In all these sets of equations, $a, b, \ldots$ can be positive integers.)
In the series cited above, $a = 1$, $b = 2$, and $c = 1$.
Another set generating six numbers:

$$x_1 = ad;\ x_2 = ac + bd;\ x_3 = bc;$$
$$y_1 = ac;\ y_2 = ad + bc;\ y_3 = bd.$$

To generate eight numbers as above:

$$x_1 = a;\ x_2 = b;\ x_3 = 3a + 3b;\ x_4 = 2a + 4b;$$
$$y_1 = 2a + b;\ y_2 = a + 3b;\ y_3 = 3a + 4b;\ y_4 = 0.$$

(F) Here is a versatile constellation:

$$1 + 6 + 7 + 17 + 18 + 23 = 2 + 3 + 11 + 13 + 21 + 22;$$
$$1^2 + 6^2 + 7^2 + 17^2 + 18^2 + 23^2 = 2^2 + 3^2 + 11^2 + 13^2 + 21^2 + 22^2;$$
$$1^3 + 6^3 + 7^3 + 17^3 + 18^3 + 23^3 = 2^3 + 3^3 + 11^3 + 13^3 + 21^3 + 22^3;$$
$$1^4 + 6^4 + 7^4 + 17^4 + 18^4 + 23^4 = 2^4 + 3^4 + 11^4 + 13^4 + 21^4 + 22^4;$$
$$1^5 + 6^5 + 7^5 + 17^5 + 18^5 + 23^5 = 2^5 + 3^5 + 11^5 + 13^5 + 21^5 + 22^5.$$

And here are the equations:

$$a^n + (a + 4b + c)^n + (a + b + 2c)^n + (a + 9b + 4c)^n + (a + 6b + 5c)^n + (a + 10b + 6c)^n =$$
$$= (a + b)^n + (a + c)^n + (a + 6b + 2c)^n + (a + 4b + 4c)^n + (a + 10b + 5c)^n + (a + 9b + 6c)^n,$$

where a, b, c are any positive integers and n can be 1, 2, 3, 4, or 5.
(G) Given:

$$4^2 + 5^2 + 6^2 = 2^2 + 3^2 + 8^2.$$

Then:

$$42^2 + 53^2 + 68^2 = 24^2 + 35^2 + 86^2.$$

But this is not the only way to join the digits of the first equation left-right, left-right, left-right, with squares summing like the squares of their reverses. Here are five more:

$$42^2 + 58^2 + 63^2 = 24^2 + 85^2 + 36^2;$$
$$43^2 + 52^2 + 68^2 = 34^2 + 25^2 + 86^2;$$
$$43^2 + 58^2 + 62^2 = 34^2 + 85^2 + 26^2;$$
$$48^2 + 52^2 + 63^2 = 84^2 + 25^2 + 36^2;$$
$$48^2 + 53^2 + 62^2 = 84^2 + 35^2 + 26^2.$$

In general, if $2n$ one-digit numbers have this relation between their squares:

$$x_1^2 + x_2^2 + \ldots + x_n^2 = y_1^2 + y_2^2 + \ldots + y_n^2,$$

then:

$$(10x_1 + y_1)^2 + (10x_2 + y_2)^2 + \ldots + (10x_n + y_n)^2$$
$$= (10y_1 + x_1)^2 + (10y_2 + x_2)^2 + \ldots + (10y_n + x_n)^2,$$

and by changing the order in the right hand of the first equation, $n! = n(n-1)(n-2)\ldots 2(1)$ such relations can be formed.

Take the digits from 1 through 8 and, on your own, determine which digits must be x_1, x_2, x_3, and x_4 and which y_1, y_2, y_3, and y_4, and form several equations of squares of two-digit numbers.

(H) Here are constellations of twelve numbers, six two-digit numbers and their reverses:

$$13 + 42 + 53 + 57 + 68 + 97 =$$
$$= 79 + 86 + 75 + 35 + 24 + 31,$$
$$13^2 + 42^2 + 53^2 + 57^2 + 68^2 + 97^2 =$$
$$= 79^2 + 86^2 + 75^2 + 35^2 + 24^2 + 31^2,$$
$$13^3 + 42^3 + 53^3 + 57^3 + 68^3 + 97^3 =$$
$$= 79^3 + 86^3 + 75^3 + 35^3 + 24^3 + 31^3;$$

$$12 + 32 + 43 + 56 + 67 + 87 =$$
$$= 78 + 76 + 65 + 34 + 23 + 21,$$
$$12^2 + 32^2 + 43^2 + 56^2 + 67^2 + 87^2 =$$
$$= 78^2 + 76^2 + 65^2 + 34^2 + 23^2 + 21^2,$$
$$12^3 + 32^3 + 43^3 + 56^3 + 67^3 + 87^3 =$$
$$= 78^3 + 76^3 + 65^3 + 34^3 + 23^3 + 21^3$$

(I)

$$145 = 1! + 4! + 5! = 1 + 24 + 120;$$
$$40{,}585 = 4! + 0! + 5! + 8! + 5!$$
$$= 24 + 1 + 120 + 40{,}320 + 120.$$

(Note that by convention $0! = 1$.)

There are no other such numbers. Can you find any of the four numbers that are 1 more or less than the sum of the factorials of their digits?

(J) Every power of 376 ends in 376, and every power of 625 ends in 625:

$$376^2 = 141{,}376, \quad 376^3 = 53{,}157{,}376, \text{ and so on;}$$
$$625^2 = 390{,}625, \quad 625^3 = 244{,}140{,}625, \text{ and so on.}$$

How can it be established that there are no other such three-digit numbers?

Prove on your own that if the square of an n-digit number ends in the same n digits, the same is true of every higher power. (For example, $76^2 = 5{,}776$, and every higher power of 76 ends in 76.)

349. A ROW OF POSITIVE INTEGERS

(A) Write the positive integers 1, 2, 3, . . . in the shape of a triangle:

```
                              .
                           .....
                        ..........
                  50 .............
               37 51 .............
            26 38 52 .............
         17 27 39 53 ...... 107 ...
      10 18 28 40 54 ...... 108 ...
    5 11 19 29 41 55 ...... 109 ...
  2 6 12 20 30 42 56 ...... 110 ...
1 3 7 13 21 31 43 57 ...... 111 ...
  4 8 14 22 32 44 58 ...... 112 ...
    9 15 23 33 45 59 ...... 113 ...
      16 24 34 46 60 ...... 114 ...
         25 35 47 61 ...... 115 ...
            36 48 62 .............
               49 63 .............
                  64 .............
```

Look closely at this triangle:

1. The lowest number in each column is the square of the column's number from left to right.

2. The product of any two adjacent numbers in a row is in the row: for example, $5 \times 11 = 55$. The product is n places to the right of the smaller multiplier n: for example, 55 is five places to the right of 5.

3. The numbers in the longest row $= n^2 - n + 1 = (n - 1)^2 + n$, where $n = 1, 2, 3, 4, 5, \ldots$ Each third number of the row after 3 is divisible by 7; each thirteenth number after 13 or 91 is divisible by 13, and so on. The numbers of each row have analogous properties.

(B) The series of positive integers can be broken into a series of equations of addition:

$$1 + 2 = 3;$$
$$4 + 5 + 6 = 7 + 8;$$
$$9 + 10 + 11 + 12 = 13 + 14 + 15;$$
$$16 + 17 + 18 + 19 + 20 = 21 + 22 + 23 + 24, \text{ and so on.}$$

1. At each step there are two more integers in the equation.

2. The first term of each equation is the square of the number of integers on the right side. Thus any equation can be written without writing all the preceding ones.

(C) The sum of the squares of the first n integers is:

$$1^2 + 2^2 + 3^2 + \ldots + n^2 = \frac{n(n+1)(2n+1)}{6}.$$

The sum of the squares of the first $\frac{1}{2}(n+1)$ odd integers has the same formula as the sum of the squares of the first $\frac{1}{2}n$ even integers:

$$1^2 + 3^2 + 5^2 + \ldots + n^2 = \frac{(n+1)^3 - (n+1)}{6};$$

$$2^2 + 4^2 + 6^2 + \ldots + n^2 = \frac{(n+1)^3 - (n+1)}{6}.$$

(D)

$$3^2 + 4^2 = 5^2;$$
$$10^2 + 11^2 + 12^2 = 13^2 + 14^2.$$

The first equation describes the simplest integral-sided triangle (legs of 3 and 4, hypotenuse of 5) that can be used to illustrate the Pythagorean theorem. The second was the basis of *The Difficult Problem,* a painting by the Russian N. P. Bogdanov-Belsky (1868-1945) which shows a group of rural students trying to solve mentally a problem chalked on the blackboard:

$$\frac{10^2 + 11^2 + 12^2 + 13^2 + 14^2}{365} = ?$$

This is not such a "difficult problem" when you know that the first three squares add to 365 and so do the last two, giving the answer: 2.

Can you find another equation of the same kind, composed of positive integers, with two terms on the left? With three? Is there a series of equations with four, five, . . . terms on the left, as in (B) of this problem?

(E) Are there two consecutive positive integers whose cubes add to the cube of the next number (as $3^2 + 4^2 = 5^2$)?

No. Proof by contradiction: Let the positive integers be $(x-1), x,$ and $(x+1)$. Then:

$$(x-1)^3 + x^3 = (x+1)^3;$$
$$2x^3 - 3x^2 + 3x - 1 = x^3 + 3x^2 + 3x + 1;$$
$$x^3 - 6x^2 = x^2(x-6) = 2.$$

But x^2 is positive; $(x-6)$ must be positive; x must be 7 or more, but then x^2 (at least 49) times $(x-6)$ (at least 1) is greater than 2, which is impossible. Therefore, there are no such three consecutive positive integers.

(F) Consider the multiplication table:

1	2	3	4	5	...	...	p	...	n
2	4	6	8	10	...	...	$2p$	...	$2n$
3	6	9	12	15	...	...	$3p$	...	$3n$
4	8	12	16	20	...	...	$4p$	...	$4n$
5	10	15	20	25	...	...	...	...	...
...	...	...	...	...	...	...	...	...	...
...	...	...	...	...	...	...	...	...	...
p	$2p$	$3p$	$4p$	...	...	...	p^2		...
...	...	...	...	...	...	...	...	...	...
n	$2n$	$3n$	$4n$	...	...	...	...	...	n^2

In such a table, of course, the product (say, 15) is at the intersection of the row and the column of its factors (3 and 5, or 5 and 3). Now that we have made corridors bending around a right angle in the table, other patterns become apparent:

1. The sum of the numbers in a square array with 1 at the upper left is a square:

$$1 = 1^2;$$
$$1 + 2 + 2 + 4 = 3^2;$$
$$1 + 2 + 3 + 2 + 4 + 6 + 3 + 6 + 9 = 6^2.$$

2. The sum of the numbers in a corridor is a cube:

$$1 = 1^3;$$
$$2 + 4 + 2 = 2^3;$$
$$3 + 6 + 9 + 6 + 3 = 3^3.$$

3. The square arrays consist of 1, 2, 3, . . ., n corridors.

A famous ancient formula follows:

$$1^3 + 2^3 + 3^3 + \ldots + n^3 = (1 + 2 + 3 + \ldots + n)^2.$$

Since $1 + 2 + 3 + \ldots + n = \frac{n(n+1)}{2}$ (the sum of an arithmetical progression):

$$1^3 + 2^3 + 3^3 + \ldots + n^3 = \frac{n(n+1)}{2}^{2}.$$

Here is an unexpected geometrical interpretation! Count the rectangles (including squares) in diagrams *a* and *b*. In *a* there are 9:

2-by-2 square	1
1-by-2 rectangles	2 + 2
1-by-1 squares	4
	9

(a)

In *b* there are 36:

3-by-3 square	1	
2-by-3 rectangles	2 + 2	
2-by-2 squares	4	
1-by-3 rectangles	3 + 3	
1-by-2 rectangles	6 + 6	
1-by-1 squares	9	(b)
	36	

The numbers on the right are those of the first corridors of the table. The square with $2^2 = 4$ cells has $1^3 + 2^3 = 9$ rectangles, and the square with $3^2 = 9$ cells has $1^3 + 2^3 + 3^3 = 36$ rectangles. How many rectangles does a square with n^2 cells have?

(G) In the ancient formula for the sum of cubes, the terms are 1, 2, 3, and so on. The French mathematician Joseph Liouville (1809-82) set himself the task of finding nonconsecutive integers (with repetitions permitted) the sum of whose cubes would equal the square of their sum:

$$a^3 + b^3 + c^3 + \ldots = (a + b + c + \ldots)^2.$$

His remarkable method can be understood through these examples:

The number 6 is divisible by 1, 2, 3, and 6; and 1 has one divisor, 2 has two (1, and 2), 3 has two (1, and 3), and 6 has four (1, 2, 3, 6). But:

$$1^3 + 2^3 + 2^3 + 4^3 = (1 + 2 + 2 + 4)^2 = 81.$$

The number 30 has the divisors 1, 2, 3, 5, 6, 10, 15, and 30. They have 1, 2, 2, 2, 4, 4, 4, and 8 divisors respectively:

$$1^3 + 2^3 + 2^3 + 2^3 + 4^3 + 4^3 + 4^3 + 8^3$$
$$= (1 + 2 + 2 + 2 + 4 + 4 + 4 + 8)^2 = 729.$$

On your own, try some other numbers.

350. A PERSISTENT DIFFERENCE

Pick a four-digit number whose digits are not all the same. From its digits form the smallest four-digit number m and the largest, M. Find $(M - m)$. Keep repeating the procedure. (Treat, say, 397 as 0397.) Eventually you will reach 6,174 – permanently, since:

$$7{,}641 - 1{,}467 = 6{,}174.$$

For example, start with 4,818:

$$8{,}841 - 1{,}488 = 7{,}353;$$
$$7{,}533 - 3{,}357 = 4{,}176;$$
$$7{,}641 - 1{,}467 = 6{,}174; \text{and so on.}$$

Can you prove that you always reach 6,174? Originally this problem was given as "a nut that has not yet been cracked." Many readers tried it. It was soon shown that a proof requires only a test of 30 four-digit numbers. (The first communication to this effect was from Y. N. Lambina of Ryazan.)

What are the 30 numbers? What number is always reached when a similar procedure is applied to two-digit numbers? Three-digit numbers? Five-digit numbers?

351. A PALINDROMIC SUM

This is a nut that has not yet been cracked.

Add its reverse to any integer. Add the sum's reverse to the sum. Continue until a sum is palindromic (same from right to left as from left to right):

38	139	48,017
83	931	71,084
121	1,070	119,101
	0,701	101,911
	1,771	221,012
		210,122
		431,134

Many steps may be necessary. (From 89 to 8,813,200,023,188 takes twenty-four steps.) It has been conjectured that every integer will eventually produce a palindrome.

P. R. Mols, an industrial worker from Riga, found the number 196, which after seventy-five steps has not produced a palindrome. Rather than continue from the 36-digit number of the seventy-fifth sum, try to prove or disprove the conjecture by reasoning.

(No answer is given in the back of the book to this problem.)

[The number 196 was independently suggested by Charles W. Trigg of California, and computers now have carried it beyond many thousands of steps without producing a palindrome. For recent information on the "palindrome conjecture," which has been proved false only for the binary system, see the Mathematical Games department of *Scientific American,* August 1970. – M. G.]

XIV
Numbers Ancient but Eternally Young

Prime Numbers

352. NUMBERS PRIME AND COMPOSITE

If the positive integer N divided by the positive integer a is a positive integer, then a is a divisor of N:

1 has one divisor (1);
2 has two divisors (1, 2);
3 has two divisors (1, 3);
4 has three divisors (1, 2, 4).

Prime numbers have two divisors; composite numbers, three or more. (One is not prime or composite.)

Two, the smallest prime, is the only even prime. There are odd prime numbers (3, 5, 7, . . .) and odd composite numbers (9, 15, 21, . . .).

Every composite is the product of a unique set of prime numbers:

$$12 = 2 \times 2 \times 3;$$
$$363 = 3 \times 11 \times 11, \text{and so on.}$$

Primes are the basic numbers from which all other numbers can be derived by multiplication. Mathematicians' great interest in prime numbers is not hard to understand.

(No answer to this problem is given in the back of the book.)

353. THE SIEVE OF ERATOSTHENES

How can we find the prime numbers? The larger the number, the harder it is to decide whether it is prime.

To separate grains from a mixture, sieves are successively used that have apertures of each grain's size. A similar method is used to separate out prime numbers.

Suppose we wish to find all the prime numbers from 2 through *N*. First write them in serial order. The first prime is 2. Underline it, then cross out the multiples of 2. The first remaining number, 3, must be a prime. Underline it and cross out its multiples. Treat 5 the same way (4 is already crossed out). Continuing, you will cross out all the composite numbers from 2 through *N*, and the underlined numbers will form a table of primes from 2 through *N*:

2, 3, ~~4~~, 5, ~~6~~, 7, ~~8~~, ~~9~~, ~~10~~,
11, ~~12~~, 13, ~~14~~, ~~15~~, ~~16~~, 17, ~~18~~, 19, ~~20~~,
~~21~~, ~~22~~, 23, ~~24~~, ~~25~~, ~~26~~, ~~27~~, ~~28~~, 29, ~~30~~,
31, ~~32~~, ~~33~~, ~~34~~, ~~35~~, ~~36~~, 37, ~~38~~, ~~39~~, ~~40~~,
. . . .

This sieve was invented more than two thousand years ago by the Greek mathematician Eratosthenes (276-196 B.C.). To this day, this tedious but completely reliable method bears his name.

Over the centuries, the prime numbers from 1 to 10,000,000 were found. The American mathematician D. H. Lehmer did much pioneering work in making, carefully checking, and finally publishing (in 1914) a table of these primes. Lehmer's book is in the Lenin Library in Moscow, and, if you wish, you can make a photocopy of it.

About twenty years earlier than Lehmer, the self-taught Russian mathematician Ivan Mikheyevich Pervushin composed a table of prime numbers up to 10,000,000 and presented it as a gift to our Academy of Sciences. Pervushin's tables have been preserved in manuscript in the archives of the Academy, but have not been published.

J. P. Kulik, a professor at the University of Prague, carried the prime numbers up to 100,000,000 (six volumes of primes and divisors of composite numbers). Since 1867, Kulik's tables have been in the library at the Vienna Academy of Sciences. One volume, containing 13,000,000 – 23,000,000 has disappeared without a trace. It is not easy to restore the lost primes or to check the numbers in the preserved volumes.

(No answer to this problem is given in the back of the book.)

[Today, with the aid of computers, many enormously large prime numbers are known that are far beyond any existing tables. The prime $2^{19,937} - 1$ was discovered in 1971 by the American mathematician Bryant Tuckerman. It has 6,002 digits! – M. G.]

354. HOW MANY PRIMES?

Euclid proved there is no largest prime. If you multiply all the primes from 2 through *n* and add 1 to the product, the result is either a prime or is a composite with a prime factor larger than *n*.

The primes are distributed irregularly but with decreasing frequency among the integers. Of integers greater than 1, there are five primes in the first ten (50%), twenty-six in the first hundred (26%), and 8% in the first million. There is a formula that gives good approximations of percent of primes if used on a large range of integers.

Fibonacci Numbers

A great expert on numbers, an artful calculator named Leonardo ("of Pisa," as he was usually called), lived in Italy in the thirteenth century. He was also known as Fibonacci, which means "son of Bonacci." In 1202 he published a book in Latin, *Liber Abaci (The Book About Abaci),* which contained all that was then known about arithmetic and algebra. It was one of the first books in Europe to teach calculation with Arabic numerals. For more than two centuries it was the definitive work on numerical calculations. As was the custom, Fibonacci participated in mathematical tournaments (public contest for the best and speediest solutions of difficult problems). His skill in solving numerical problems was astonishing.

355. A PUBLIC TEST

In 1225, Holy Roman Emperor Frederick II came to Pisa with a group of mathematicians to test Leonardo publicly – so high was Fibonacci's reputation. One problem given at the tournament was:

To find a square which remains a square if it is decreased by 5 or increased by 5.

Obviously, the answer is not an integer. After thinking for a while, Fibonacci found the number:

$$\frac{1{,}681}{144} \text{ or } \left(\frac{41}{12}\right)^2.$$

When 5 is subtracted it remains a square:

$$\frac{961}{144} = \left(\frac{31}{12}\right)^2;$$

and when 5 is added it remains a square:

$$\frac{2{,}401}{144} = \left(\frac{49}{12}\right)^2.$$

In G. N. Popov's *Historical Problems* (1932) a method of solution is offered:

$$x^2 + 5 = u^2 \text{ and } x^2 - 5 = v^2.$$

Then:

$$u^2 - v^2 = 10.$$

But $10 = \dfrac{80 \times 18}{12^2}$.

Therefore: $(u+v)(u-v) = \frac{80 \times 18}{12^2}$.

Let: $u + v = \frac{80}{12}$ and $u - v = \frac{18}{12}$,

and we obtain Fibonacci's answer.

Perhaps this was how Fibonacci solved the problem at the tournament. If so, what power of imagination he must have had to think of replacing 10 by the fraction given.

(No answer to this problem is given in the back of the book.)

356. THE FIBONACCI SERIES

$$1, 1, 2, 3, 5, 8, 13, 21, 34, 55, \ldots$$

Each number equals the sum of the preceding two: $1 + 1 = 2$, $1 + 2 = 3$, and so on.

If two consecutive numbers in the series are y and x, then either:

$$x^2 - xy - y^2 = 1, \text{ or } x^2 - xy - y^2 = -1.$$

For example:

$$x = 2, \quad y = 1;$$
$$x = 5, \quad y = 3;$$
$$x = 13, \quad y = 8;$$

are roots of the first equation, and:

$$x = 3, \quad y = 2;$$
$$x = 8, \quad y = 5;$$
$$x = 21, \quad y = 13;$$

are the roots of the second.

The Fibonacci series is important not only to mathematicians but to botanists. Leaves on a branch are often placed helically around the stalk; that is, each leaf is a little higher than and to one side of the preceding one. Different plants have characteristic angles of divergence of adjacent leaves. The angle is usually expressed by a fraction of 360 degrees. For the linden and elm the fraction is 1/2; for the beech, 1/3; oak and cherry 2/5; poplar and pear, 3/8; willow, 5/13, and so on. The same angle is preserved in the arrangement of each tree's branches, buds, and flowers. The fractions are composed of Fibonacci numbers.

(No answer to this problem is given in the back of the book.)

357. A PARADOX

When you cut up a figure and reorder the parts, the shape may change but, clearly, the area cannot.

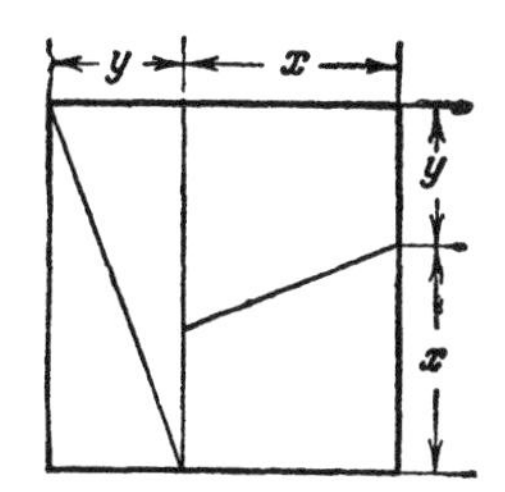

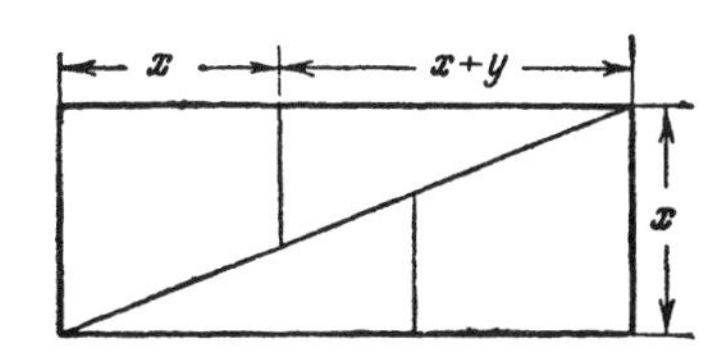

But consider the diagram above. The square is cut into two congruent triangles and two congruent trapezoids. Can we choose x and y so the square can be transformed into a rectangle as shown?

A young friend wrote me: "Using graph paper I tried some values of x and y but the pieces would not form a rectangle. When I tried $x = 5, y = 3$, the rectangle had an area of $5 \times 13 = 65$." (See diagrams below.)

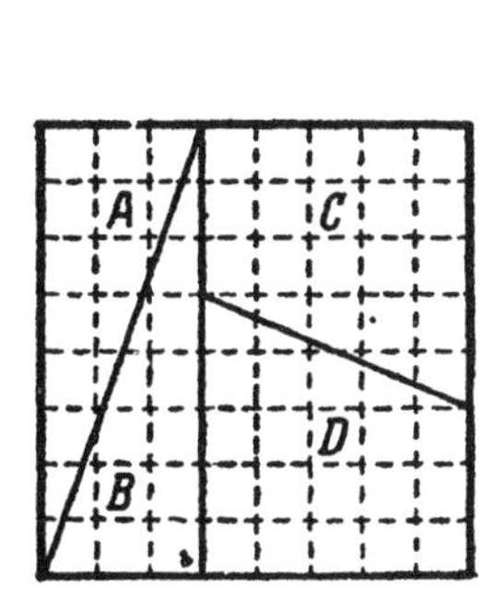

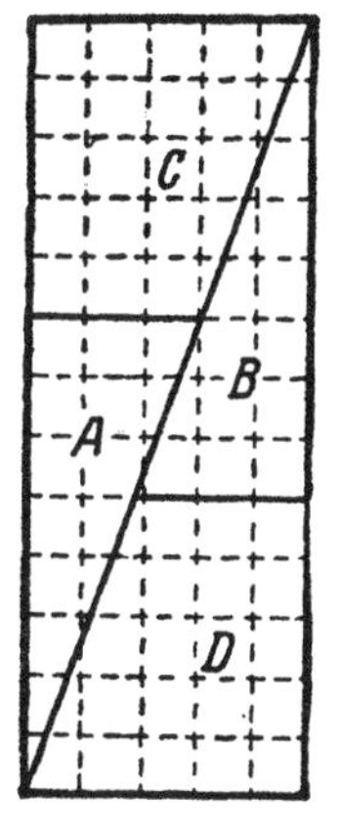

But the square's area is only 64!

"With a 13-by-13 square ($x = 8, y = 5$), my rectangle's area was 168 instead of 169; with a 21-by-21 square ($x = 13, y = 8$), 442 instead of 441. What is wrong?"

And what part do Fibonacci numbers play in the paradox?

358. PROPERTIES OF THE FIBONACCI SERIES

Here are the first twenty numbers of the series:

1	8	89	987
1	13	144	1,597
2	21	233	2,584
3	34	377	4,181
5	55	610	6,765

1. Successive terms are formed thus:

$$S_{n-2} + S_{n-1} = S_n.$$

2. As in any series, we want to be able to derive any number S_n of the series directly from n. It is natural to expect that the formula will contain only integers, or perhaps fractions. This is not so. Two irrational numbers are necessary, namely:

$$a_1 = \frac{1+\sqrt{5}}{2}; \quad a_2 = \frac{1-\sqrt{5}}{2}.$$

a_1, the ratio of the golden section, appeared in the last problem; a_2 is its negative reciprocal.

Here is the formula for S_n:

$$S_n = \frac{\left(\frac{1+\sqrt{5}}{2}\right)^n - \left(\frac{1-\sqrt{5}}{2}\right)^n}{\sqrt{5}} = \frac{a_1{}^n - a_2{}^n}{\sqrt{5}}.$$

When $n = 1$:

$$S_1 = \frac{\frac{1+\sqrt{5}}{2} - \frac{1-\sqrt{5}}{2}}{\sqrt{5}} = \frac{\frac{2\sqrt{5}}{2}}{\sqrt{5}} = 1.$$

When $n = 2$:

$$S_2 = \frac{\left(\frac{1+\sqrt{5}}{2}\right)^2 - \left(\frac{1-\sqrt{5}}{2}\right)^2}{\sqrt{5}} = 1, \quad \frac{6+2\sqrt{5}-(6-2\sqrt{5})}{4\sqrt{5}} = 1.$$

We can prove that, for S as defined by the formula, the Fibonacci relation $S_{n+1} = S_{n\ 1} + S_n$ holds:

$$S_{n-1} + S_n = \frac{a_1{}^{n-1} + a_1{}^n - a_2{}^{n-1} - a_2{}^n}{\sqrt{5}}$$

$$= \frac{\left(\frac{1+\sqrt{5}}{2}\right)^{n+1}\left[\frac{2^2}{(1+\sqrt{5})^2} + \frac{2}{1+\sqrt{5}}\right] - \left(\frac{1-\sqrt{5}}{2}\right)^{n+1}\left[\frac{2^2}{(1-\sqrt{5})^2} + \frac{2}{1-\sqrt{5}}\right]}{\sqrt{5}}$$

But both expressions in brackets = 1 (as you can easily show), so the whole expression $= S_{n+1}$. This completes a proof by induction: the formula will generate the Fibonacci series from two correct terms, and since it has been shown correct for the first two terms, it must be correct for all terms.

3. The formula for the sum of the first n numbers of a Fibonacci series has an amusing appearance:

$$S_1 + S_2 + \ldots + S_n = S_{n+2} - 1.$$

Thus the sum of the first six terms is $1 + 1 + 2 + 3 + 5 + 8 = 20$, and the eighth (not seventh) term is 21, or 1 more.

4. The sum of the squares of the first n numbers of a Fibonacci series is the product of two adjacent numbers:

$$S_1^2 + S_2^2 + \ldots + S_n^2 = S_n \times S_{n+1}.$$

For example:

$$1^2 + 1^2 = 1 \times 2;$$
$$1^2 + 1^2 + 2^2 = 2 \times 3;$$
$$1^2 + 1^2 + 2^2 + 3^2 = 3 \times 5.$$

5. The square of each Fibonacci number, reduced by the product of the preceding and following numbers, is alternately + 1 and − 1 :

$$2^2 - 1 \times 3 = +1;$$
$$3^2 - 2 \times 5 = -1;$$
$$5^2 - 3 \times 8 = +1.$$

6. $S_1 + S_3 + \ldots + S_{2n-1} = S_{2n}$.
7. $S_2 + S_4 + \ldots + S_{2n} = S_{2n+1} - 1$.
8. $S_n^2 + S_{n+1}^2 = S_{2n+1}$.
9. Each third number of the series is even, each fourth number is divisible by 3, each fifth by 5, and each fifteenth by 10.
10. It is impossible to construct a triangle whose sides are three different Fibonacci numbers. Can you see why?
(No answer to this problem is given in the back of the book.)

Figurate Numbers

359. PROPERTIES OF FIGURATE NUMBERS

1. The ancient Greeks were fascinated by numbers that could be arranged in a series and given geometrical interpretations. Consider, for example, arithmetical progressions in which the difference (d) between two consecutive integers is constant:

$$1, 2, 3, 4, 5, \ldots \quad (d = 1);$$
$$1, 3, 5, 7, 9, \ldots \quad (d = 2);$$
$$1, 4, 7, 10, 13, \ldots \quad (d = 3).$$

Or generally:

$$1, \; 1 + d, \; 1 + 2d, \; 1 + 3d, \; 1 + 4d, \ldots$$

Each element of each row has its position, n. In order to obtain the nth element (call it a_n), we add to the first element of the row the product of the difference and the number of steps from 1 to n which is $(n - 1)$:

$$a_n = 1 + d(n - 1).$$

The elements of every such series are called linear figurate numbers, or figurate numbers of the first order.

2. Let us form consecutive sums of a row of linear figurate numbers. The first "sum" is simply the first element of the series, the second is the sum of the first two elements, and the nth is the sum of the first n elements.

The first series of linear figurate numbers 1, 2, 3, 4, 5, . . . produces this series of sums: 1, 3, 6, 10, 15, . . . They are called triangular numbers.

The second row 1, 3, 5, 7, 9, . . . produces square numbers:

$$1, 4, 9, 16, 25, \ldots$$

The third row 1, 4, 7, 10, 13, . . . produces pentagonal numbers:

$$1, 5, 12, 22, 35, \ldots$$

Hexagonal, heptagonal, and higher polygonal numbers are produced by later rows.

Polygonal numbers are called planar figurate numbers, or figurate numbers of the second order.

3. The geometric names of polygonal numbers are explained by geometric interpretations the ancient Greeks gave them. The diagram (four polygons) shows a method of forming polygonal arrays of dots having 1, 2, or any number of dots radiating out from the lower left corner. Counting the dots at each step will give you the first four planar series.

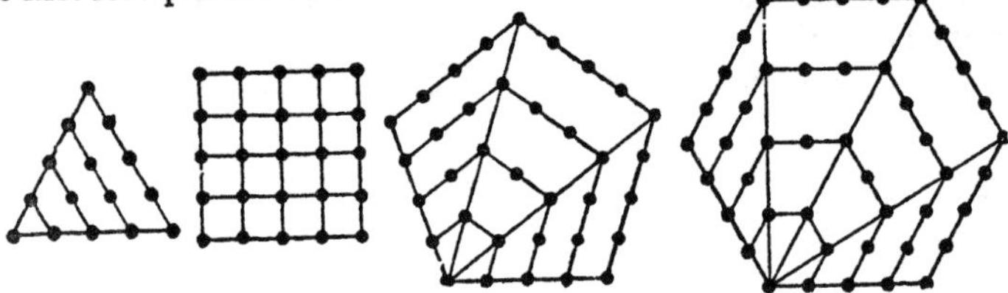

4. Let us tabulate facts about the planar figurate numbers, including the formulas giving the nth terms of each series:

		Numbers						
d	*Figures*	S_1	S_2	S_3	S_4	S_5		*Formulas*
1	*Triangle*	1	3	6	10	15	. . .	$\frac{n(n+1)}{2}$
2	*Square*	1	4	9	16	25	. . .	n^2
3	*Pentagon*	1	5	12	22	35	. . .	$\frac{n(3n-1)}{2}$
4	*Hexagon*	1	6	15	28	45	. . .	$n(2n-1)$
.	. . . .	.	.	.	.	.	. . .	. . . .
d		1	$2+d$	$3+3d$	$4+4d$	$5+10d$		$\frac{n[dn-(d-2)]}{2}$

The last line generalizes the planar series with a constant difference *d;* the general formula for the *n*th number is at bottom right.

5. Many interesting relations hold between the set of positive integers and that of planar figurate numbers, also among the planar figurate numbers.

Pierre de Fermat of Toulouse, France, a jurist and civic leader, made a hobby of mathematics. In his lifetime (1601-1665) he made many significant discoveries in the theory of numbers, for example:

Any positive integer is triangular or is the sum of 2 or 3 triangular numbers.

Any positive integer is square or is the sum of 2, 3, or 4 square numbers.

Generally, any positive integer is the sum of not more than k k-gonal numbers.

Euler proved this for some cases; the French mathematician Augustin Cauchy found a general proof in 1815.

6. The ancient Greek mathematician Diophantus, who lived in the third century B. C., found a simple connection between triangular numbers T and square numbers K:

$$8T + 1 = K.$$

You can exemplify this formula by considering the triangular number 21.

The diagram below shows 169 dots in a square array. Let $13 \times 13 = 169$ be our square number K. One dot occupies the square's center and the other 168 are grouped in 8 triangular numbers T in the shape of 8 right triangles with sawtooth hypotenuses.

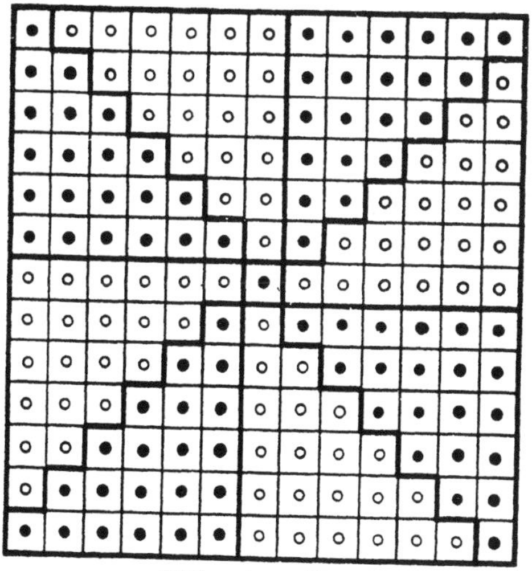

$8T + 1 = K$

7. On your own, prove algebraically the validity of Diophantus's formula. Also, that no triangular number can end in 2, 4, 7, or 9; and that every hexagonal number is a triangular number with an odd position in the triangular series.

8. Making consecutive sums from the planar figurate numbers $V_1 = S_1$, $V_2 = S_1 + S_2$, $V_3 = S_1 + S_2 + S_3$, and so on, we obtain spatial figurate numbers, or figurate numbers of the third order.

For example, the row of triangular numbers (1, 3, 6, 10, 15, . . .) produces this row of figurate numbers of the third order:

$$1, 4, 10, 20, 35, \ldots$$

These are called pyramidal numbers, because they can be represented by building tetrahedral pyramids of balls as in the last diagram (showing pyramids of 4 and 10 balls).

9. Here is a table of facts about figurate numbers of the third order, including the formulas giving the *n*th terms of each series. As in the previous table, the general series and formula are on the last line.

	Numbers						
d	V_1	V_2	V_3	V_4	V_5		Formulas
1	1	4	10	20	35	. . .	$\frac{1}{6}n(n+1)(n+2)$
2	1	5	14	30	55	. . .	$\frac{1}{6}n(n+1)(2n+1)$
3	1	6	18	40	75	. . .	$\frac{1}{2}n^2(n+1)$
4	1	7	22	50	95	. . .	$\frac{1}{6}n(n+1)(4n-1)$
.	.	.	.	.	.	. . .	. . . .
d	1	$3+d$	$6+4d$	$10+10d$	$15+20d$		$\frac{1}{6}n(n+1)[dn-(d-3)]$

(No answer to this problem is given in the back of the book.)

Answers

Answers

I. Amusing Problems

1. OBSERVANT CHILDREN

Note the smoke coming out of the locomotive chimney. If the locomotive was standing, its smoke would be going in the wind's direction. If it was moving forward in the absence of a wind, its smoke would be inclined backward. As shown on opposite page 1, the smoke of the moving locomotive is straight up. Therefore, the train is moving at a speed equal to the wind's speed: 20 miles per hour.

2. THE STONE FLOWER

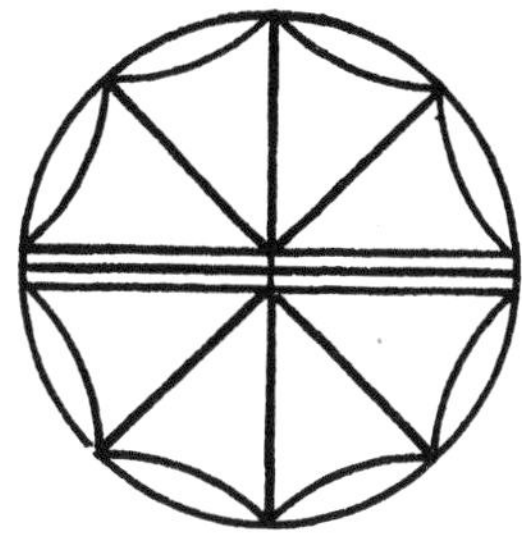

3. MOVING CHECKERS

Number the checkers from left to right, as shown. If the open space is on the left, move checkers 2 and 3 to the left (move I in the picture). In the space now vacant place checkers 5 and 6 (move II). Now move checkers 6 and 4 to the left (move III).

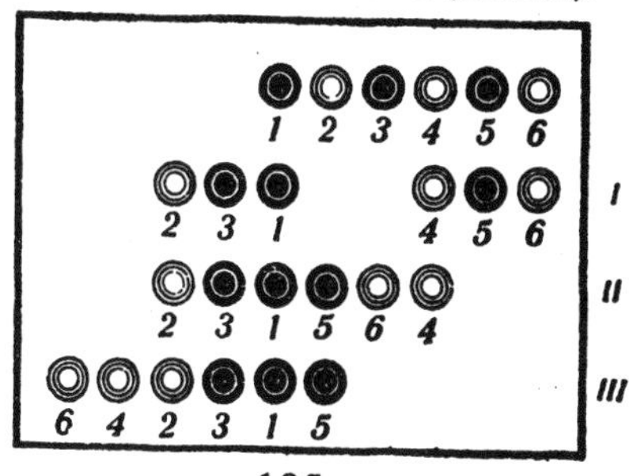

4. THREE MOVES

First pile to second; second to third; third to first:

Pile	*Initial number*	*First move*	*Second move*	*Third move*
First	11	11 − 7 = 4	4	4 + 4 = 8
Second	7	7 + 7 = 14	14 − 6 = 8	8
Third	6	6	6 + 6 = 12	12 − 4 = 8

5. COUNT!

Thirty-five.

6. THE GARDENER'S ROUTE

A possible route is shown in the diagram.

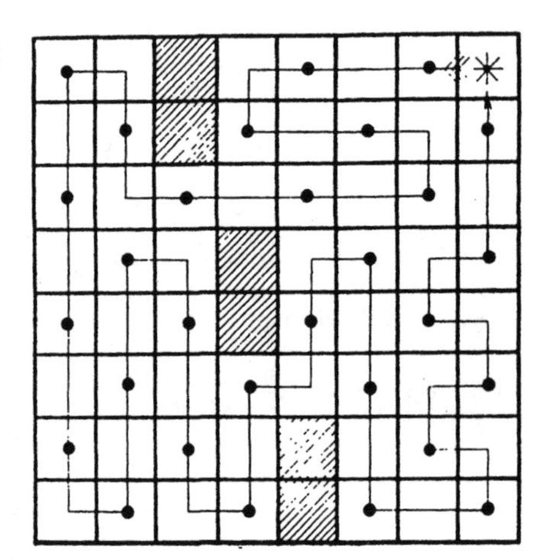

7. FIVE APPLES

Give the fifth girl her apple in the basket.

8. DON'T THINK TOO LONG

Four cats, each near the tail of a cat in an adjacent corner.

9. DOWN AND UP

At the start, 1 inch of the yellow pencil gets smeared with wet paint. As the blue pencil is moved downward, a second inch of the blue pencil's length is smeared. After the next upward movement the second inch of the blue pencil smears a second inch of the yellow pencil.

Each pair of down-up moves of the blue pencil smears 1 more inch of each pencil. Five pairs of moves will smear 5 inches. This, together with the initial inch, makes 6 inches for each pencil.

(Looking at his boots, Leonid Mikhailovich noticed that their entire lengths were muddied where they usually rub each other while he walks.

"How puzzling," he thought. "I didn't walk in any deep mud, yet my boots are muddied up to the knees."

Now you understand the origin of the puzzle.)

10. CROSSING A RIVER

First the boys cross the river. One stays ashore while the other brings the boat to the soldiers and gets out. A soldier gets in the boat and crosses. The boy who has stayed there brings the boat back to the soldiers, then takes the other boy to the other shore. Again a boy brings the boat back, gets out, a second soldier crosses over . . . until all the soldiers have crossed.

11. WOLF, GOAT, AND CABBAGE

A wolf does not eat cabbage, so the crossing can start with the goat.

The man leaves the goat and returns, puts the cabbage in the boat and takes it across. On the other bank, he leaves the cabbage but takes the goat.

He leaves the goat on the first bank and takes the wolf across. He leaves the cabbage with the wolf and rows back alone.

He takes the goat across.

12. ROLL THEM OUT

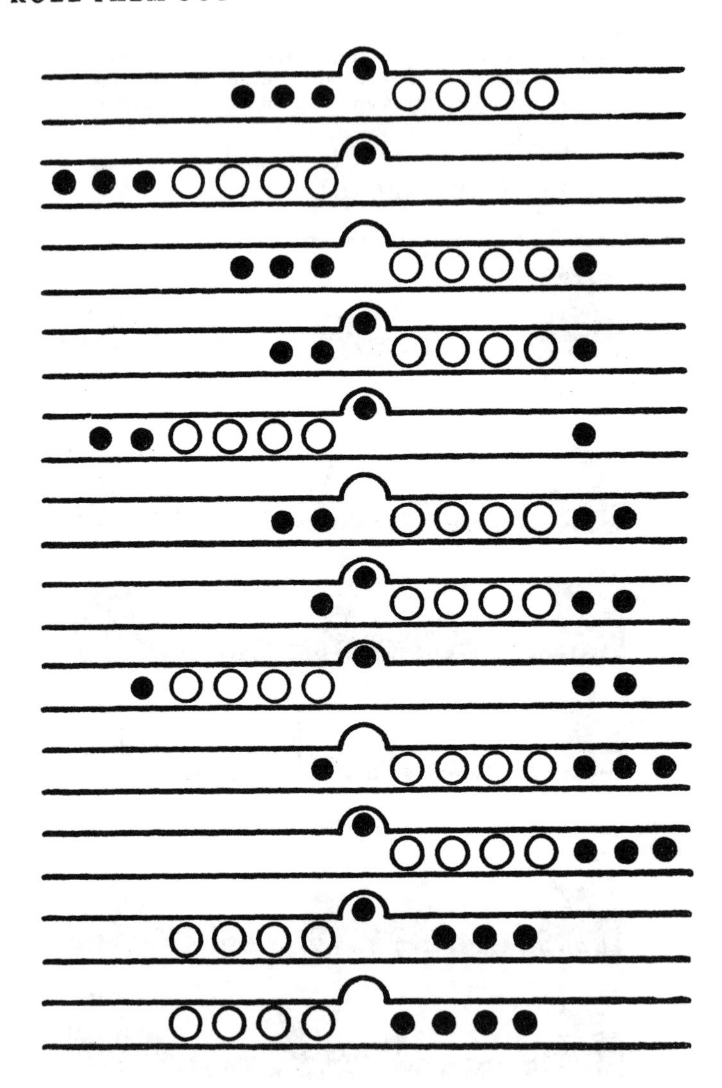

13. REPAIRING A CHAIN

He opens all 3 rings of 1 piece (3 operations). With these he links the other 4 pieces together. Total: 6 operations.

14. CORRECT THE ERROR

Two solutions are shown:

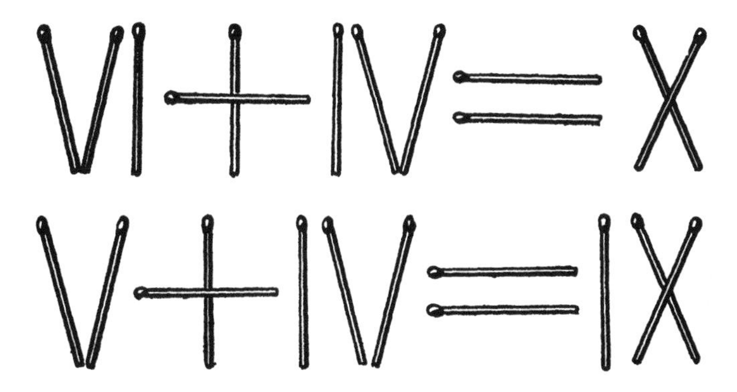

15. FOUR OUT OF THREE (A JOKE)

III=3; IV=4

16. THREE AND 2 IS 8 (ANOTHER JOKE)

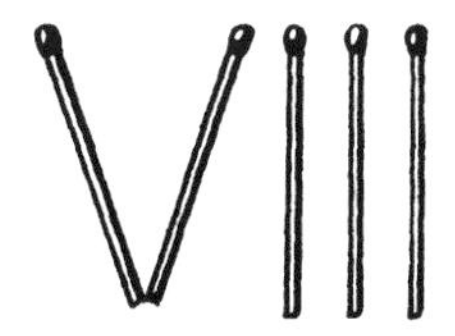

17. THREE SQUARES

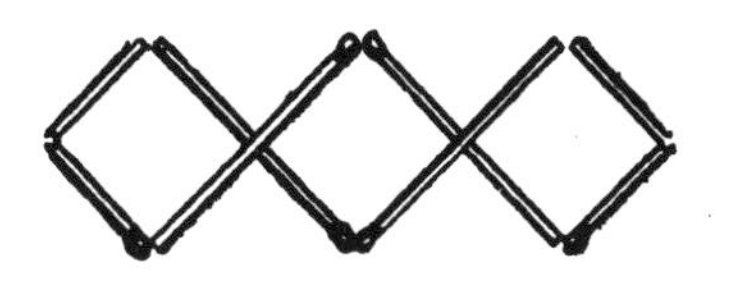

18. HOW MANY ITEMS?

From 36 blanks, 36 items are made. The lead shavings are enough to make 6 blanks, which make 6 more items. But don't stop here. The new shavings are good for 1 more item. Total: 43.

19. ARRANGING FLAGS

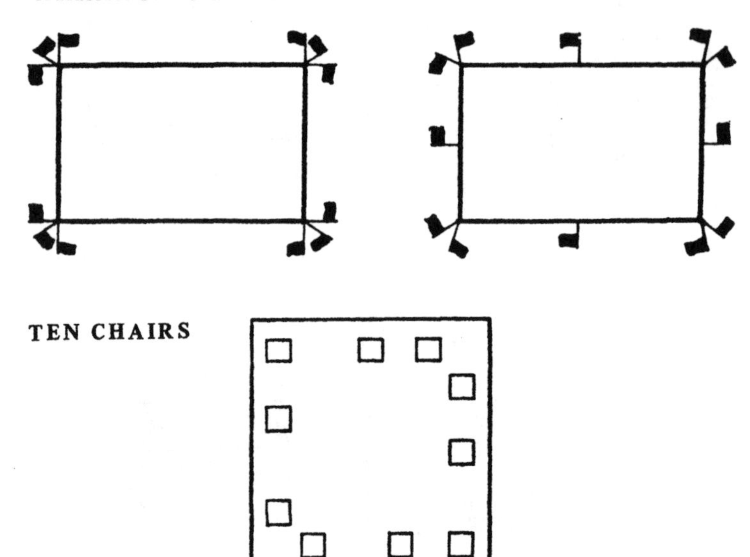

20. TEN CHAIRS

21. KEEP IT EVEN

Two solutions are shown.

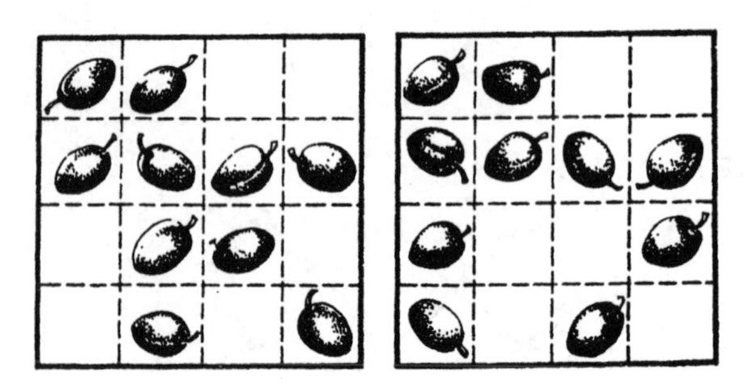

22. A MAGIC TRIANGLE

The triangles show a 17-solution and then two 20-solutions.

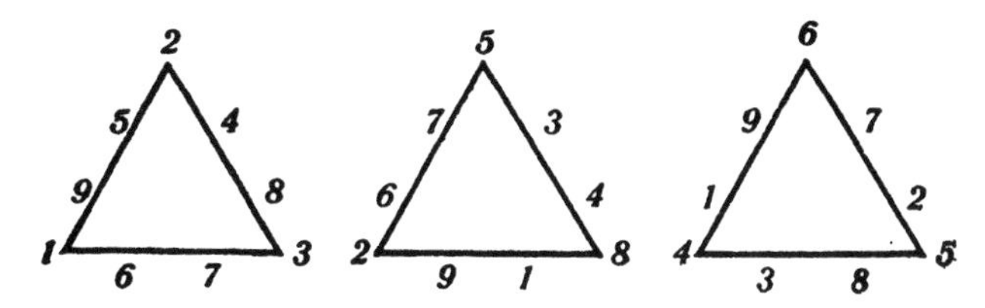

23. GIRLS PLAYING BALL

The diagram shows that all 13 girls play ball when they play by skipping 5. Skipping 6 gives the same order, but in the other direction.

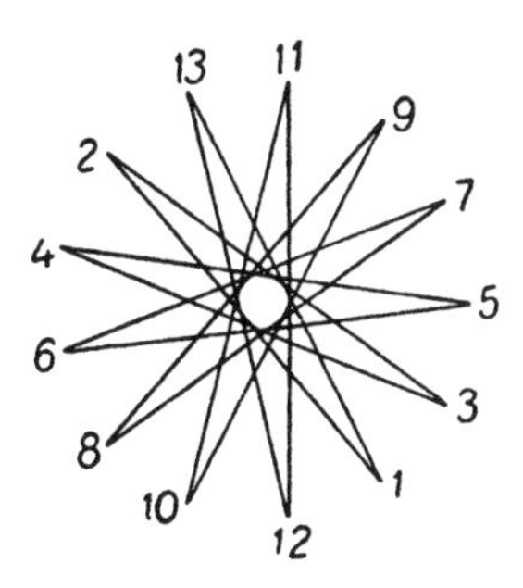

24. FOUR STRAIGHT LINES

The figure shows one solution.

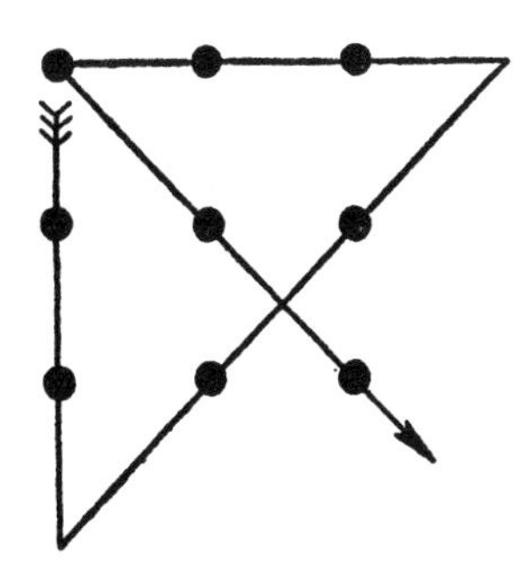

25. GOATS FROM CABBAGE

The picture shows the solution.

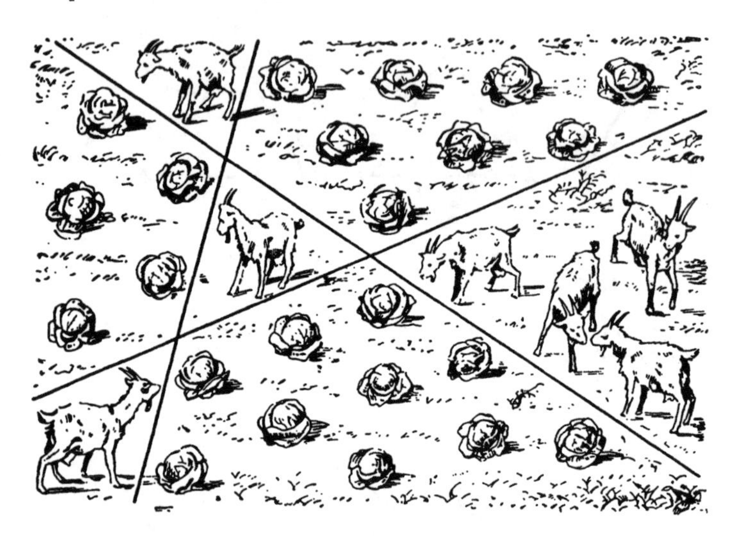

26. TWO TRAINS

100 miles (60 + 40).

27. THE TIDE COMES IN (A JOKE)

When a problem deals with a physical phenomenon, the phenomenon should be considered as well as the numbers given. As the water rises, so does the rope ladder. The water will never cover the rung.

28. A WATCH FACE

The sum of the numbers on the watch face is 78. If the two lines cross, there must be 4 equal parts, but 78 is not divisible by 4. Then the lines do not cross, giving 3 parts with sums of 26

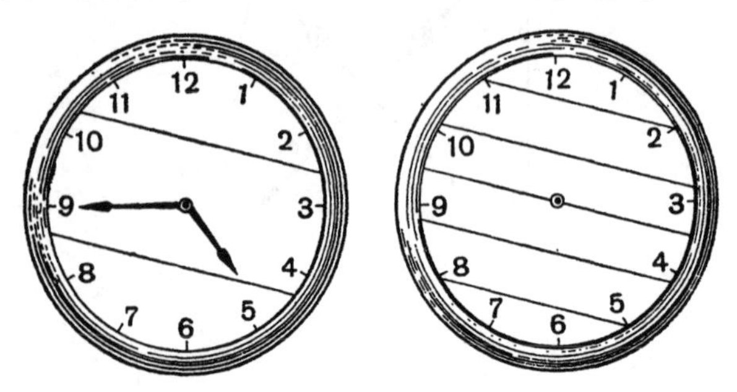

each. Once you see the pairs on the face that add to 13 (12 + 1, 11 + 2, and so on), the answer shown to the first question is easy to find. And the answer shown to the second question immediately follows.

29. A BROKEN CLOCK FACE

There are three adjacents Xs in IX, X, and XI, and two of them must be in one part. The crack must split IX, not XI, so that the numbers add to 80.

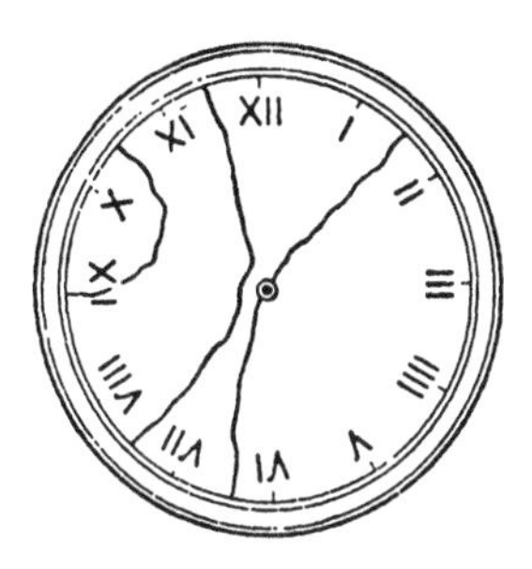

30. THE WONDROUS CLOCK

As the problem says, the apprentice mixed up the clock hands so that the minute hand was short and the hour hand long.

The first time the apprentice returned to the client was about 2 hours and 10 minutes after he had set the clock at six. The long hand moved only from 12 to a little past 2. The short hand made 2 full circles and an additional 10 minutes. Thus the clock showed the correct time.

Next day around 7:05 a.m. he came a second time, 13 hours and 5 minutes after he had set the clock for six. The long hand, acting as hour hand, covered 13 hours to reach 1. The short hand made 13 full circles and 5 minutes, reaching 7. So the clock showed the correct time again.

31. THREE IN A ROW

20: 8 rows of 3 buttons (see *a*) and 12 rows of 2 buttons (*b*).

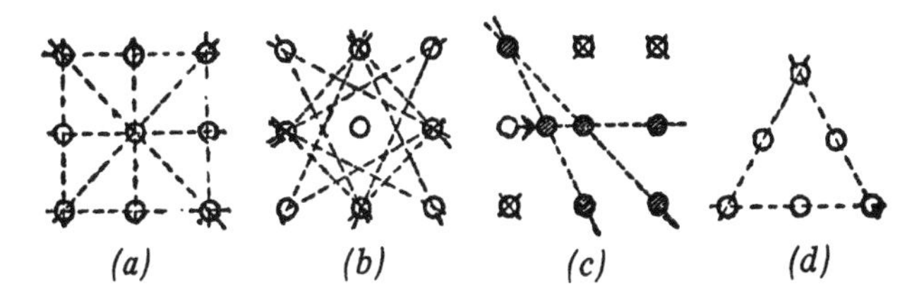

In *c*, the x-ed-out buttons are removed. The dotted-line circle moves slightly to the right, as the arrow shows.

A second arrangement of 6 buttons in 3 rows is shown in *d*.

32. TEN ROWS

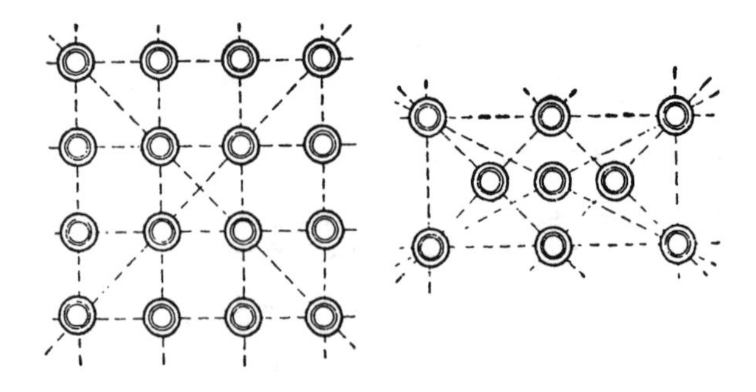

33. PATTERN OF COINS

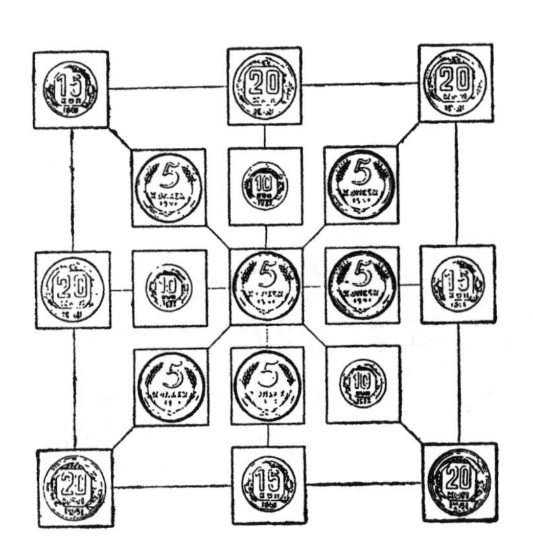

34. FROM 1 THROUGH 19

There are nine pairs of numbers that total 20 (1 + 19, 2 + 18, and so on). The remaining number, 10, goes in the center to make 30.

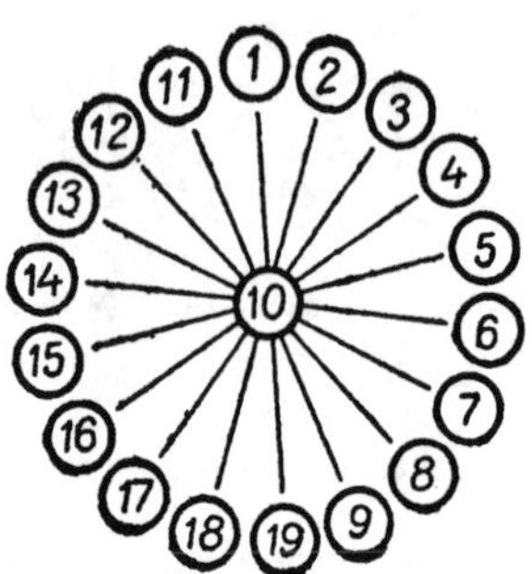

35. SPEEDILY YET CAUTIOUSLY

(A) Neither.

(B) A pound of metal is always worth more than half a pound of the same metal.

(C) Six strokes took 30 seconds, therefore 12 strokes will take 60 seconds – such is the usual train of thought. But when the clock struck six, there were only 5 intervals between strokes, and each interval was $30 \div 5 = 6$ seconds. Between the first and twelfth strokes there will be 11 intervals of 6 seconds each; therefore, 12 strokes will take 66 seconds.

(D) There is always a plane that contains any given 3 points.

36. A CRAYFISH FULL OF FIGURES

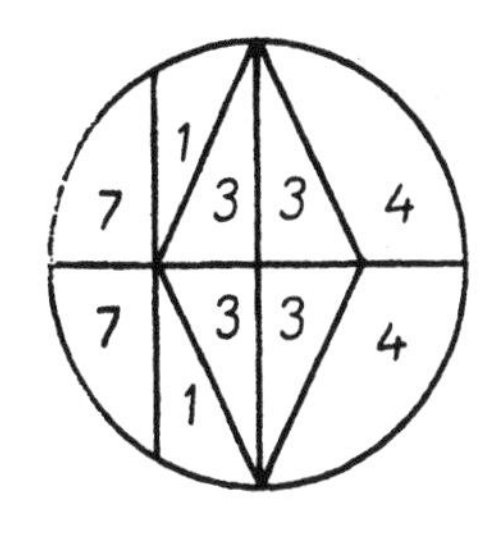

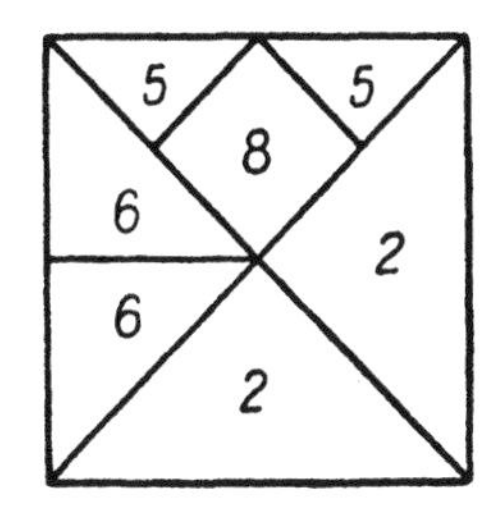

37. THE PRICE OF A BOOK

$2.

38. THE RESTLESS FLY

The problem is simpler than it looks. The cyclists took 6 hours to meet. The fly traveled $6 \times 30 = 180$ miles.

39. UPSIDE-DOWN YEAR

1961.

40. TWO JOKES

(A) $4. She had read 86 upside down.

(B) Turn 9 upside down and exchange it with the 8. Both columns will add to 18.

41. HOW OLD AM I?

The difference in age is still 23 years, so I must be 23 if my father is twice as old.

42. TELL "AT A GLANCE"

The columns don't look like they have the same sums, but look closer: comparing digits, nine 1s match one 9; comparing tens, eight 2s match two 8s, and so on. Check by adding – the sums are equal.

43. A QUICK ADDITION

(A) In the first and fifth lines, the last digits add to 10, and the other corresponding pairs of digits add to 9. The sum of the numbers is therefore 1,000,000.

The second and sixth numbers, the third and seventh, and the fourth and eighth add to 1,000,000. The sum of all eight is 4,000,000.

(B) The eight numbers are:

7,621
3,057
2,794
4,518
5,481
7,205
6,942
2,378

To do the quick adding, you need only multiply 9,999 by 4, that is multiply 10,000 by 4 and subtract 4. The answer is 39,996.

(C) 48,726,918, total 172,603,293. Every digit of your third number plus the corresponding digit of the second number equals 9. The sum is simply the first number plus 100,000,000 and minus 1.

44. WHICH HAND?

	Dime ("even" coin) is in:	
	Right hand	Left hand
Right hand (× 3)	odd × even = even	odd × odd = odd
Left hand (× 2)	even × odd = even	even × even = even
	Sum: even	Sum: odd

The trick still works if you ask your friend to multiply by other odd and even numbers than 3 and 2.

45. HOW MANY?

Four brothers and three sisters.

46. WITH THE SAME FIGURES

$$22 + 2 + 2 + 2 = 28; \quad 888 + 88 + 8 + 8 + 8 = 1{,}000.$$

47. ONE HUNDRED

$111 - 11 = 100$; $(5 \times 5 \times 5) - (5 \times 5) = 100$; $(5 + 5 + 5 + 5) \times 5 = 100$; $(5 \times 5)\,[5 - (5 \div 5)] = 100$.

48. A DUEL IN ARITHMETIC

There are two ways to get 1,111 by replacing 10 digits with zeros, five ways by using 9 zeros, six ways by using 8 zeros, three ways with 7 zeros, one way with 6 zeros, and one with 5 zeros—a total of eighteen in all. The last variant is

$$111 + 333 + 500 + 077 + 090 = 1{,}111.$$

Try to find the other seventeen variants on your own.

49. TWENTY

As in packing a suitcase, we will start with the large items and work down. All solutions will be in descending order.

We can't use 19, 17, or 15 because they don't leave room for seven more addends. To 13, it is necessary and sufficient to add seven 1s:

$$13 + 1 + 1 + 1 + 1 + 1 + 1 + 1 = 20.$$

After 11, we can't use 9, 7, or 5. Trying 3:

$$11 + 3 + 1 + 1 + 1 + 1 + 1 + 1 = 20.$$

No other solution with 11 is possible.

Now start with 9. We can't use 7 next ($9 + 7 = 16$ and there are no six odd numbers that add to 4). We get one solution with 5 next:

$$9 + 5 + 1 + 1 + 1 + 1 + 1 + 1 = 20.$$

And another with 3 next:

$$9 + 3 + 3 + 1 + 1 + 1 + 1 + 1 = 20.$$

The method is clear, and the seven other solutions (of eleven) follow:

$$7 + 7 + 1 + 1 + 1 + 1 + 1 + 1 = 20;$$
$$7 + 5 + 3 + 1 + 1 + 1 + 1 + 1 = 20;$$
$$7 + 3 + 3 + 3 + 1 + 1 + 1 + 1 = 20;$$
$$5 + 5 + 5 + 1 + 1 + 1 + 1 + 1 = 20;$$
$$5 + 5 + 3 + 3 + 1 + 1 + 1 + 1 = 20;$$
$$5 + 3 + 3 + 3 + 3 + 1 + 1 + 1 = 20;$$
$$3 + 3 + 3 + 3 + 3 + 3 + 1 + 1 = 20.$$

Note: Only the sixth-to-last solution has as many as four different addends.

50. HOW MANY ROUTES?

Yes. You will get mixed up if you try to draw each route from A to C – it is too complicated. It is simpler to solve the problem for points near A and progress point by point to C. The diagram labels all the points from 1a (which is A) to 5e (which is C).

It is evident that there is only 1 route each from A to the nearest points on AB and AD (2*a* and 1*b*). You can get to 2*b* through either of these points (2 routes). Now the crossing 2*c* can be reached from 2*b* (2 routes) or from 1*c* (1 more route, making 3 in all). Analogously, there are 3 routes to 3*b*.

Now it becomes clear that the number of routes shown in each crossing is the sum of the number of routes shown immediately to the left and immediately below – which is logical, because all moves are up or to the right.

We keep adding, working from point to point, until we reach C with its 70 different routes from A.

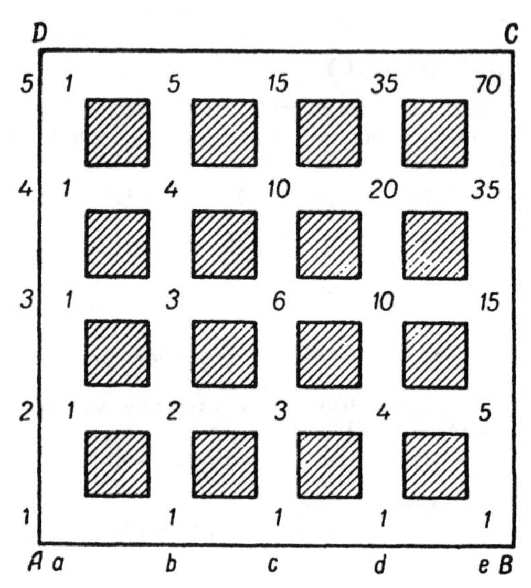

51. ORDER THE NUMBERS

We are to place, for example, A and a at the ends of one diameter, and B and b at the same ends, respectively, of an adjacent diameter, so that $A + B = a + b$. Therefore, $A - a = b - B$. It follows that all differences between opposite pairs of numbers must be equal.

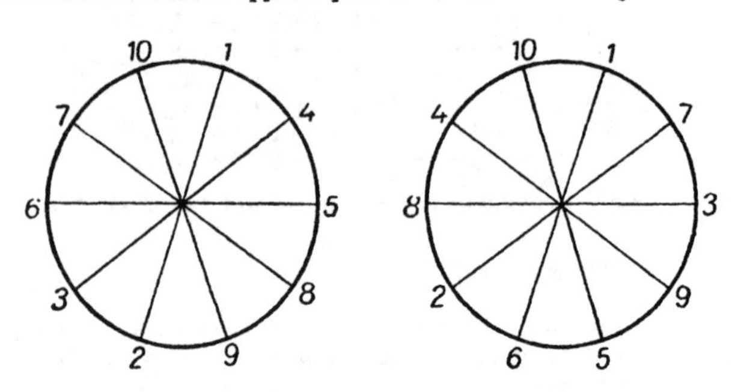

The numbers from 1 through 10 must be grouped in 5 pairs, each with the same difference. There are only two ways to do it, with differences of 1 and 5:

1 – 2	1 – 6
4 – 3	7 – 2
5 – 6	3 – 8
8 – 7	9 – 4
9 – 10	5 – 10

Both solutions are shown in the diagram. Variants can be formed by moving pairs of opposite numbers.

To eliminate rotations, fix 1 where it is in the circle on the left, with 2 opposite. Now, not only (4 – 3) can be clockwise from (1 – 2): the equality will hold with 6 – 5, 8 – 7, or 10 – 9 on the second diameter. This gives 4 pairs for the second diameter: there are 3 pairs to choose from for the third diameter, 2 for the fourth, and 1 for the fifth. Thus there are 24 variants (including the basic one) for the circle on the left. With 24 more for the circle on the right, there are 48 in all.

52. DIFFERENT ACTIONS, SAME RESULT

The only solution for 4 numbers:

$$1 + 1 + 2 + 4 = 1 \times 1 \times 2 \times 4.$$

For 5 numbers, three solutions:

$$1 + 1 + 1 + 2 + 5 = 1 \times 1 \times 1 \times 2 \times 5;$$
$$1 + 1 + 1 + 3 + 3 = 1 \times 1 \times 1 \times 3 \times 3;$$
$$1 + 1 + 2 + 2 + 2 = 1 \times 1 \times 2 \times 2 \times 2.$$

(On your own, see if you can find multiple solutions for 6, 7, and more numbers.)

53. NINETY-NINE AND ONE HUNDRED

$$9 + 8 + 7 + 65 + 4 + 3 + 2 + 1 = 99;$$
$$9 + 8 + 7 + 6 + 5 + 43 + 21 = 99.$$

$$1 + 2 + 34 + 56 + 7 = 100;$$
$$1 + 23 + 4 + 5 + 67 = 100.$$

54. A CUT-UP CHESSBOARD

One solution is shown.

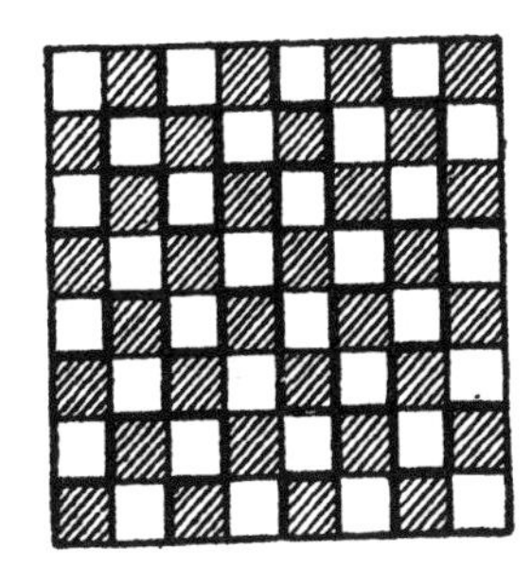

55. LOOKING FOR A LAND MINE

One solution is shown: one route is a solid line; the other, dotted.

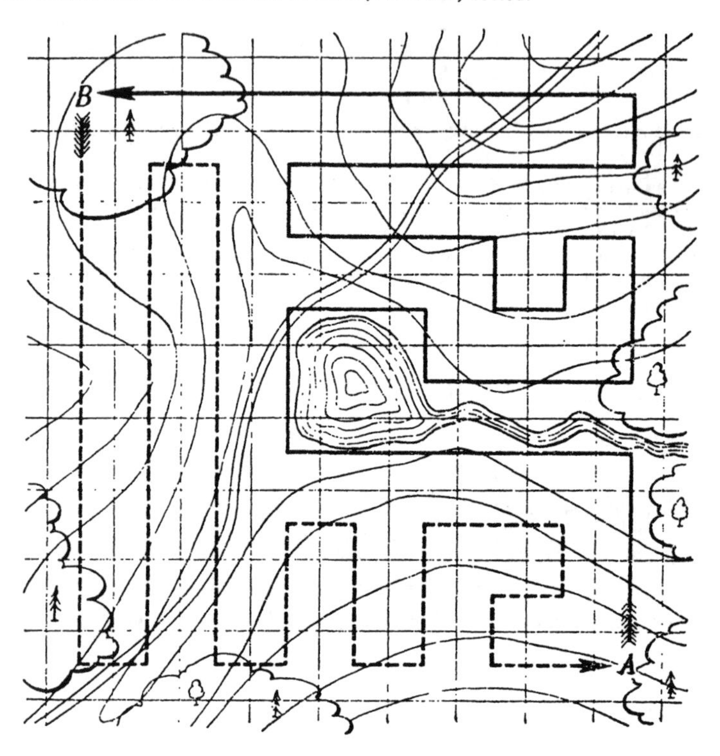

56. GROUPS OF TWO

Other solutions are variants of the one given. The basic procedure is: Make a crossed pair at either end of the row. In the next two moves, make a crossed pair second from each end of the remaining row of 8 uncrossed matches. The last two moves then become obvious.

57. GROUPS OF THREE

Move match 5 to 1, then 6 – 1, 9 – 3, 10 – 3, 8 – 14, 7 – 14, 4 – 2, 11 – 2, 13 – 15, 12 – 15.

The basic procedure is parallel to that used in the last problem: Form a triplet at either end of the row, then triplets second from both ends of the remaining row. Again, there are variants.

It can be shown that at least 8 matches are needed to make pairs, at least 12 to make triplets, and at least $4k$ to make k-lets.

58. THE STOPPED CLOCK

Before I left, I wound the wall clock. When I returned, the change in time it showed equaled the time it took to go to my friend's and return, plus the time I spent there. But I knew the latter, because I looked at my friend's watch both when I arrived and when I left.

Subtracting the time of the visit from the time I was absent from my house, and dividing by 2, I obtained the time it took me to return home. I added this time to what my friend's watch showed when I left, and set the sum on my wall clock.

59. PLUS AND MINUS SIGNS

Again, there is only one way:

$$123 - 45 - 67 + 89 = 100$$

60. THE PUZZLED DRIVER

The first digit of 15,951 could not change in 2 hours. Therefore, 1 is the first and last digit of the new number. The second and fourth digits changed to 6. If the middle digit is 0, 1, 2, . . ., then the car traveled 110, 210, 310, . . ., miles in 2 hours. Clearly the first alternative is the correct one, and the car traveled 55 miles per hour.

61. FOR THE TSIMLYANSK POWER INSTALLATION

Distributing among the nine young workers the 9 extra sets produced by the chief, the daily average for all ten men is $15 + 1 = 16$ sets. Then the chief turns out $16 + 9 = 25$ sets daily, and the entire brigade, $(15 \times 9) + 25 = 160$ sets.

Those who know algebra can solve the problem by composing an equation with one unknown.

62. DELIVERING GRAIN ON TIME

At 30 miles per hour a truck travels a mile in 2 minutes; at 20 miles per hour, in 3 minutes. At the latter speed the truck is 1 minute slower per mile. To lose 2 hours, or 120 minutes, takes 120 miles, which is how far the kolhoz is from the city.

You might assume the speed desired is halfway between 20 and 30 miles per hour, or 25 miles per hour, but this is wrong.

At 30 mph, the truck would cover 120 miles in 4 hours. The trip is to take 1 hour longer, or to arrive at 11:00 a.m., and calls for a speed of $(120 \div 5) = 24$ miles per hour.

63. RIDING THE TRAIN TO A DACHA

If the girls had been on a standing train, the first girl's calculations would have been correct, but their train was moving. It took 5 minutes to meet a second train, but then it took the second train 5 more minutes to reach where the girls met the first train. So the time between trains is 10 minutes, not 5, and only 6 trains per hour arrive in the city.

64. FROM 1 TO 1,000,000,000

The numbers can be grouped by pairs:

999,999,999 and 0;
999,999,998 and 1;
999,999,997 and 2, and so on.

There are half a billion pairs, and the sum of the digits in each pair is 81. The digits in the unpaired number, 1,000,000,000, add to 1. Then:

$$(500{,}000{,}000 \times 81) + 1 = 40{,}500{,}000{,}001.$$

65. A SOCCER FAN'S NIGHTMARE

If the ping-pong ball rolls flush against the wall, the cast-iron ball cannot crush it.

Those who know geometry can determine that if the diameter of a large ball is at least 5.83 $(3 + 2\sqrt{2})$ times as large as the diameter of a little ball, then the little ball will be safe if it hugs the wall.

A cast-iron ball that is larger than a soccer ball is more than 4.83 times as large in diameter as a ping-pong ball.

66. MY WATCH

In 24 hours the watch gained $1/2 - 1/3 = 1/6$ minute. It would seem it would be 5 minutes fast in $5 \times 6 = 30$ days; that is, the morning of May 31. But already on the morning of May 28 the watch was $27 \div 6 = 4\frac{1}{2}$ minutes fast. At the close of that day the watch gained 1/2 minute more, so it was 5 minutes fast on May 28.

67. STAIRS

$2\frac{1}{2}$ times ($5 \div 2$, not $6 \div 3$).

68. A DIGITAL PUZZLE

A decimal point.

69. INTERESTING FRACTIONS

1/5; 1/7. Any fraction with 1 for numerator and any odd number $(2n - 1)$ for denominator increases to n times its value when its denominator is added to its numerator and to its denominator.

70. WHAT IS IT?

$1\frac{1}{2}$.

71. THE SCHOOLBOY'S ROUTE

From the machine and tractor station to the railroad station is $1/3 - 1/4 = 1/12$ of the way. Boris walks it in 5 minutes, so the whole trip takes 1 hour. A quarter of this is 15 minutes. So he leaves his house at 7:15 and reaches school at 8:15.

72. AT THE STADIUM

Not 12 seconds. There are 7 segments from the first flag to the eighth, and 11 from the first to the twelfth. He runs each segment in 8/7 seconds; therefore, 11 segments take $\frac{88}{7} = 12\frac{4}{7}$ seconds.

73. WOULD HE HAVE SAVED TIME?

Yes. He took as much time for the second half of his trip as the whole trip would have taken on foot. So no matter how fast the train was, he lost exactly as much time as he spent on the train.

He would have saved 1/30 by walking all the way.

74. THE ALARM CLOCK

In 3½ hours the alarm clock has become 14 minutes slow. At noon the alarm clock will fall behind approximately an additional minute. Its hands will show noon in 15 minutes.

75. LARGE SEGMENTS INSTEAD OF SMALL

He noticed that $7/12 = 1/3 - 1/4$, so he cut 4 sheets into 12 thirds, and 3 sheets into 12 fourths. Each worker got one third and one fourth, or 7/12.

For the other distributions he used:

$$\frac{5}{6} = \frac{1}{2} + \frac{1}{3};\ \frac{13}{12} = \frac{1}{3} + \frac{3}{4};\ \frac{13}{36} = \frac{1}{4} + \frac{1}{9};\ \frac{26}{21} = \frac{2}{3} + \frac{4}{7}\text{, and so on.}$$

76. A CAKE OF SOAP

Since 1/4 cake weighs 3/4 pound, an entire cake weighs 3 pounds.

77. ARITHMETICAL NUTS TO CRACK

(A) $1 \times 1; \frac{1}{1}, \frac{2}{2}, \ldots; 1 - 0.\ 2 - 1, \ldots; 1°, 2°, \ldots; 01$, and many others.

(B) $37 = \frac{333}{3 \times 3}$; $37 = 3 \times 3 \times 3 + \frac{3}{.3}$.

(C) $99 + \frac{99}{99}$; $55 + 55 - 5 - 5$; $\frac{666 - 66}{6}$; or, in a general form: $\frac{(100a + 10a + a) - (10a + a)}{a}$, where a is any digit.

(D) $44 + \frac{44}{4} = 55$.

(E) $9 + \frac{99}{9} = 20$.

(F)

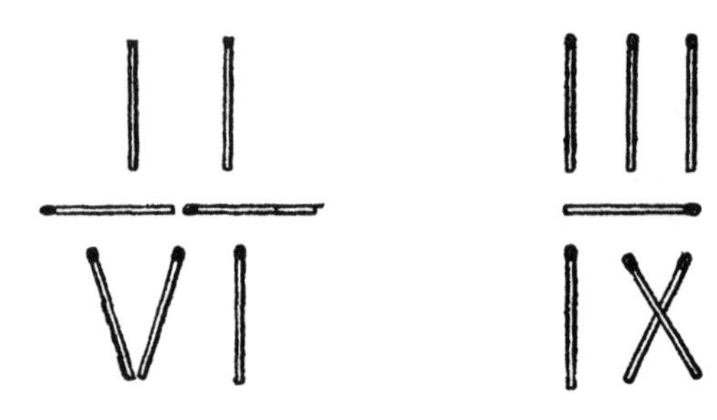

Can you find another solution?

(G) $$1 + 3 + 5 + 7 + \frac{75}{75} + \frac{33}{11} = 20.$$

(H) $$79\tfrac{1}{3} + 5 = 84 + \tfrac{2}{6}\ ;\ 75\tfrac{1}{3} + 9 = 84 + \tfrac{2}{6}\,.$$

(I) 1 and $\frac{1}{2}$; $\frac{1}{3}$ and $\frac{1}{4}$; in general, $\frac{1}{n-1}$ and $\frac{1}{n}$, where $(n - 1)$ is an integer greater than 1. Also x and $\frac{x}{x+1}$, where x is a positive integer.

(J) $\frac{35}{70} + \frac{148}{296}$; $\frac{45}{90} + \frac{138}{276}$; $\frac{15}{30} + \frac{486}{972}$; $0.5 + \frac{1}{2}(9 - 8)(7 - 6)(4 - 3)$, and so on.

(K) $78\frac{3}{6} + 21\frac{45}{90}$; $50\frac{1}{2} + 49\frac{38}{76}$; $29\frac{1}{3} + 70\frac{56}{84}$, and so on.

78. DOMINO FRACTIONS

One of the possible solutions:

$$\frac{1}{3} + \frac{6}{1} + \frac{3}{4} + \frac{5}{3} + \frac{5}{4} = 10;$$
$$\frac{2}{1} + \frac{5}{1} + \frac{2}{6} + \frac{6}{3} + \frac{4}{6} = 10;$$
$$\frac{4}{1} + \frac{2}{3} + \frac{4}{2} + \frac{5}{2} + \frac{5}{6} = 10.$$

Try some other sums.

79. MISHA'S KITTENS

Three-quarters of a kitten is one-quarter of Misha's kittens. He has $4 \times \frac{3}{4} = 3$ kittens.

80. AVERAGE SPEED

Without thinking you might answer 8 miles per hour (half of 12 + 4). But if you consider the whole route as 1, the horse takes $\frac{1}{2} \div 12 = \frac{1}{24}$ unit of time for the first half, and $\frac{1}{2} \div 4 = \frac{1}{8}$ unit for the second half. The sum is $\frac{1}{6}$ unit, so the average speed is 6 miles per hour.

81. THE SLEEPING PASSENGER

Two-thirds of one-half the whole trip, that is, one-third.

82. HOW LONG IS THE TRAIN

The speed of the passenger in the first train, in relation to the movement of the second train, is 45 + 36 = 81 miles per hour, or:

$$\frac{5{,}280 \times 81}{60 \times 60} = 118.8 \text{ feet per second.}$$

Therefore, the length of the second train is 6 × 118.8 = 712.8 feet.

83. A CYCLIST

He walks one-third of the way, or half as far as he rides, but it takes him twice as long. Therefore, he rides four times as fast as he walks.

84. A CONTEST

Volodya has done 2/3 of his assignment; 1/3 remains. Kostya has done 1/6 of his assignment; 5/6 remains.

Kostya must increase his output to $\frac{5}{6} \div \frac{1}{3} = 2\frac{1}{2}$ times as fast as Volodya.

85. WHO IS RIGHT?

Masha's girl friend was right. Masha was multiplying two-thirds of a number by four-thirds of it. But $\frac{2}{3} \times \frac{4}{3} = \frac{8}{9}$, or the correct answer minus one-ninth of itself. One-ninth of the correct volume is 20 cubic yards, and the correct answer is 180 cubic yards.

86. THREE SLICES OF TOAST

She puts two slices in the pan; after 30 seconds she has toasted one side of each. She turns over the first slice, takes the second from the pan, and in its place puts the third slice. After the second half minute the first slice is done and the other two are half done. In the last 30 seconds she finishes the second and third slices.

II. Difficult Problems

87. BLACKSMITH KHECHO'S INGENUITY

The prisoners placed 1 piece of chain (10 pounds) in a basket and sent it down. Into the empty basket that came up they put 2 pieces of chain (20 pounds). They kept adding 2 pieces to each basket that came up until they sent a 70-pound load down, getting back a 60-pound load.

Khecho replaced the 6 pieces of chain (60 pounds) with the servant (80 pounds). The girl descended as 7 pieces of chain came up. He unloaded 6 pieces and signaled the girl below to climb out. He lowered the remaining piece of chain, bringing the empty basket up.

The servant got in the basket again (total weight 80 + 10 = 90 pounds) and Daridjan (100 pounds) descended. They both got out, Daridjan on the ground, the servant in the tower. Down went the basket still with 1 piece of chain in it, and up came the other basket, now empty.

Khecho repeated his first set of actions and soon lowered the servant to the ground again. He signaled Daridjan and the servant (100 + 80 = 180 pounds) to get in, allowing Khecho (180 pounds) to descend with 1 piece of chain. Now the two women were in the tower and Khecho on the ground.

The servant was brought down as before, then Daridjan replaced her on the ground. In due time, the servant made her fourth and last trip down, bringing up 7 pieces of chain. As she stepped out, Khecho fastened the basket to keep the chain in the top basket from falling.

88. CAT AND MICE

Start from the cross in the diagram (position 13) and go clockwise through positions 1, 2, 3, . . . , crossing out each thirteenth dot: 13, 1, 3, 6, 10, 5, 2, 4, 9, 11, 12, 7, and 8. Call position 8 the white mouse, and Purrer starts clockwise from the fifth mouse clockwise from the white mouse (i.e., position 13 relative to position 8). Or he starts counterclockwise from the fifth mouse counterclockwise from the white mouse.

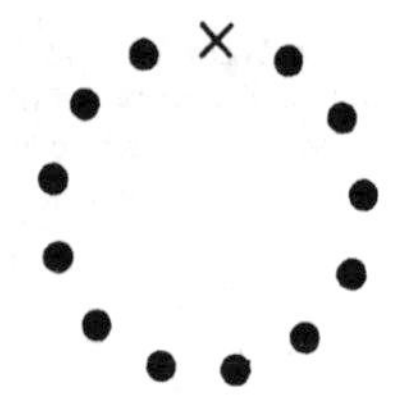

89. SISKIN AND THRUSH

From left to right, the seventh and fourteenth cages.

90. MATCHES AND COINS

A good method is to aim at the match you just started from. Say you start with the fifth match and put a coin by the seventh. Now start from the third so you can put a coin by the fifth, start from the first so you can put a coin by the third, and so on (see diagram).

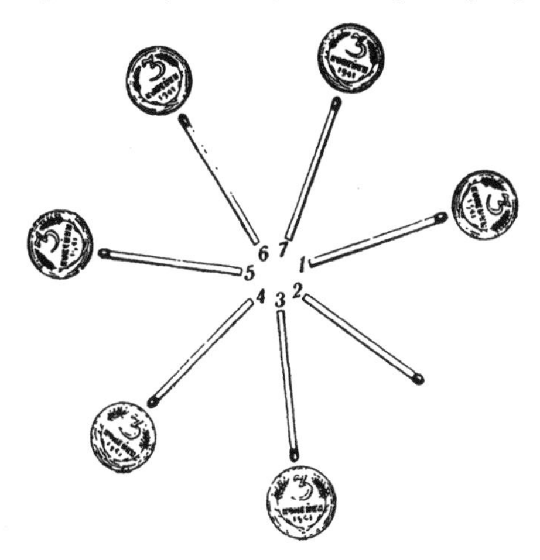

91. LET THE PASSENGER TRAIN THROUGH!

The work train backs into the siding, which can hold its rear 3 cars.

Uncoupling them in the siding, the rest of the work train goes forward a sufficient distance.

The passenger train comes up and couples on the 3 cars left by the work train. It backs up on the main track.

The work train backs up into the siding, which will now hold its engine and the remaining 2 cars.

The passenger train uncouples the 3 cars it took from the siding and goes through.

92. THE WHIM OF THREE GIRLS

Call the three fathers A, B, C, and their daughters correspondingly a, b, c.

First shore	Second shore
A B C	. . .
a b c	. . .

1. First two girls go:

First shore	Second shore
A B C	. . .
a . .	. b c

2. A girl returns and rows the third girl across:

First shore	Second shore
A B C	. . .
. . .	a b c

3. One of the three girls returns and remains ashore with her father. The other two fathers row across:

A . .	. *B C*
a . .	. *b c*

4. A father returns to the first shore with his daughter; the girl remains there, and two fathers go:

. . .	*A B C*
a b .	. . *c*

5. The last girl returns to the first shore and crosses with the second girl:

. . .	*A B C*
a . .	. *b c*

6. The girl on the first shore is fetched by her father (or one of the other two girls):

. . .	*A B C*
. . .	*a b c*

93. AN EXPANSION OF PROBLEM 92

(A) Crossing in a boat that holds three: Call the fathers *A, B, C, D,* and their daughters correspondingly *a, b, c, d.*

First shore	In the boat	Second shore
A B C D		. . . .
a b c d		. . . .

1. Three girls go:

First shore	In the boat	Second shore
A B C D		. . . .
a . . .	*b c d* →	. *b c d*

Two return:

First shore	In the boat	Second shore
A B C D		. . . .
a b c .	← *b c*	. . . *d*

2. A father goes with his daughter and the father whose daughter is on the other shore:

First shore	In the boat	Second shore
A B . .	*C D* } →	. . *C D*
a b . .	*c*	. . *c d*

A father and his daughter return:

First shore	In the boat	Second shore
A B C .	← { *C*	. . . *D*
a b c .	*c*	. . . *D*

3. Three fathers go:

First shore	In the boat	Second shore
. . . .	*A B C* →	*A B C D*
a b c .		. . . *d*

A girl returns:

First shore	In the boat	Second shore
. . . .		*A B C D*
a b c d	← *d*	. . . .

4. The girl who has just returned takes two with her:

First shore	In the boat	Second shore
. . . .		*A B C D*
a . . .	*b c d* →	. *b c d*

Father A returns for his daughter (or one of the other 3 girls does):

A . . . a . . .	$\leftarrow A$	. B C D . b c d

5. The last pair go:

. . . .	$\left.\begin{matrix}A\\a\end{matrix}\right\}\rightarrow$	A B C D a b c d

(B) Crossing in a boat that holds two persons (solution by Y. V. Morozova):

	First shore	Island	Second shore
	A B C D a b c d	. . . .	. . . .
1.	A B C D a b . .	. . . .	. . . . . . c d
2.	A B C D a . . .	. . . .	. . . . . b c d
3.	A B . . a b . .	. . . .	. . C D . . c d
4.	A B C . a b . .	. . . . . . c .	. . . D . . . d

(C took c to the island, returned to the first shore, and gave the boat to two girls.)

	First shore	Island	Second shore
5.	A B C	. . . . a b c .	. . . D . . . d
6.	A . . . a . . .	. . . . . b c .	. B C D . . . d
7.	A . . . a . . .	. . . . . b . .	. B C D . . c d
8.	. . . . a . . .	. . . . . b . .	A B C D . . c d

(B went to fetch A and took him directly to the second shore.)

	First shore	Island	Second shore
9.	. . . . a . . .	. . . .	A B C D . b c d
10.	. . . .	. . . .	A B C D a b c d

94. JUMPING CHECKERS

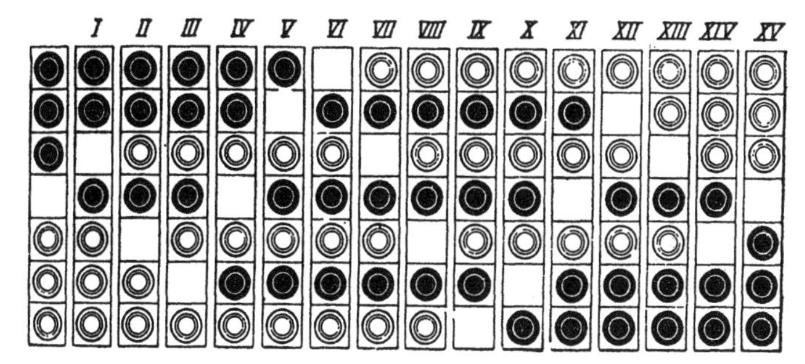

95. WHITE AND BLACK

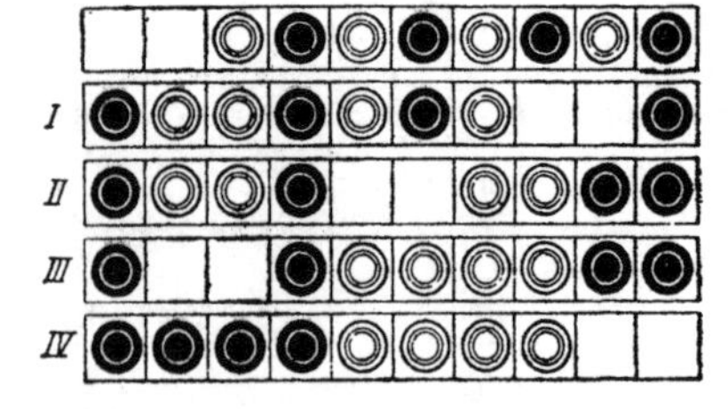

96. COMPLICATING THE PROBLEM

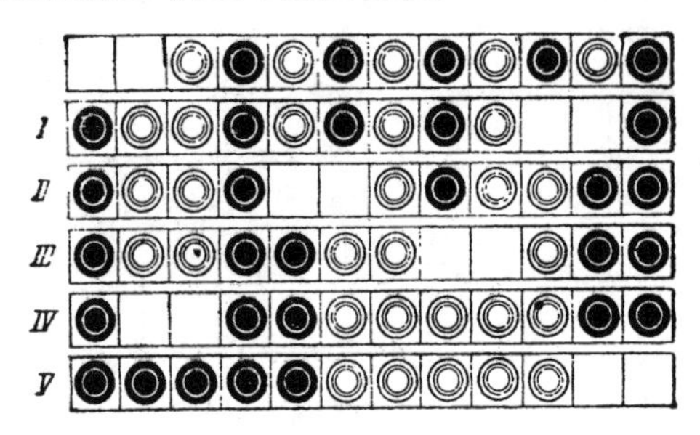

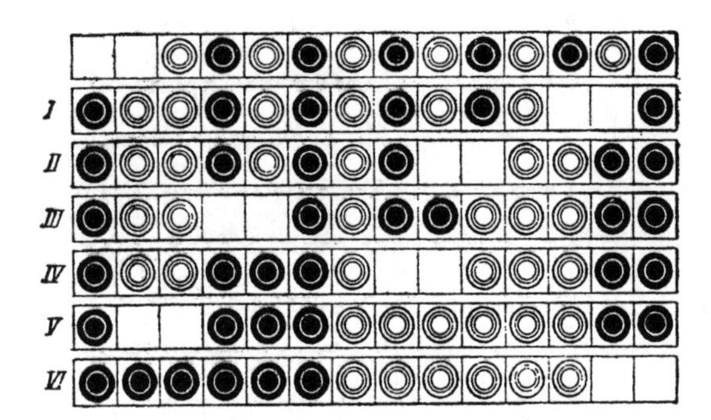

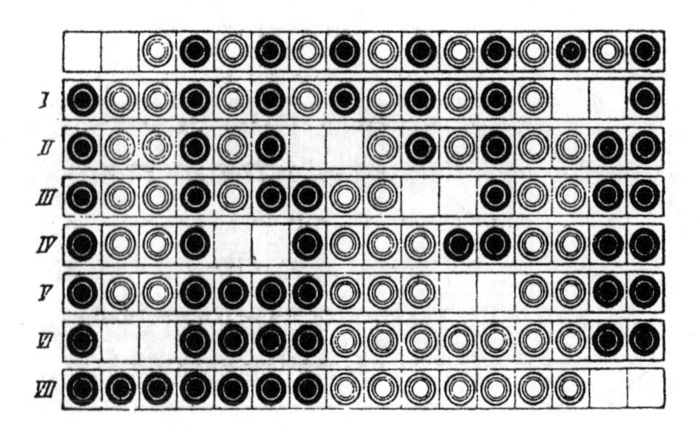

97. THE GENERAL PROBLEM

In the diagram, vertical strokes help us concentrate on the 2 pairs at left and 2 pairs at right.

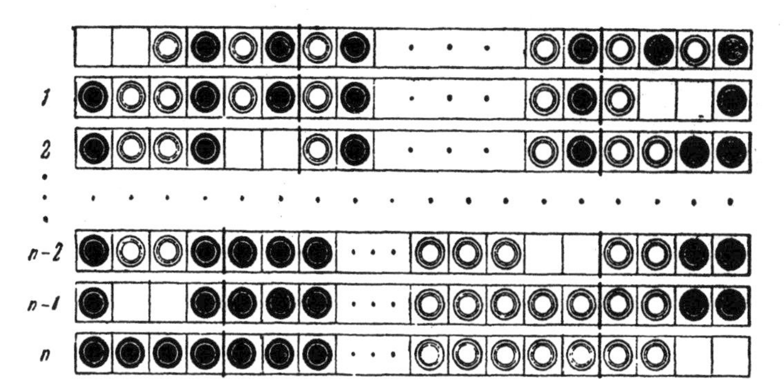

In the first 2 moves leave the inner ($n - 4$) pairs alone and arrive at the position shown for the outer 4 pairs, leaving the vacancy to the right of the 2 left pairs.

In the next ($n - 4$) moves you will be able to put the inner pairs in order, black left, white right. At move ($n - 2$) leave the vacancy to the left of the two right pairs.

The last 2 moves put the outer pairs in order, completing the solution.

98. SMALL CARDS PLACED IN ORDER

The first deal described produces this sequence on the table, with the 4 on top:

1, 3, 5, 7, 9, 2, 6, 10, 8, 4.

Since the 4 is tenth, put the 10 fourth from the top as you form a new pile of cards. The 8 is in ninth, so put the 9 eighth, and so on. The new pile, from top down, will now be in the desired order:

1, 6, 2, 10, 3, 7, 4, 9, 5, 8.

99. TWO ARRANGEMENT PUZZLES

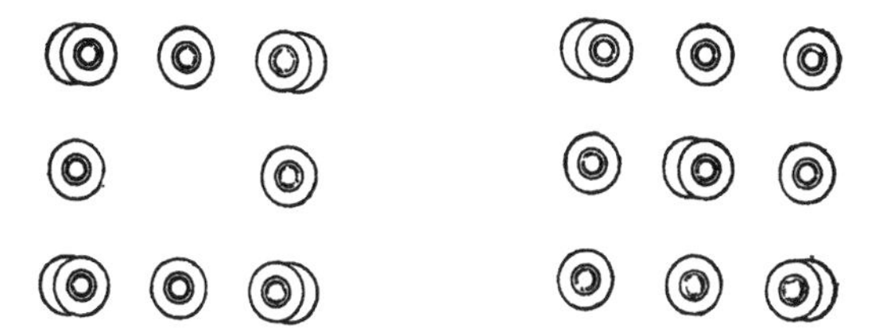

100. A MYSTERIOUS BOX

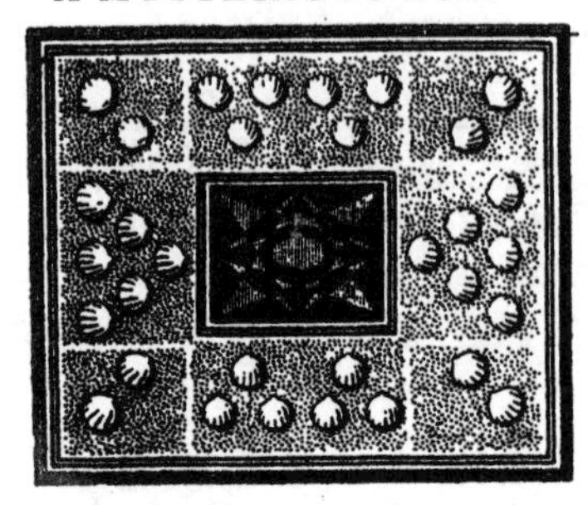

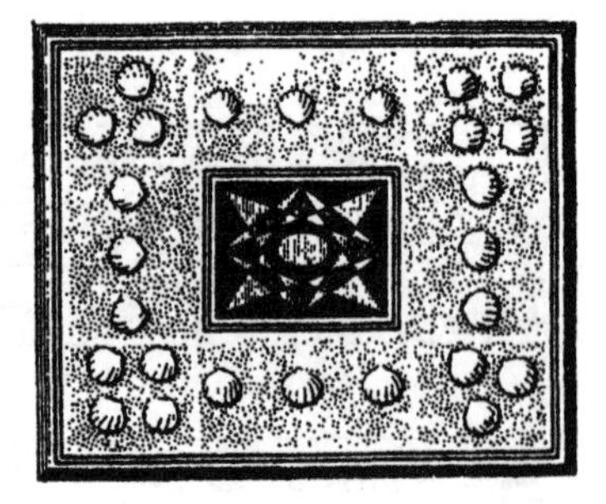

101. THE COURAGEOUS GARRISON

2	7	2
7	**36**	7
2	7	2

3	5	3
5	**32**	5
3	5	3

4	3	4
3	**28**	3
4	3	4

5	1	5
1	**24**	1
5	1	5

6	—	5
—	**22**	—
5	—	6

102. DAYLIGHT LAMPS

He can use any number of lamps from 18 through 36, in some cases with some loss of symmetry. The maximum case is shown in the last diagram.

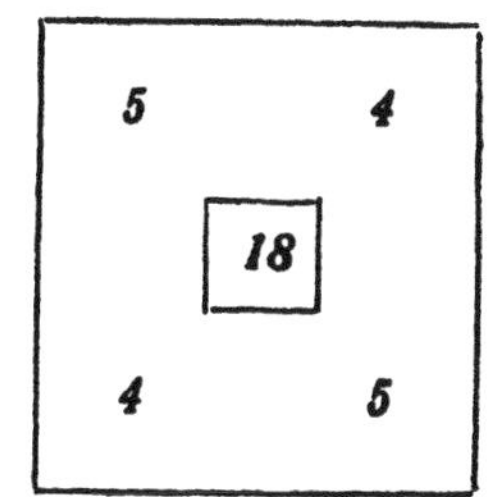

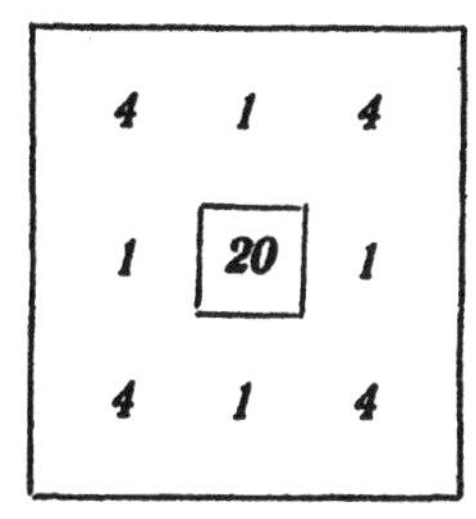

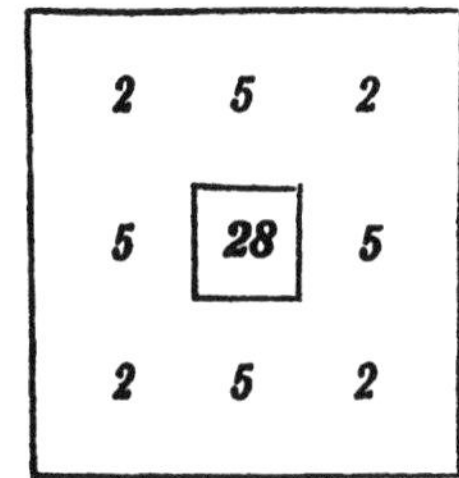

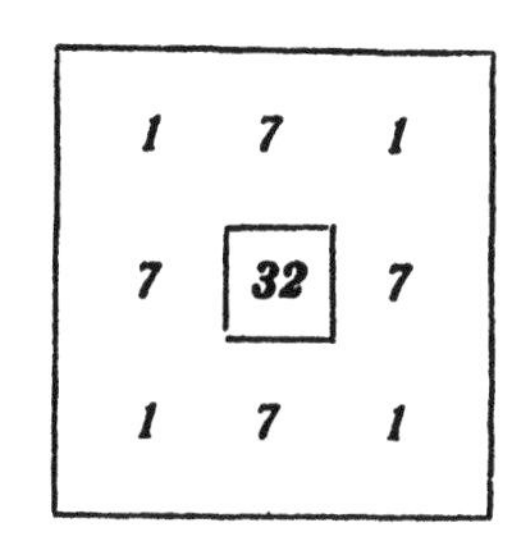

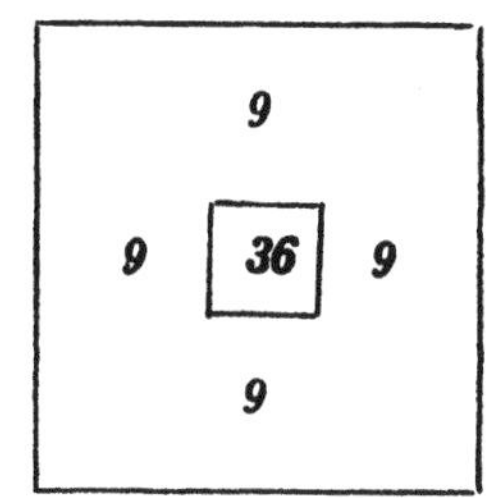

103. ARRANGEMENT OF EXPERIMENTAL RABBITS

It follows from condition 3 that from 22 through 44 rabbits can be housed (see solution of Problem 102).

But the number of rabbits must be a multiple of 3 (condition 4). Thus the number can be 24, 27, 30, 33, 36, 39, or 42. Further, trial shows 24 rabbits cannot be housed 11 to a side (condition 3) without leaving empty sections (condition 1), and that 33, 36, 39, or 42 rabbits cannot be housed 11 to a side without placing more than 3 rabbits in some sections (condition 2).

By elimination, 30 rabbits were expected, and 27 arrived. The diagrams show how they were housed. (In both pairs of diagrams, the second floor is on the left, and the first floor on the right.)

2	3	3
3	20	2
3	2	2

1	1	1
1	10	2
1	2	1

3	1	3
1	18	2
3	2	3

2	1	1
1	9	1
1	1	1

104. PREPARING FOR A FESTIVAL

(A) Four solutions are shown below. The third and fourth were found by fourth-grader Batyr Erdniyev of Stavropol. He also arranged 12 lamps in 7 rows as shown in the fifth diagram, the one that looks like a dunce cap.

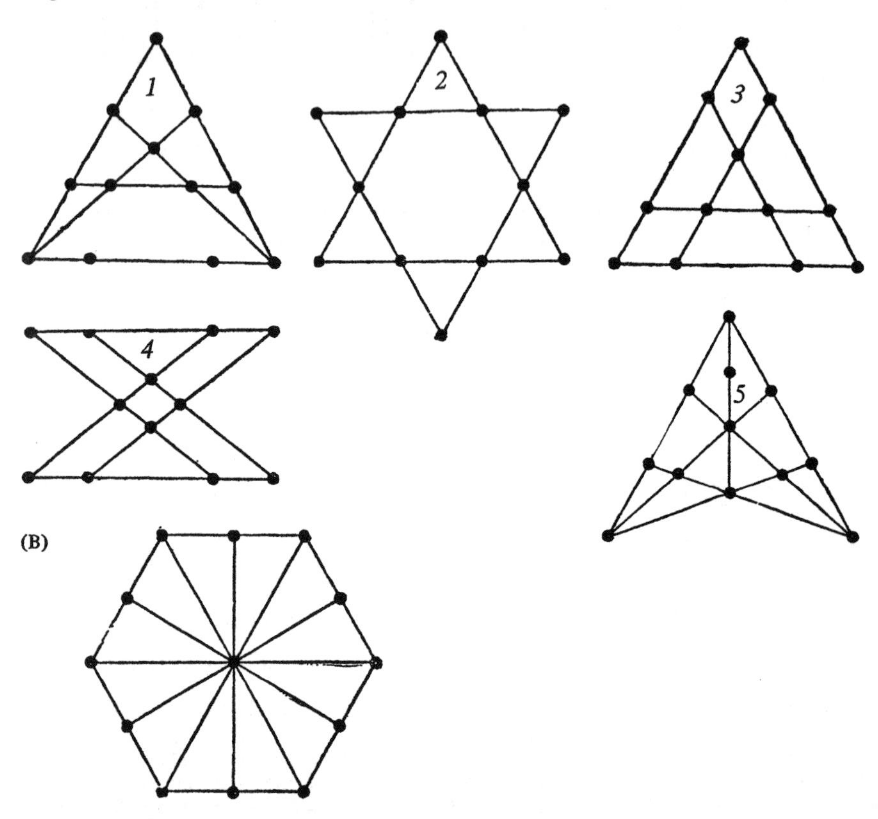

(C) The problem's basic condition is satisfied by the five-pointed star (left) but it is better not to have intersections where there are no objects. The gardener, therefore, chose the irregular star at right (found by V. I. Lebedev, a Moscow engineer).

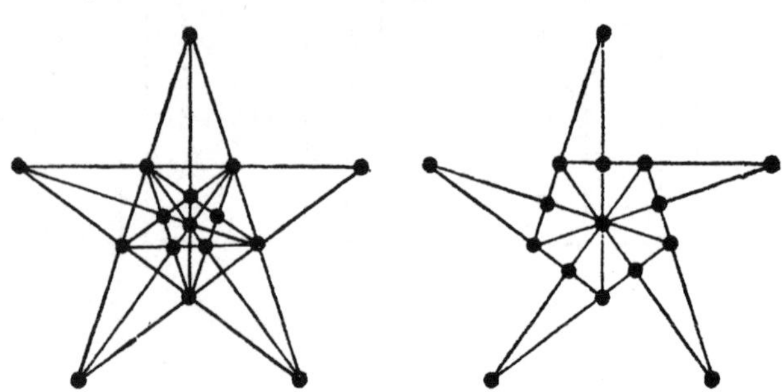

(D) A square superimposed on a square. A simpler solution: form a 5-by-5 square array!

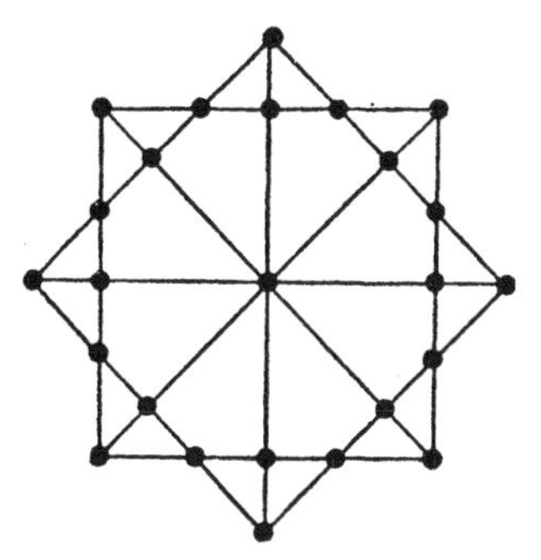

105. PLANTING OAKS

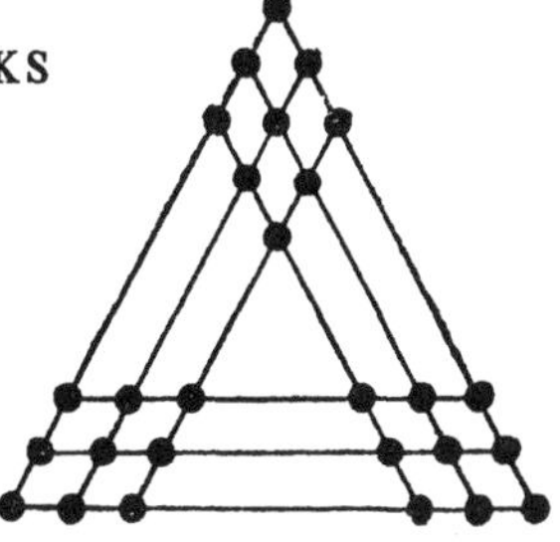

106. GEOMETRICAL GAMES

(A) All the possible solutions can be quickly and easily obtained by simple geometrical constructions. Represent the checkers by dots on a piece of paper. Strike out any 3 dots and 1 bottom dot. Connect one of the remaining 2 top dots to any 2 remaining dots, then connect the second top dot with the other 2 remaining bottom dots (diagram below). Discard combinations that result in parallel lines. Place 4 checkers corresponding to the crossed-out dots at the 4 intersection points of the lines drawn.

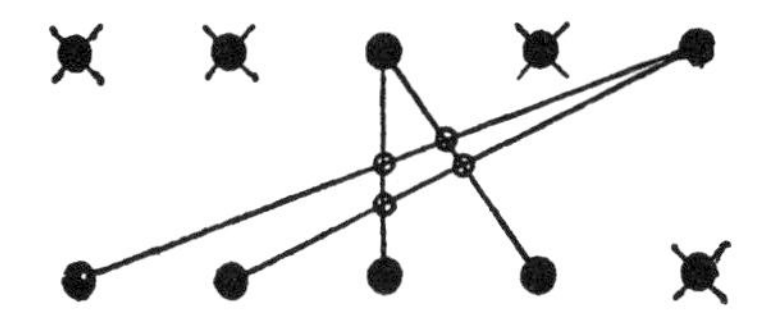

(B)

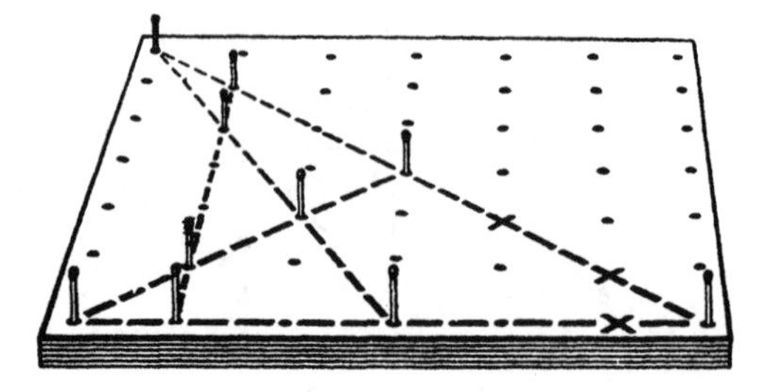

107. EVEN AND ODD

The minimum is 24 moves (in the first move, for example, checker 1 is shifted to circle *A*):

1. 1 – *A*;	7. 3 – *B*;	13. 3 – *C*;	19. 6 – *C*;
2. 2 – *B*;	8. 1 – *B*;	14. 1 – *C*;	20. 8 – *B*;
3. 3 – *C*;	9. 6 – *C*;	15. 5 – *A*;	21. 6 – *B*;
4. 4 – *D*;	10. 7 – *A*;	16. 1 – *B*;	22. 2 – *E* (or *C*);
5. 2 – *D*;	11. 1 – *A*;	17. 3 – *A*;	23. 4 – *B*;
6. 5 – *B*;	12. 6 – *E*;	18. 1 – *A*;	24. 2 – *B*.

108. A PATTERN

The best approach is to select moves in short chains. Thus, after exchanging checkers 1 and 7, move checker 7 into position 7 (where checker 20 is), then move checker 20 into position 20 (where checker 16 is), and so on. On the sixth move, the exchange puts both checkers in the correct square, so start a new chain.

The quickest solution is 19 moves in five chains:

1 – 7, 7 – 20, 20 – 16, 16 – 11, 11 – 2, 2 – 24;
3 – 10, 10 – 23, 23 – 14, 14 – 18, 18 – 5;
4 – 19, 19 – 9, 9 – 22;
6 – 12, 12 – 15, 15 – 13, 13 – 25;
17 – 21.

109. A PUZZLE GIFT

Take one of the 9 candies from the outer, largest box and move it to the smallest box. Then there are 5 candies (2 pairs plus 1) in this inner box; these 5 must be included in the number of candies in the second inner box. This box contains 5 + 4 = 9 candies (4 pairs plus 1).

The third box contains 9 + 4 = 13 candies (6 pairs plus 1), and the largest box holds 13 + 8 = 21 candies (10 pairs plus 1). Other solutions are possible.

110. KNIGHT'S MOVE

The first capture may be any pawn except pawns *c*4, *d*3, *d*4, *e*5, *e*6, and *f*5. Place the knight, for example, on *a*3, with the first capture pawn *c*2, then pawns *b*4, *d*3, *b*2, *c*4, *d*2, *b*3, *d*4, *e*6, *g*7, *f*5, *e*7, *g*6, *e*5, *f*7, and *g*5.

111. SHIFTING THE CHECKERS

(A) The first move below is checker 2 to cell 1, and so on:

1. 2 – 1;	11. 7 – *B;*	21. 1 – *C;*	31. 7 6;	41. 6 – *C;*	51. 1 – 4;
2. 3 – 2;	12. 8 – 7;	22. 9 – 7;	32. 7 7;	42. 5 – 4;	52. 1 – 3;
3. 4 – 3;	13. 8 – 6;	23. 9 – 8;	33. 7 8;	43. 5 – 5;	53. 1 – *A;*
4. 4 – *A;*	14. 8 – 5;	24. 9 – 9;	34. 1 7;	44. 5 – 6;	
5. 5 – 4;	15. 9 – 8;	25. 9 – 10;	35. 1 6;	45. 5 – 7;	
6. 5 – 3;	16. 9 – 7;	26. 8 – 6;	36. 1 5;	46. 4 – 3;	
7. 6 – 5;	17. 9 – 6;	27. 8 – 7;	37. 1 *B;*	47. 4 – 4;	
8. 6 – 4;	18. 1 – 9;	28. 8 – 8;	38. 6 5;	48. 4 – 5;	
9. 7 – 6;	19. 1 – 8;	29. 8 – 9;	39. 6 6;	49. 4 – 6;	
10. 7 – 5;	20. 1 – 7;	30. 7 – 5;	40. 6 7;	50. 1 – 5;	

and the other 22 moves are obvious.

(B) 1. A step east;
2. A jump west;
3. A step west;
4. A jump east;
5. A step north;
6. A jump south.

Continuing with abbreviations:

7. s. S;	21. j. S;	35. s. E;
8. j. N;	22. s. W;	36. j. W;
9. j. E;	23. j. N;	37. j. S;
10. s. W;	24. j. N;	38. j. E;
11. j. W;	25. s. S;	39. j. E;
12. s. N;	26. j. S;	40. s. W;
13. s. E;	27. j. E;	41. j. W;
14. j. W;	28. s. N;	42. s. E;
15. s. S;	29. j. S;	43. j. N;
16. j. E;	30. j. W;	44. s. S;
17. s.N;	31. j. N;	45. j. S;
18. j. S;	32. s. E;	46. s. N.
19. s. E;	33. j. W;	
20. s. N;	334. j. N;	

112. A GROUPING OF INTEGERS 1 THROUGH 15

$$\left.\begin{matrix}1\\8\\15\end{matrix}\right\} d=7;\quad \left.\begin{matrix}2\\7\\12\end{matrix}\right\} d=5;\quad \left.\begin{matrix}6\\10\\14\end{matrix}\right\} d=4;\quad \left.\begin{matrix}9\\11\\13\end{matrix}\right\} d=2;\quad \left.\begin{matrix}3\\4\\5\end{matrix}\right\} d=1.$$

113. EIGHT STARS

The only solution is shown. There is a slow but sure method of solving the problem. The first star was placed in the lowest possible white square in the column 1. Now place the second star

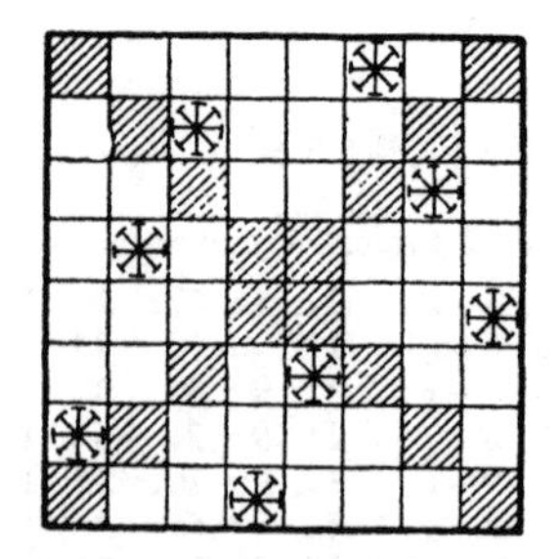

in the lowest possible white square in column 2. The first and second white squares from the bottom are out because they are diagonally in line with the first star. So use the third white square from the bottom (for the moment).

In column 3 use the bottom square, in column 4 use the third square from the bottom, and so on.

When you reach an "impossible" column, move the preceding star up, as little as possible. If you can't find a permissible square for the preceding star, or if all the permissible squares still make the next column "impossible," remove the preceding star, and repeat the process one column to the left.

114. TWO PROBLEMS IN PLACING LETTERS

(A) With identical letters, place one in any cell on diagonal *AC* of the diagram below. Of the 4 cells on diagonal *BD*, 1 is in the same column. The second letter can go in one of the 2 other cells.

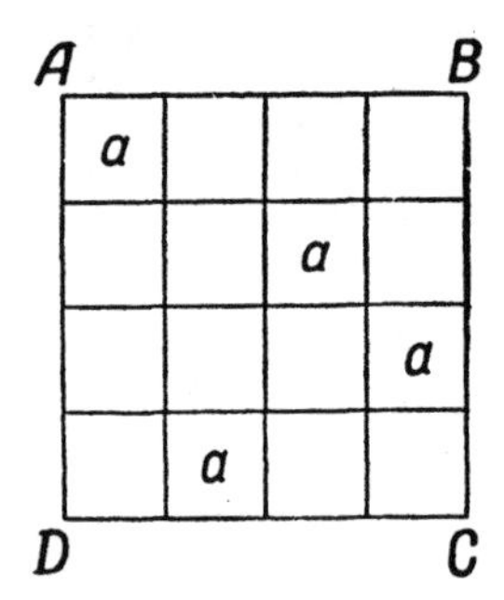

It is not hard to see that the last 2 letters can only be placed as shown.

Since the first letter can be placed in any of 4 cells on *AC*, and the second in any of 2 cells on *BD*, there are $4 \times 2 = 8$ solutions. However, all 8 can be obtained from the first by rotating and reflecting the square.

Four different letters lead to the same 8 choices of cells, but for each choice there are 24 ways to put the letters in: *a, b, c, d; a, b, d, c;* . . . ; *d, c, b, a*. So there are $8 \times 24 = 192$ solutions.

(B) It follows from the problem's conditions that the letters in the 4 corner cells must be different. We write them in, in an arbitrary order (see diagram *a*, facing page). In the middle cells of the diagonal containing *a* and *d* there must be *b* and *c*, but they can be placed in two ways (diagrams *b* and *c*).

After these 6 cells are filled, there is only one way to fill the others. (First fill in the outer

a			b
c			d

(a)

a			b
	b		
		c	
c			d

(b)

a			b
	c		
		b	
c			d

(c)

cells, then those on the second diagonal.) The results are shown in the two layouts below.

a	c	d	b
d	b	a	c
b	d	c	a
c	a	b	d

a	d	c	b
b	c	d	a
d	a	b	c
c	b	a	d

Since the 4 letters in the corners can be placed $4 \times 3 \times 2 \times 1 = 24$ ways, with 2 solutions each, there are 48 solutions.

115. CELLS OF DIFFERENT COLORS

One solution is shown below.

1 *red*	4 *black*	2 *green*	3 *white*
2 *white*	3 *green*	1 *black*	4 *red*
3 *black*	2 *red*	4 *white*	1 *green*
4 *green*	1 *white*	3 *red*	2 *black*

Call the colors *A, B, C,* and *D,* and the numbers *a, b, c, d.* (The second diagram represents the first, with red, black, green, white equal to *A, B, C, D,* and so on.) As in Problem 114(B), there

Aa	Bd	Cb	Dc
Db	Cc	Ba	Ad
Bc	Ab	Dd	Ca
Cd	Da	Ac	Bb

are 48 correct ways to place the colors – and 48 correct ways to place the numbers. Since the two steps are independent of each other in this problem, the total number of solutions is 48 × 48 = 2,304.

116. THE LAST PIECE

Each move shows the circle jumped from and the circle jumped to:

1. 9 – 1;	9. 1 – 9;	17. 28 – 30;	25. 25 – 11;
2. 7 – 9;	10. 18 – 6;	18. 33 – 25;	26. 6 – 18;
3. 10 – 8;	11. 3 – 11;	19. 18 – 30;	27. 9 – 11;
4. 21 – 7;	12. 16 – 18;	20. 31 – 33;	28. 18 – 6;
5. 7 – 9;	13. 18 – 6;	21. 33 – 25;	29. 13 – 11;
6. 22 – 8;	14. 30 – 18;	22. 26 – 24;	30. 11 – 3;
7. 8 – 10;	15. 27 – 25;	23. 20 – 18;	31. 3 – 1.
8. 6 – 4;	16. 24 – 26;	24. 23 – 25;	

117. A RING OF DISKS

1.	1–2, 3	2–6, 5	6–1, 3	1–6, 2
2.	1–2, 3	4–1, 3	3–6, 5	5–3, 4
3.	1–4, 5	3–4, 1	4–2, 6	2–3, 4
4.	1–4, 5	5–2, 6	6–4, 1	1–6, 5
5.	2–3, 4	3–1, 6, 5	6–2, 4	2–1, 6
6.	2–3, 4	5–2, 3	3–1, 6	1–3, 5
7.	2–4, 5	5–1, 3, 6	6–2, 4	2–1, 6
8.	2–4, 5	3–2, 5	5–1, 6	1–5, 3
9.	3–1, 2	5–3, 2	2–6, 4	4–5, 2
10.	3–1, 2	4–3, 1	1–6, 5	5–1, 4
11.	3–1, 2	1–2, 6, 4	6–2, 3	3–6, 5
12.	3–1, 2	2–1, 6, 5	6–3, 1	3–6, 4
13.	3–4, 5	2–3, 5	5–1, 6	1–2, 5
14.	3–4, 5	1–3, 4	4–2, 6	2–1, 4
15.	3–4, 5	4–1, 6, 5	6–5, 3	3–2, 6
16.	3–4, 5	5–2, 6, 4	6–3, 4	3–1, 6
17.	4–3, 2	3–1, 6, 5	6–2, 4	4–5, 6
18.	4–3, 2	1–4, 3	3–5, 6	5–3, 1
19.	4–1, 2	1–3, 6, 5	6–2, 4	4–6, 5
20.	4–1, 2	3–1, 4	1–6, 5	5–1, 3
21.	5–3, 4	4–1, 6	6–3, 5	5–6, 4
22.	5–3, 4	2–3, 5	3–1, 6	1–2, 3
23.	5–1, 2	3–2, 5	2–6, 4	4–3, 2
24.	5–1, 2	1–4, 6	6–2, 5	5–1, 6

118. FIGURE SKATERS

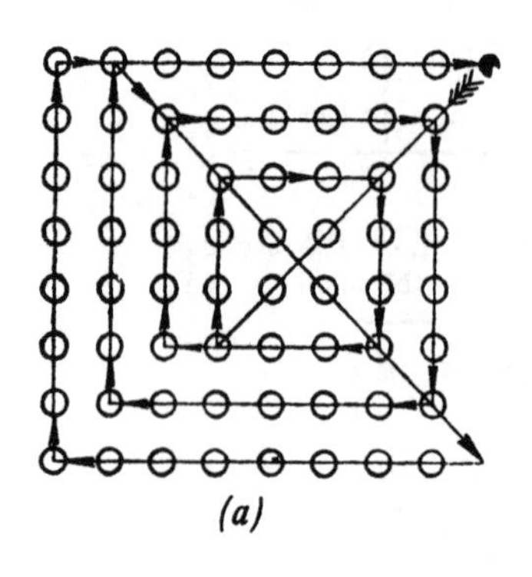
(a)

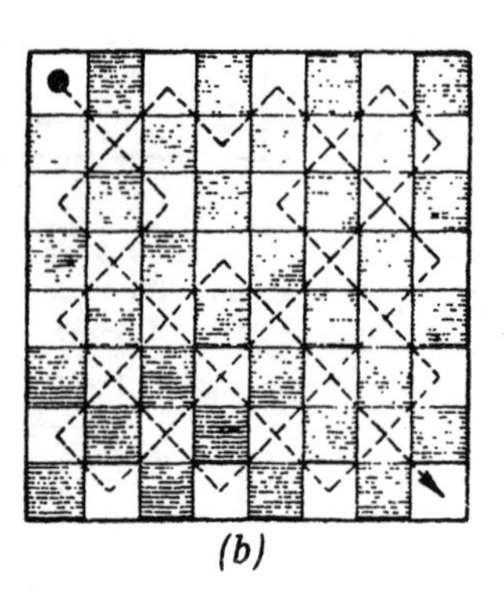
(b)

119. A KNIGHT PROBLEM

No. A knight moves from a black square to a white one and vice versa. From *a*1 (a black square) after 1, 3, 5, . . . , 61, 63 moves the knight is on a white square. Since it takes him 63 moves to visit all 64 squares minus the one he starts on, he should end on white. But *h*8 is black.

120. ONE HUNDRED AND FORTY-FIVE DOORS

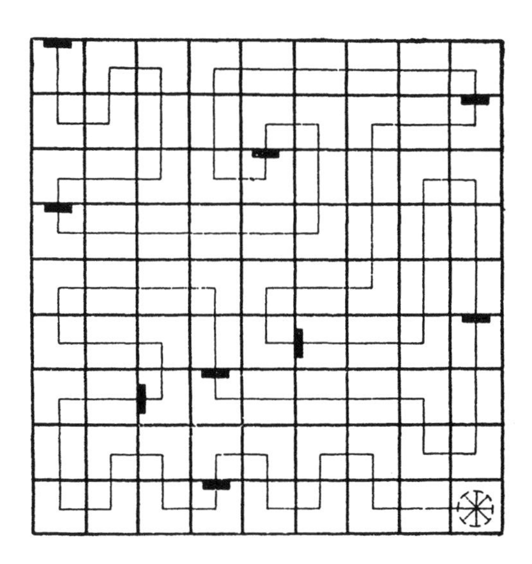

121. HOW DOES THE PRISONER ESCAPE?

He gets keys *d* and *e* and unlocks the doors of cells *E* and *D* (see the diagram on p. 49). He gets key *c* and unlocks the door of cell *C*, and gets key *a*, enabling him to unlock the door of cell *A* and get key *b*. He goes through *E* and *D* again to unlock the door of cell *B*, gets key *f*, again passes through *E* and unlocks the door of cell *F*, and gets key *g* and leaves the dungeon through cell *G*.

The path to freedom is not easy – through 85 doors.

III. Geometry with Matches

122. FIVE PUZZLES

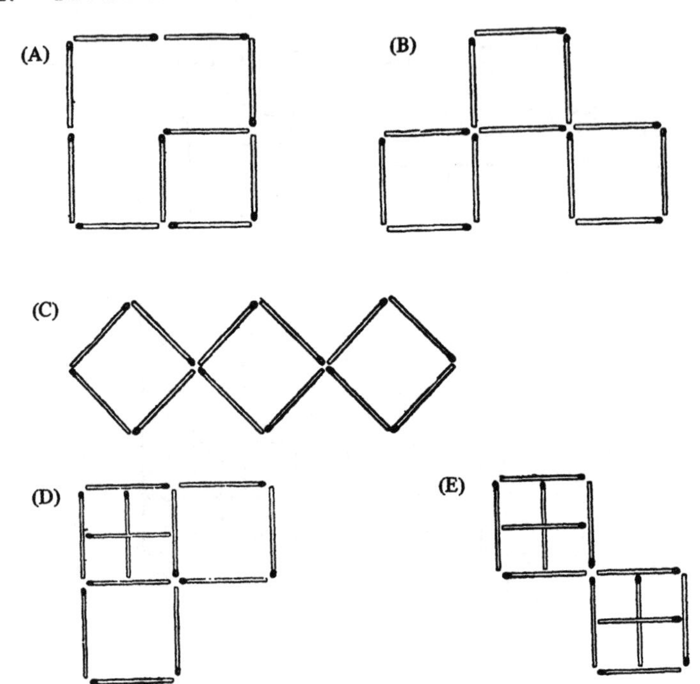

123. EIGHT MORE PUZZLES

(A) Take out the 12 matches inside the large square, and use them to make another large square.

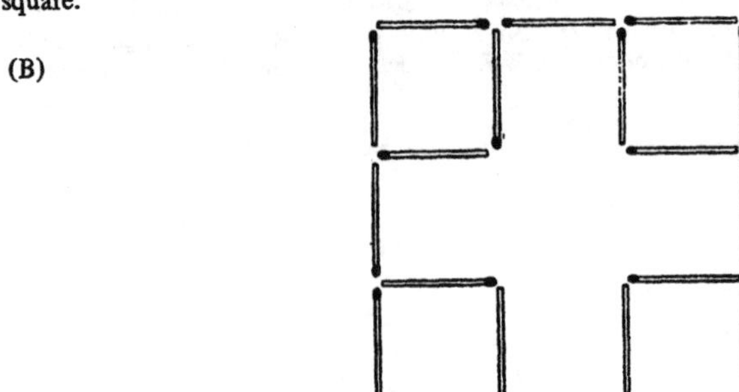

(B)

(C) See *a* below for 4 matches; *b* for 6 matches (there is another solution); and *c* for 8 matches (there are two other solutions).

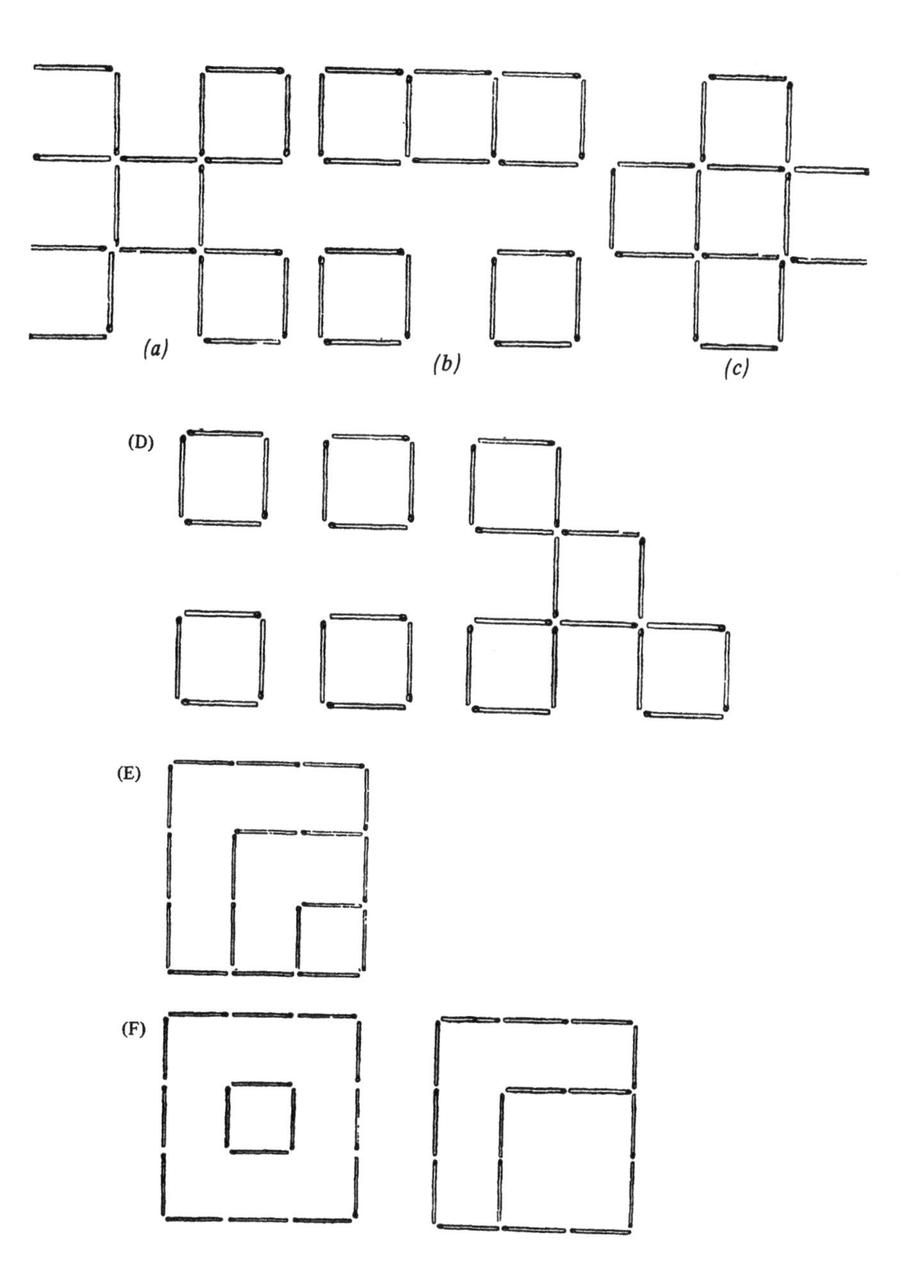

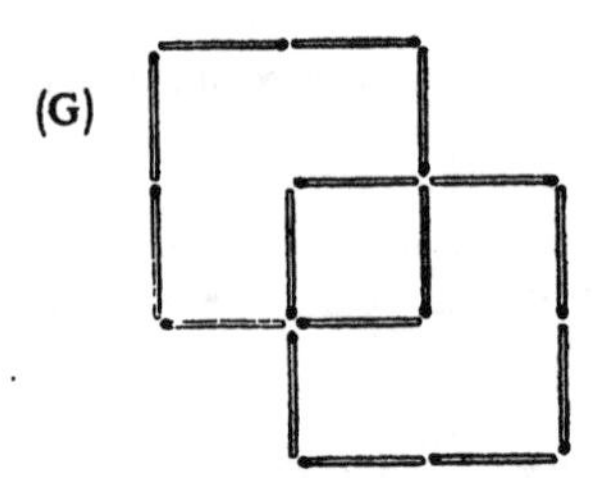

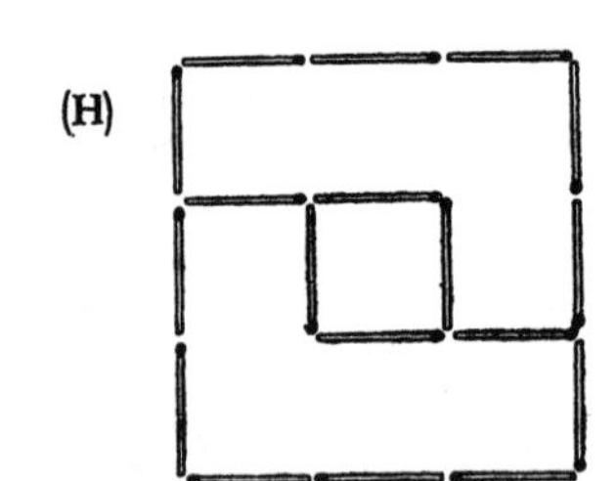

124. NINE MATCHES

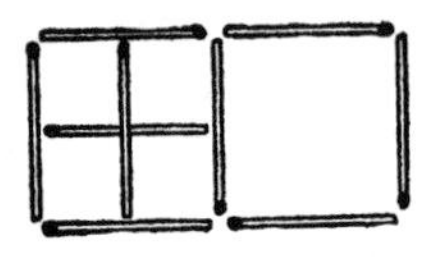

125. A SPIRAL

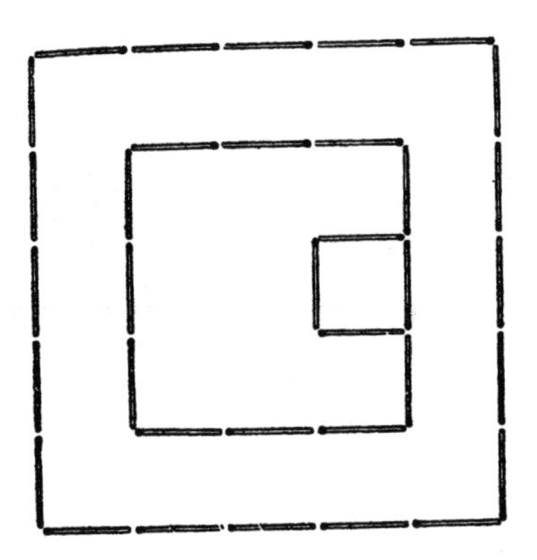

126. CROSSING THE MOAT

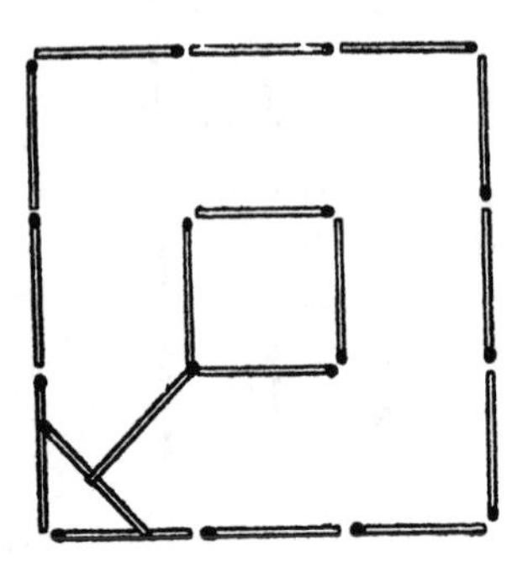

Answers

27. REMOVE TWO MATCHES

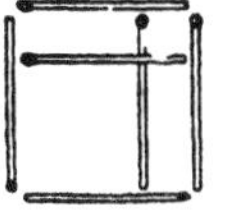

128. THE FAÇADE OF A HOUSE

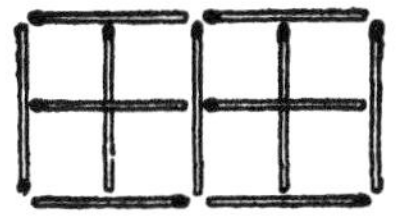

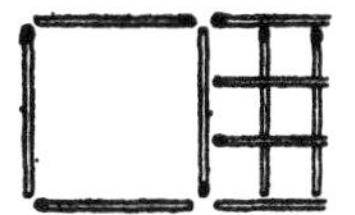

129. A JOKE

Bend matches around 2 corners, as shown.

130. TRIANGLES

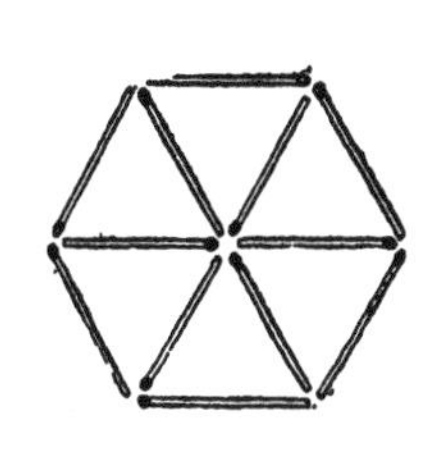

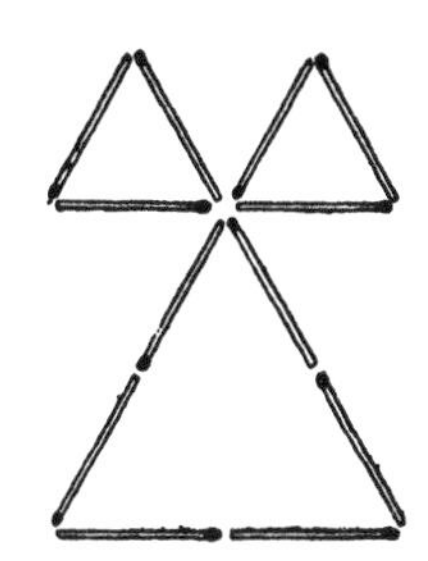

131. HOW MANY MATCHES MUST WE REMOVE?

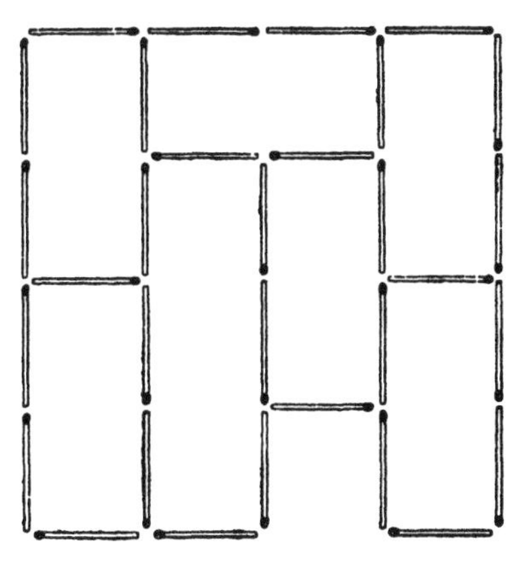

132. ANOTHER JOKE

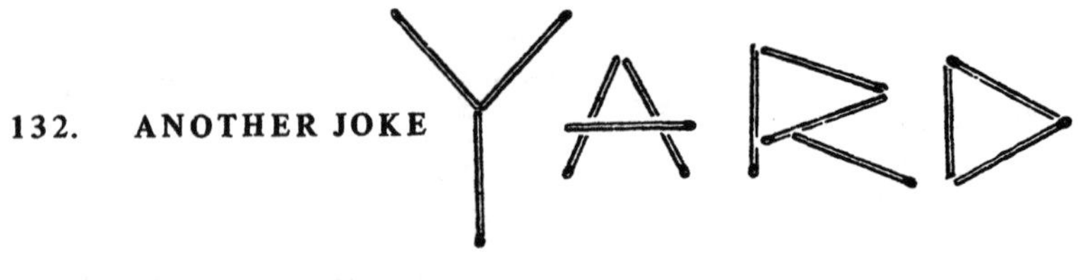

133. A FENCE

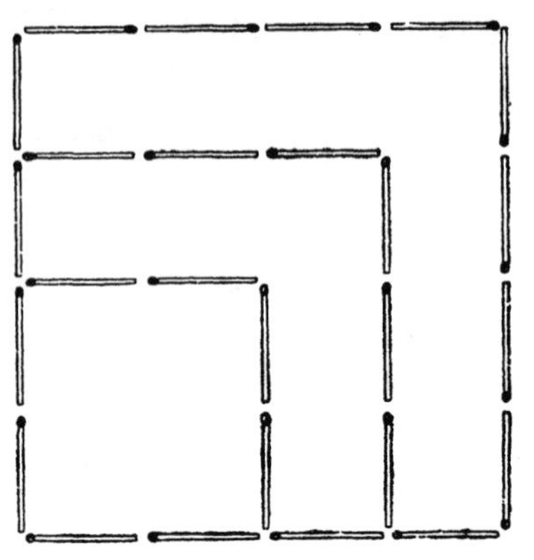

134. A THIRD JOKE

Place 2 matches at the corner of a table so that 2 edges of the table form the other sides of the square.

135. AN ARROW

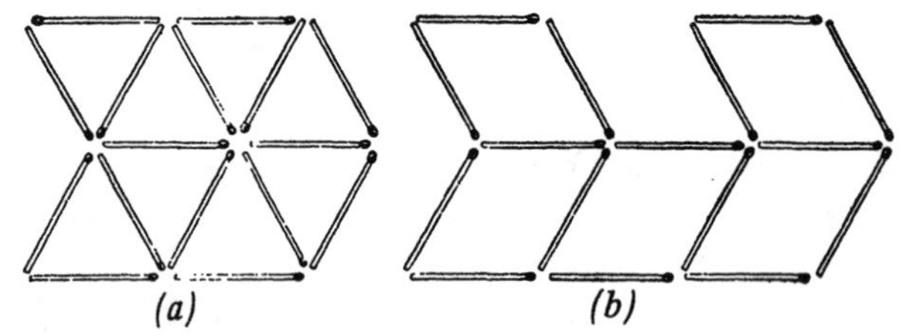

136. SQUARES AND RHOMBUSES

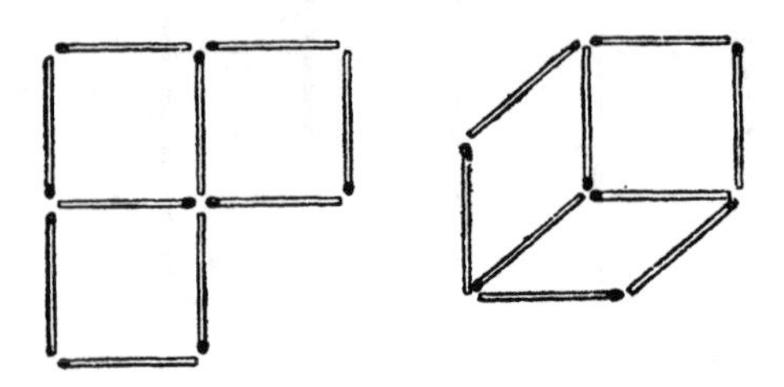

137. ASSORTED POLYGONS

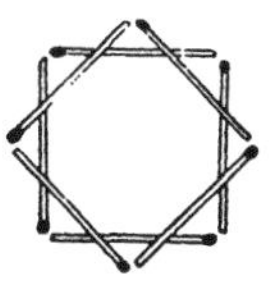

138. PLANNING A GARDEN

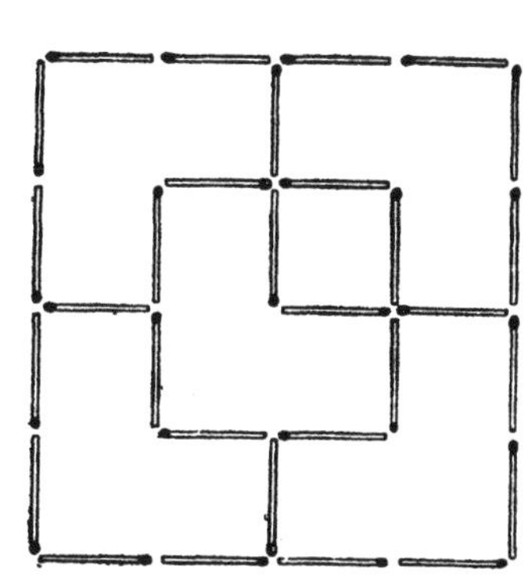

139. PARTS OF EQUAL AREAS

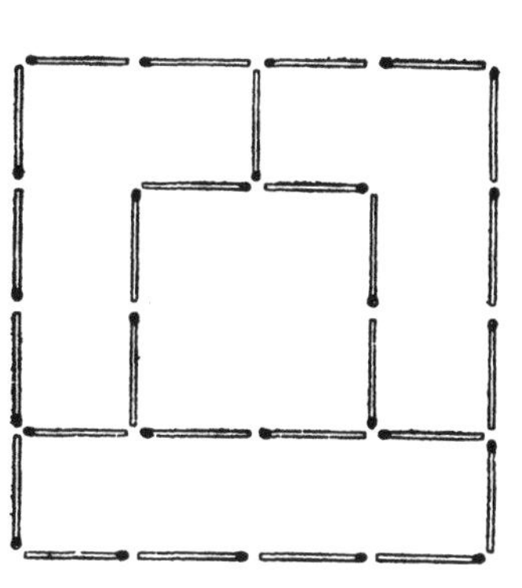

140. GARDEN AND WELL

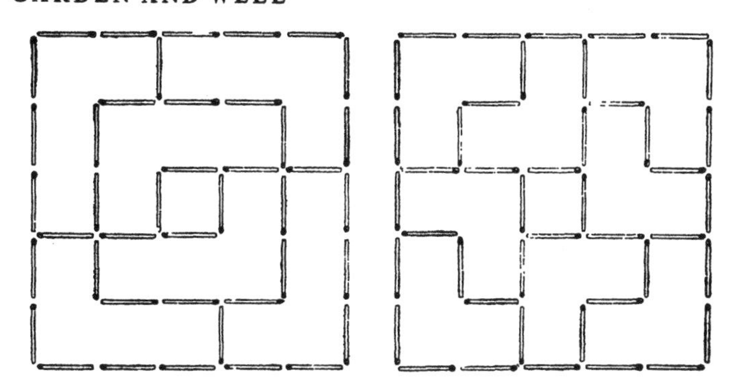

141. PARQUET

684.

142. RATIO OF AREAS

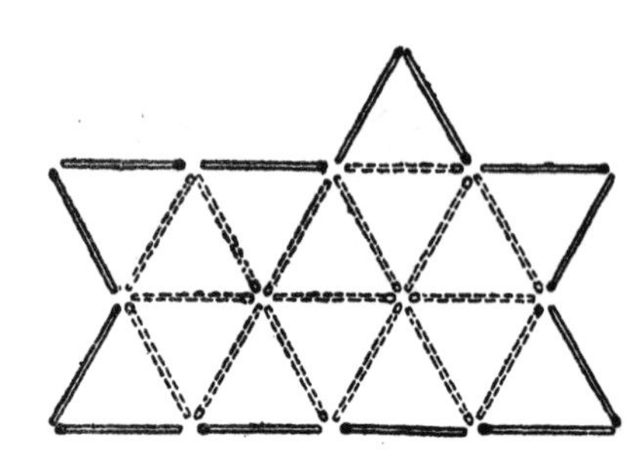

143. OUTLINE OF A POLYGON

We give three solutions:

1. First make a 3-by-4-by-5 right triangle. (Its area is 6 square units.) Then move 4 matches into the triangle, losing 3 square units as shown in the first diagram. The shaded area is 3 square units made out of 12 matches.

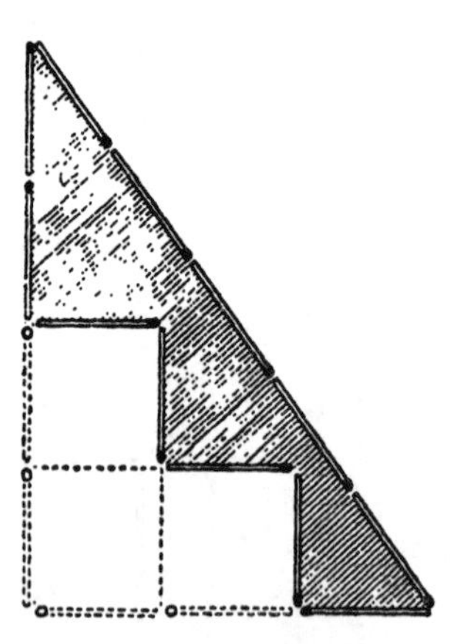

2. Start with a 4-square-unit square as shown. In the first transformation, the area remains the same. In the second, 1 square unit is lost, leaving 3 square units made out of 12 matches. (Solution by N. I. Arzhanov, a Leningrad engineer.)

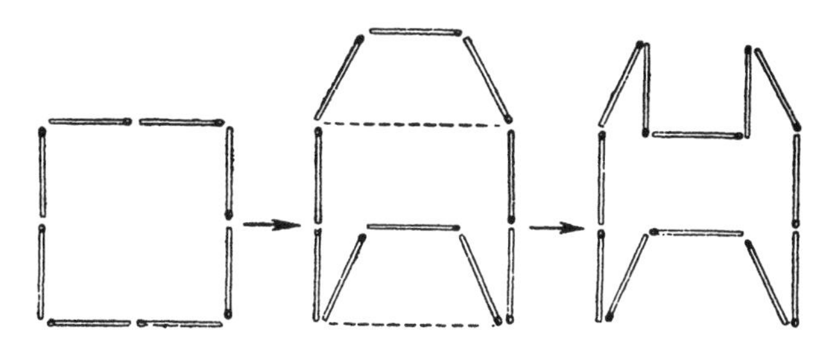

3. Make a parallelogram with base = 1 unit and altitude = 3 units. Its area is $1 \times 3 = 3$ square units made out of 12 matches. (Solution by V. I. Lebedev, a Moscow engineer.)

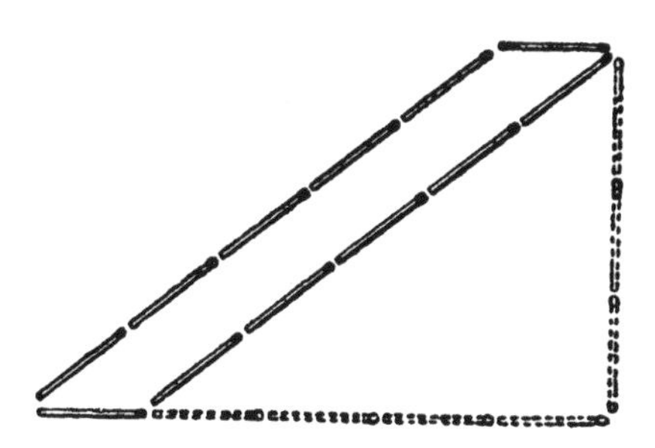

144. FINDING A PROOF

First solution: Build three adjacent equilateral triangles as shown in *a*. The two solid matches make an angle of $3 \times 60^\circ = 180^\circ$, so they make a straight line.

Second solution: See *b!*

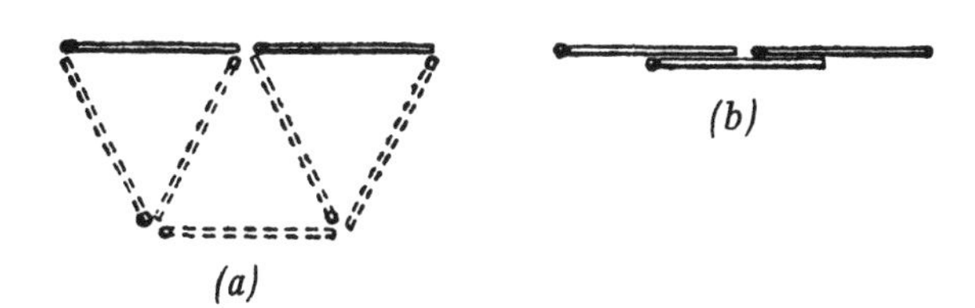

(a) (b)

145. IDENTICAL PARTS

(A)

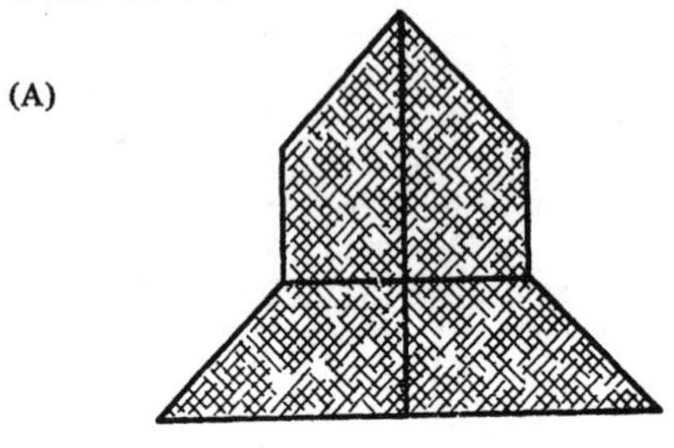

(B)

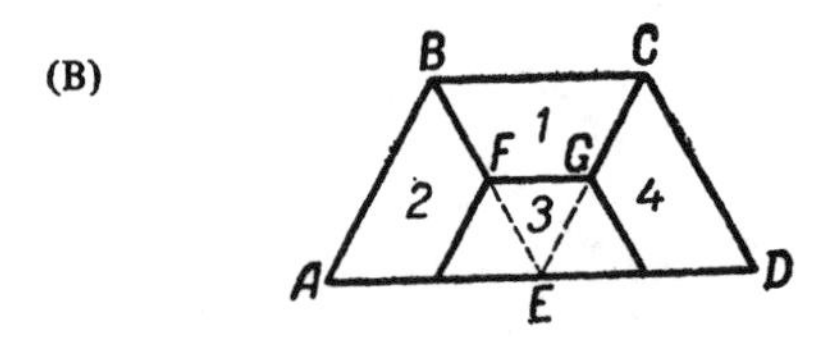

(C)

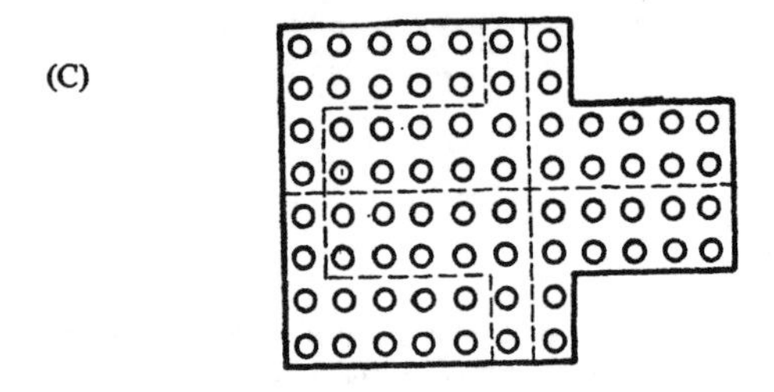

(D)

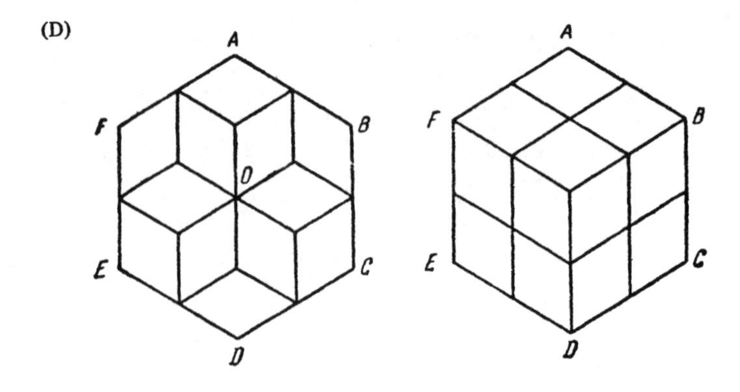

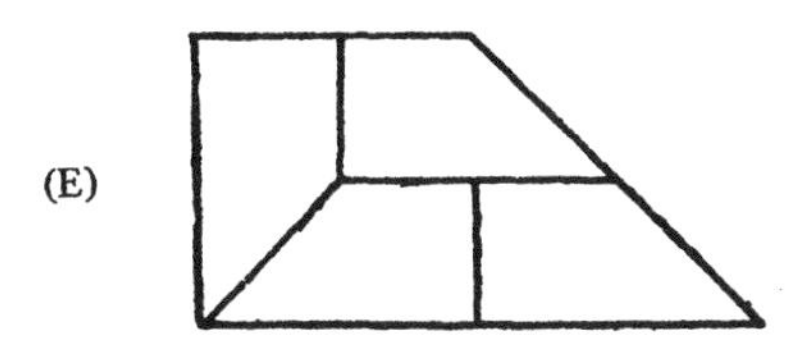

146. SEVEN ROSES ON A CAKE

147. LOST CUTTING LINES

1. First make 1-cell cuts between adjacent identical numbers (*a* in diagram).

			3		1	1	
			3	4			
				2			
	1		4	2			
	1						
		3	3				
					4	2	2
					4		

(a)

2. For symmetry, repeat each 1-cell cut at three other places on the square *(b)*.

			3		1	1	
			3	4			
				2			
	1		4	2			
	1						
		3	3				
					4	2	2
					4		

(b)

3. Separate the 4 central cells, since no 2 can belong to the same configuration. It is then possible to complete the cuts, bearing in mind that each corner cell belongs to a different configuration, and that each configuration must include one of each digit.

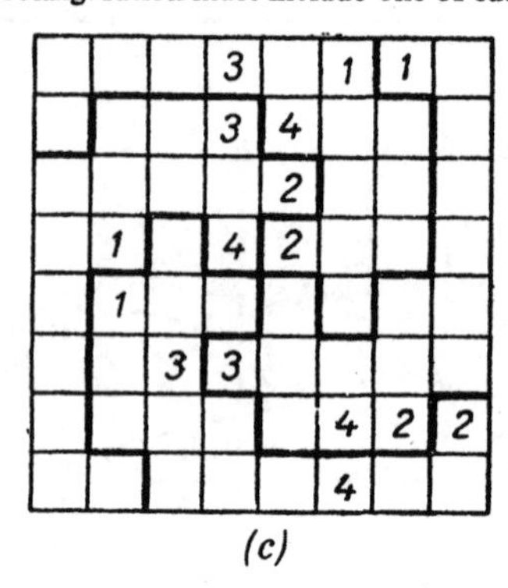

(c)

148. GIVE US ADVICE

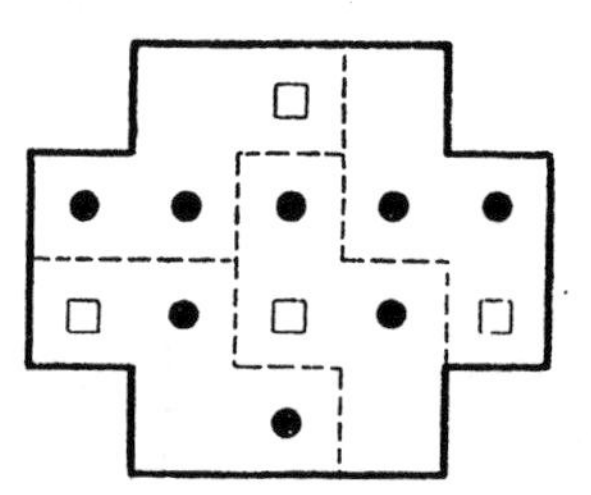

149. NO LOSSES!

A rectangle can be made out of 6 no. 1 plates, as shown below.

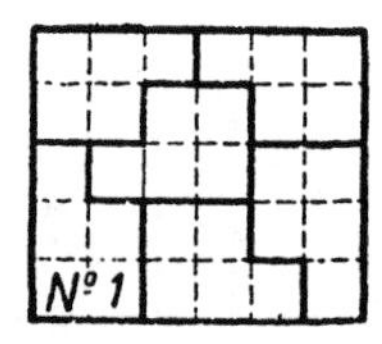

The cuts for the other six patterns are shown in diagrams *I* to *VI*.

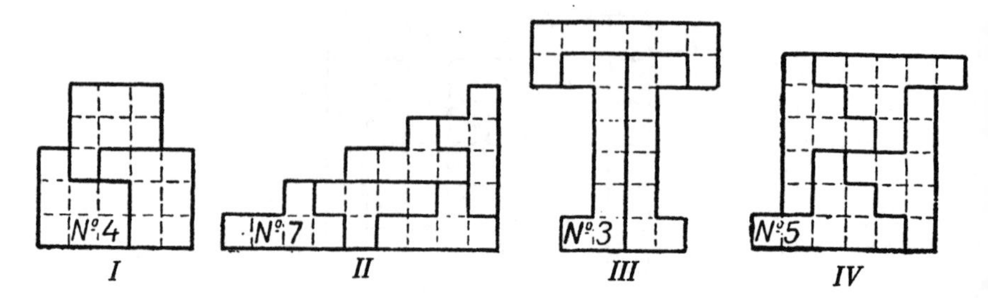

I *II* *III* *IV*

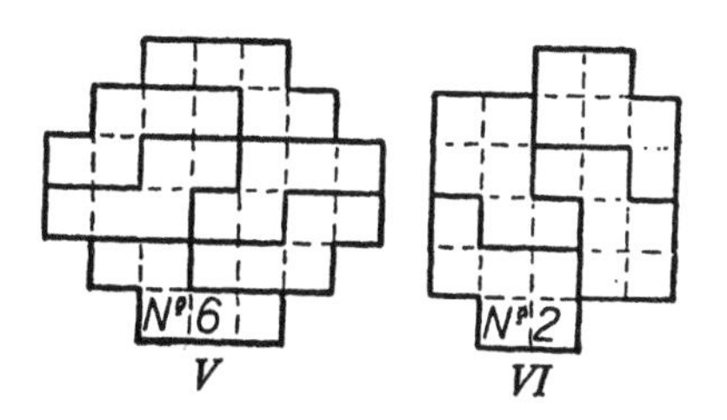

150. WHEN THE FASCISTS ATTACKED OUR COUNTRY

Vasya proved how clever he was by making a stair-step cut, as shown.

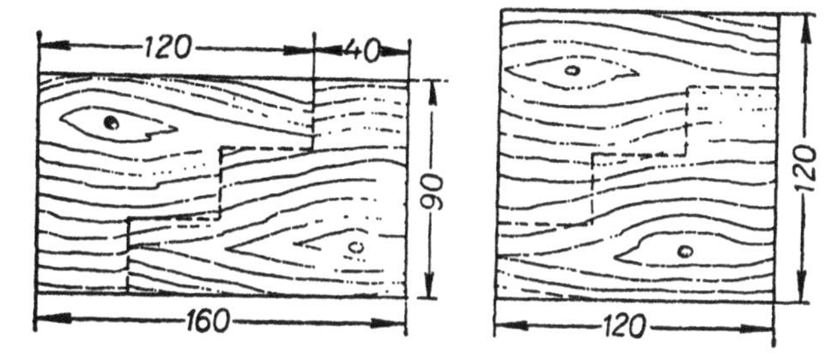

151. AN ELECTRICIAN'S REMINISCENCES

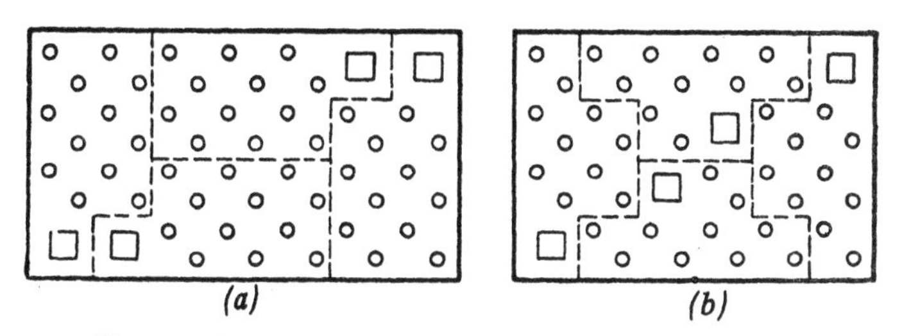

On your own, see if you can find a second solution for the second board.

152. NOTHING WASTED

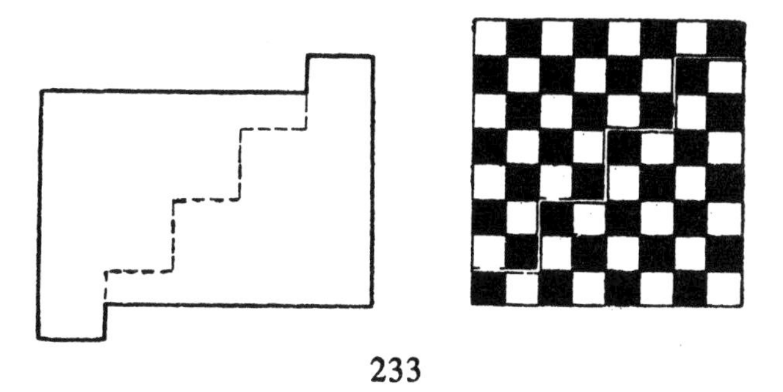

153. A DISSECTION PUZZLE

Cut along *abcde,* where *b, c,* and *d* are the centers of the squares making up the figure at left. Move the piece to complete the frame (right). On your own, see if you can find a second solution.

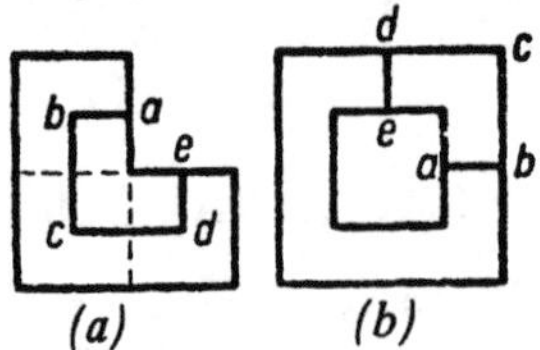

(a) (b)

154. HOW TO CUT A HORSESHOE

The lines must intersect on the horseshoe.

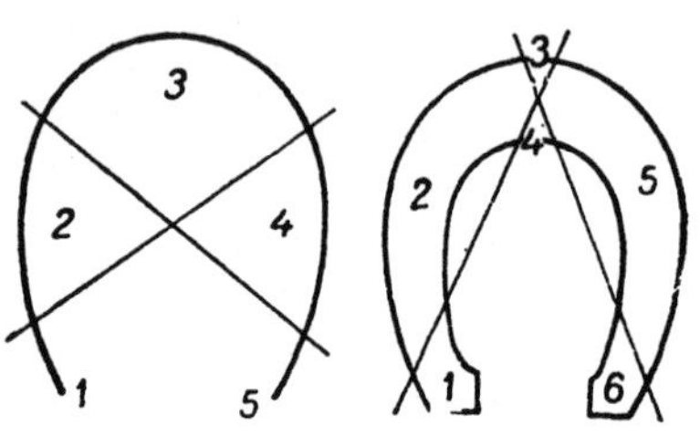

155. A HOLE IN EACH PART

This problem, unlike the preceding one, does not prohibit rearranging parts after the first cut.
Cut off the 2 holes at the top of the horseshoe. Place the top on one of the sides so that the holes align. Then another transverse cut will divide the horseshoe as required.

156. MAKE A SQUARE OUT OF A JUG

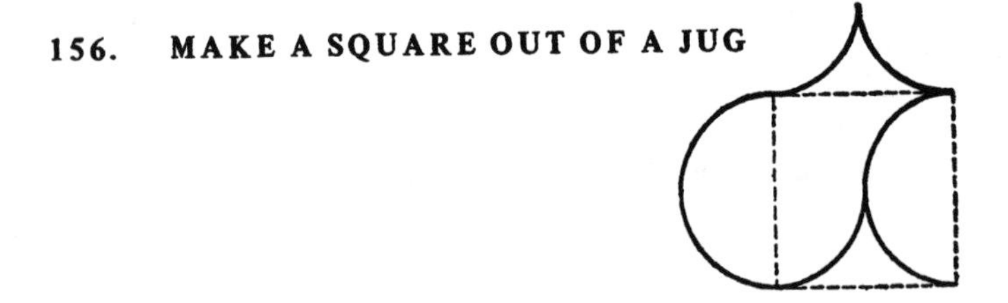

157. SQUARE THE LETTER *E*

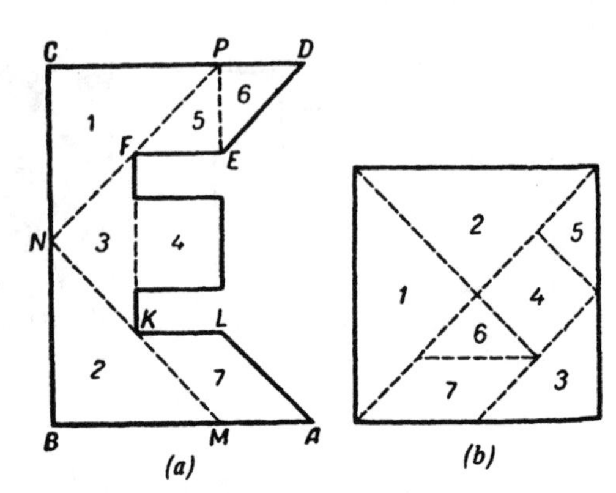

(a) (b)

158. AN OCTAGON TRANSFORMATION

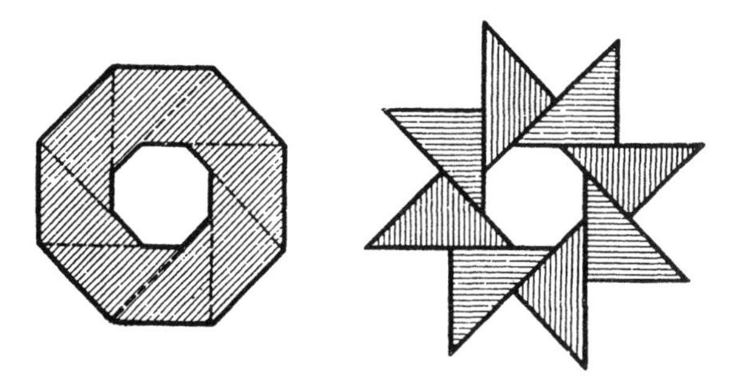

159. RESTORING A RUG

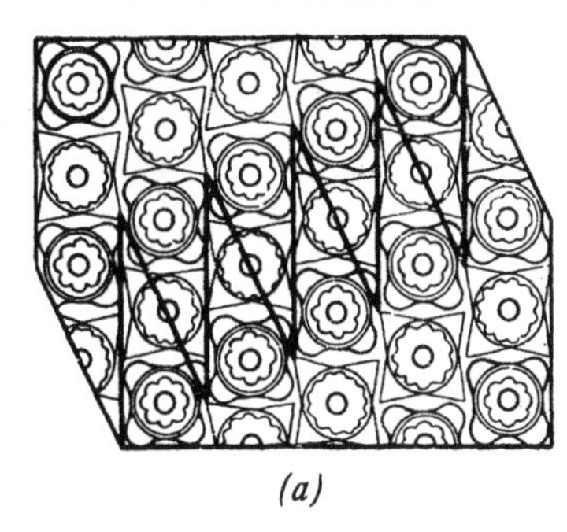

(a)

(b)

160. A CHERISHED REWARD

She made two stair-step cuts, then put the 2 parts together as shown.

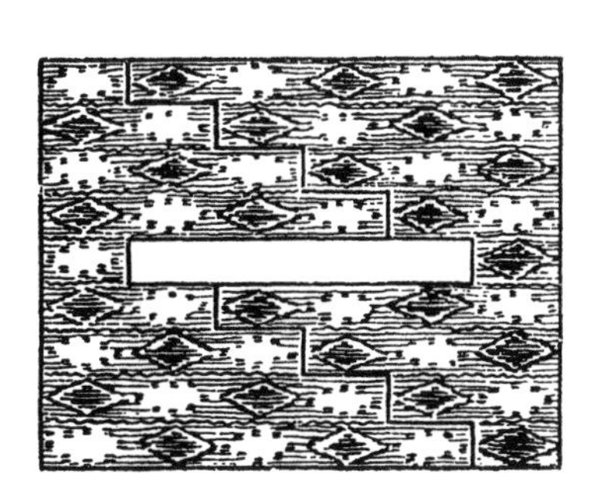

161. RESCUE THE PLAYER

Consider the 8-by-8 board with black and white columns, (*c*, next page). No matter how we cut *a* from this board it will contain an odd number (1 or 3) of white squares and an odd number (3

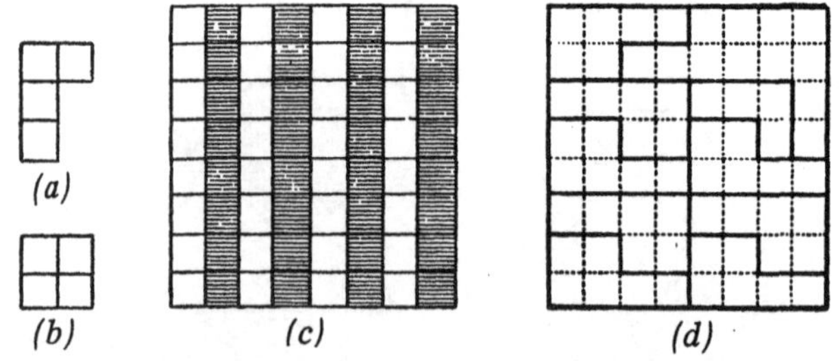

or 1) of black squares. Consequently, 15 such shapes will also contain an odd number of white and of black squares. But *b* has 2 white and 2 black squares, so the 16 parts will contain an odd number of white and of black squares. But the board has 32 white and 32 black squares. It follows that it is impossible to cut the board into 15 parts like *a* and 1 like *b*.

A solution of the other cut is shown in *d*.

162. A GIFT FOR GRANDMOTHER

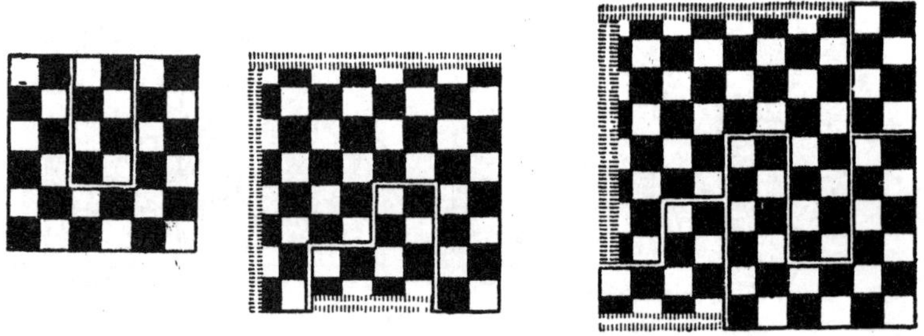

163. THE CABINETMAKER'S PROBLEM

The cabinetmaker saws each board along the lines BA, CA, B_1A_1, and C_1A_1. From the resulting 8 parts he glues together a circular tabletop as shown.

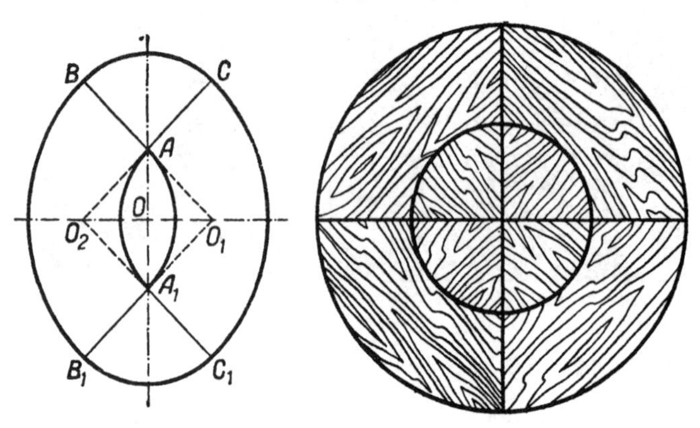

164. A FUR DRESSER NEEDS GEOMETRY!

Let ABC be the patch to be turned over. The fur dresser cut along DE and DF (which bisect BC and AB), and turned the 3 resulting parts over – the triangles around vertical axes and the quadrilateral around EF. When he sewed them together, patch ABC was reversed but preserved its shape.

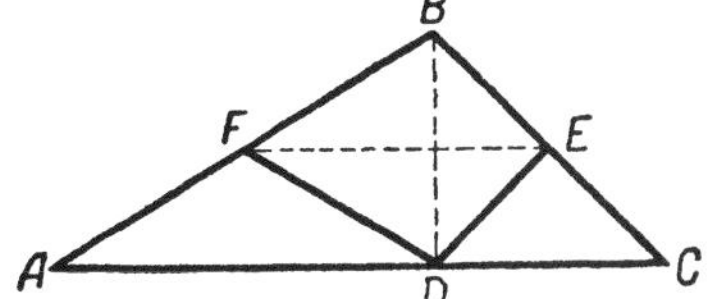

165. THE FOUR KNIGHTS

166. CUTTING A CIRCLE

For maximum results, each line should intersect with all the others, and no more than 2 lines should intersect at any point.

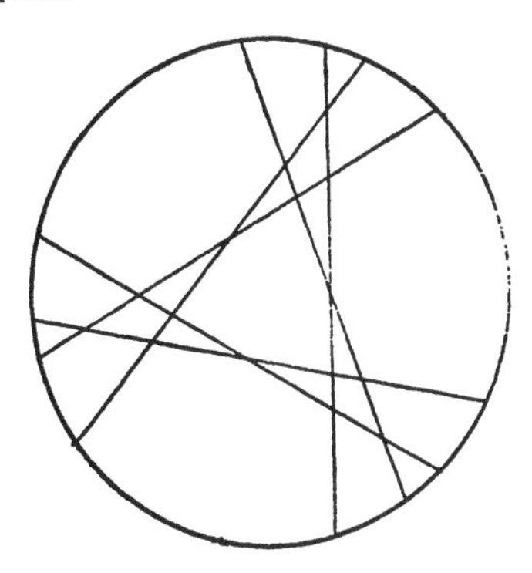

167. TRANSFORMING A POLYGON TO A SQUARE

Bisect AC at K. Let $FQ = AK = DP$. Cut along BQ and QE. The square $BPEQ$ is formed from the resulting parts.

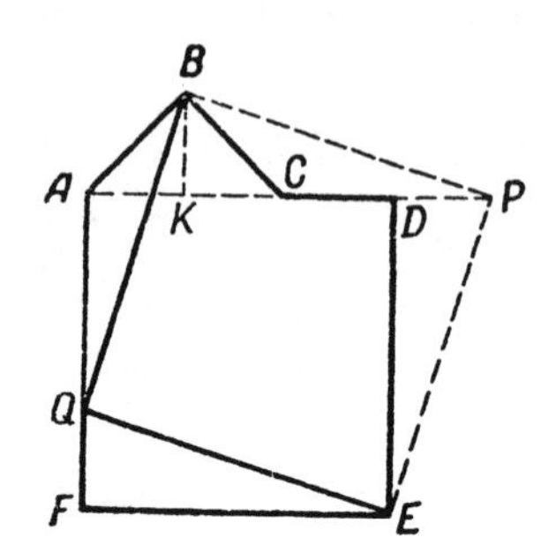

168. DISSECTION OF A REGULAR HEXAGON INTO AN EQUILATERAL TRIANGLE

In *a*, draw AC, EL perpendicular to AF, LM equal to EL. Draw EM. Use EM to build equilateral triangle EMN, and draw KP continuing KN until it meets CD at P. If the construction is done correctly, CP will equal CK.

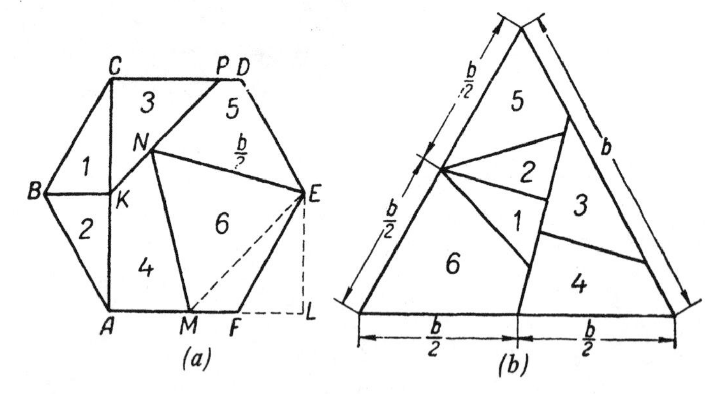

Solid lines show where to cut into 6 parts, which are used in *b* to form an equilateral triangle.

V. Skill Will Find Its Application Everywhere

169. WHERE IS THE TARGET?

The target is 75 miles from *A* and 90 miles from *B*. Look at the diagram on page 69: in the middle is a scale of miles. Pick up a 75-mile length from the scale with a compass and draw a 75-mile arc from *A*. Pick up a 90-mile length with the compass and draw a 90-mile arc from *B*. The target is where the 2 arcs intersect at sea.

170. THE SLICED CUBE

6; 27; 0; 8 (number of corners of large cube); 12 (number of edges); 6 (number of faces); 1.

171. A TRAIN ENCOUNTER

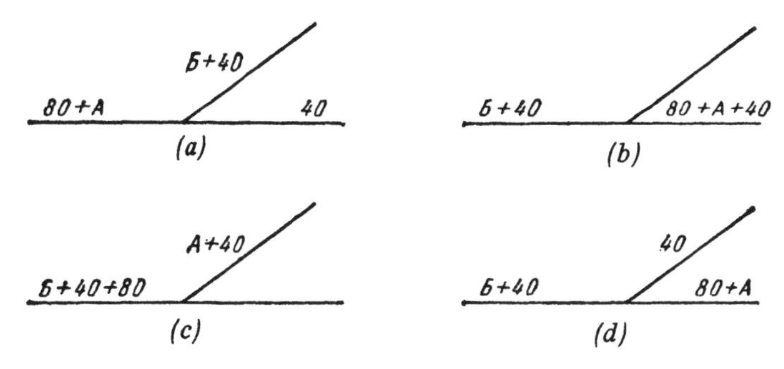

1. Locomotive *B* takes 40 cars into the siding (leaving its 40 rear cars on the right). See *a*.
2. *A* takes its 80 cars past the siding and couples on the 40 cars *B* left behind; *B* leaves the siding. See *b*.
3. *A* passes the siding from right to left with 120 cars, leaves its 80 own cars to the left but keeps *B's* 40 cars as it enters the siding. See *c*.
4. *A* leaves *B*'s 40 cars in the siding, picks up its own cars, and leaves to the right. *B*, still with 40 cars, picks up the other 40 from the siding and leaves to the left. See *d*.

172. A RAILROAD TRIANGLE

(A) The diagram below shows a 10-move solution.

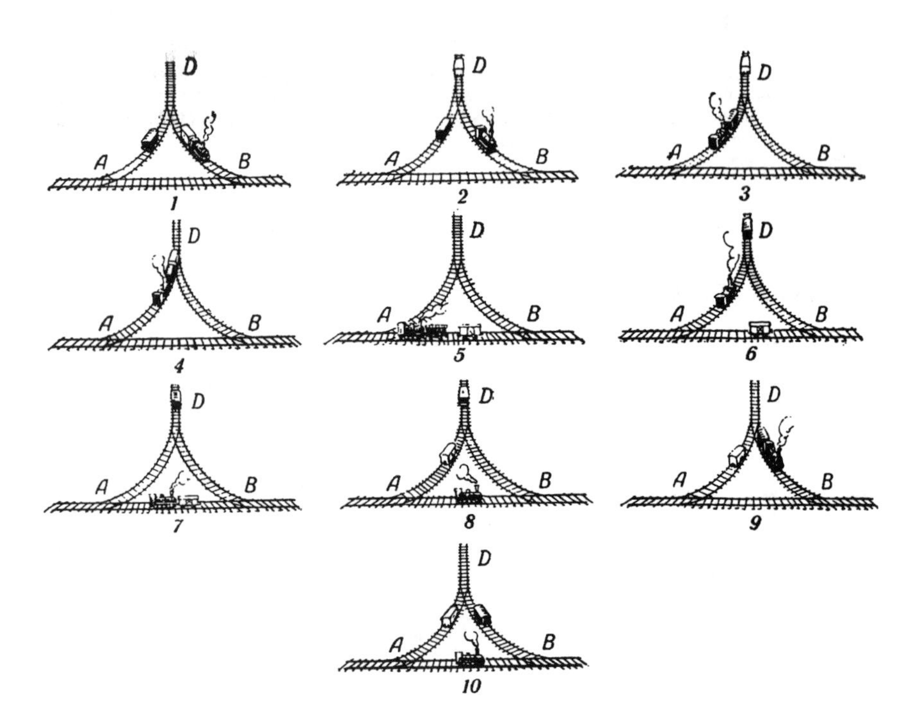

1. The engineer backs into *BD* and couples on the white car.
2. He backs the white car into *D* and uncouples, then moves out along *DB*.
3. He passes *B*, backs up past *A*, enters *AD*, and couples on the black car.
4. He pushes the black car forward, couples it on the white car, and starts backing out of *AD*, pulling both cars.
5. He backs past *A*, goes partway forward to *B*, and uncouples the white car.
6. Leaving the white car on *AB*, he goes backward past *A* pulling the black car, then pushes it up *AD* into *D* and uncouples. He starts backing out of *AD*.
7. He backs past *A* and moves forward on *AB* till he can couple on the white car.
8. He backs past *A*, pushes the white car onto *AD* and uncouples, then backs past *A* again and moves forward partway to *B*.
9. He passes *B*, backs into *BD* and couples on the black car, pulling it forward along *DB*.
10. Uncoupling the black car on *BD*, he passes *B* and backs up till he is again halfway between *A* and *B*, facing right.

(There are at least two other solutions in 10 moves.)

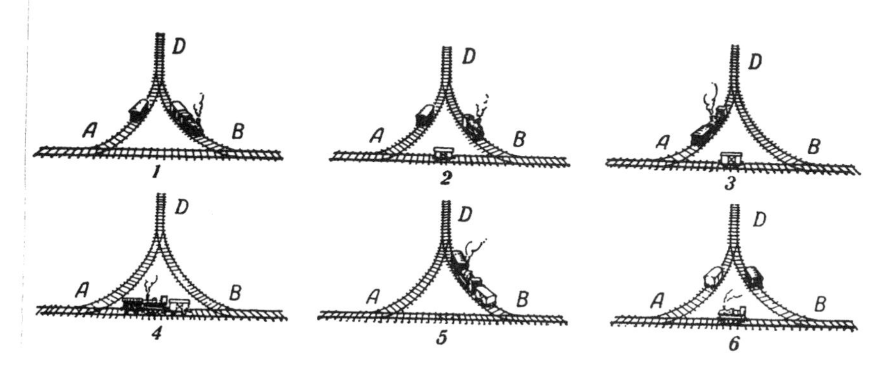

(B) The diagram above shows the 6-move solution.

1. The engineer backs into *BD* and couples on the white car.
2. He pulls the white car past *B*, backs up along *BA*, and uncouples the white car. Leaving it on *AB*, he moves forward past *B*, then backs into *BD*.
3. Moving in and out of *D*, he moves forward along *DA* and couples on the black car.
4. He pushes the black car past *A*, then backs up along *AB* and couples on the white car.
5. Sandwiched between the cars, he backs past *B* and enters *BD*. He uncouples the black car.
6. Leaving the black car on *BD*, he backs out pushing the white car past *B*, pulls it along *AB* past *A*, pushes it back into *AD* where he uncouples the white car, moves out of *AD* past *A*, then backs up until he is again halfway between *A* and *B*, but facing left.

173. A WEIGHING PUZZLE

1. Split the 180 ounces between the two pans of the balance. Each will hold 90 ounces.
2. Split one 90-ounce batch into 45 and 45 ounces, again using the balance without the weights.
3. Remove 5 ounces from a 45-ounce batch by balancing against the two weights. Now you have 40 ounces for one package, and the rest of the grits is 140 ounces for the second package.

174. BELTS

Yes. *C* and *D* turn clockwise and *B* turns counterclockwise. The wheels can also turn if all 4 belts are crossed, but not if 1 or 3 belts are.

175. SEVEN TRIANGLES

The problem can only be solved in three dimensions, as shown.

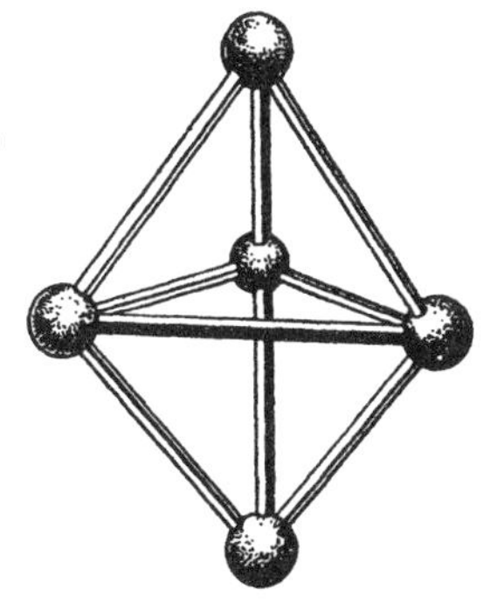

176. THE ARTIST'S CANVASES

Her proof is geometric. Take any rectangle with integral sides and divide it into unit squares *(a)*. Consider the shaded border. It contains 4 less unit squares than the perimeter. If and only if the unshaded heart of the rectangle contains 4 unit squares, the total number of unit squares – that

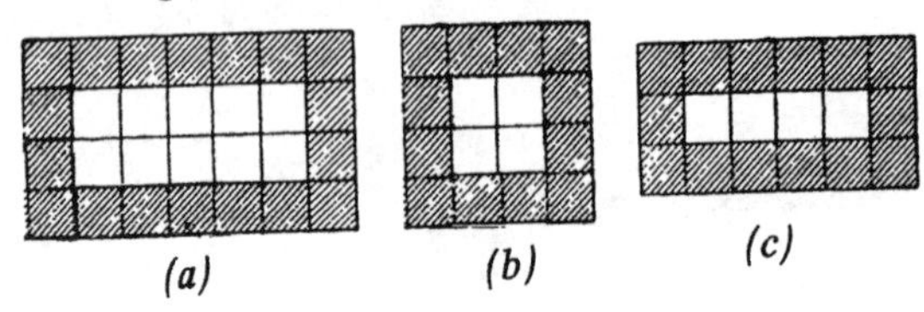

is, the area – will equal the perimeter. But 4 unit squares can be arranged in a rectangle in only two ways *(b* and *c)*. The two solutions are a 4-by-4 square and a 6-by-3 rectangle.

It is easy to find these solutions from the equation $xy = 2x + 2y$ (where x and y are the sides of the rectangle), but not to prove that they are the only solutions.

177. HOW MUCH DOES THE BOTTLE WEIGH?

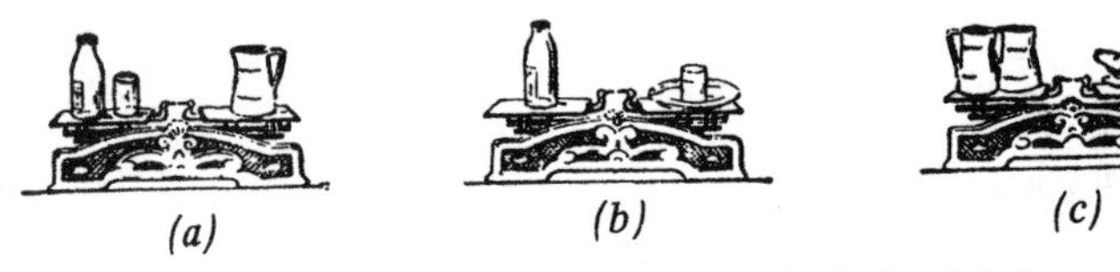

For convenience, we repeat the diagrams on page 72. In *b*, the bottle balances the plate and the glass. If a glass is added to each scale it will not disturb the balance. Therefore the bottle and 1 glass balance the plate and 2 glasses (*d* below). Comparing *a* with *d*, we see that the jug weighs

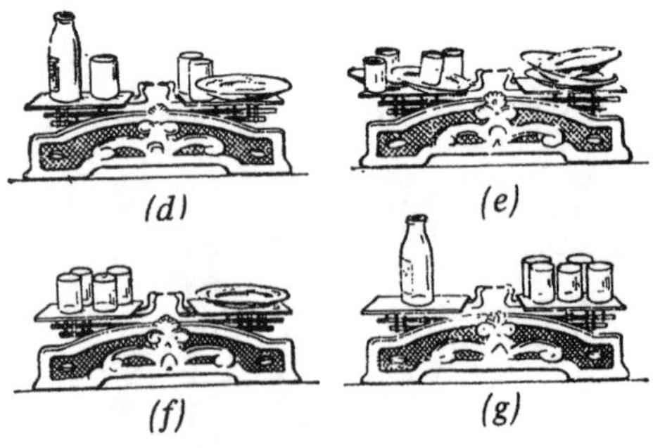

as much as a plate and 2 glasses. Also, 2 jugs balance 3 plates *(c)*. Therefore, 3 plates weigh the same as 2 plates and 4 glasses *(e)*.

Removing 2 plates from each pan in *e* leaves 1 plate balancing 4 glasses *(f)*. Now in *b*, instead of 1 plate put 4 glasses; the 5 glasses balance the bottle *(g)*.

178. SMALL CUBES

He cut each cube into 8 smaller cubes, as shown. The surface of each small cube is clearly one-fourth that of the large cube, so the total surface area is doubled.

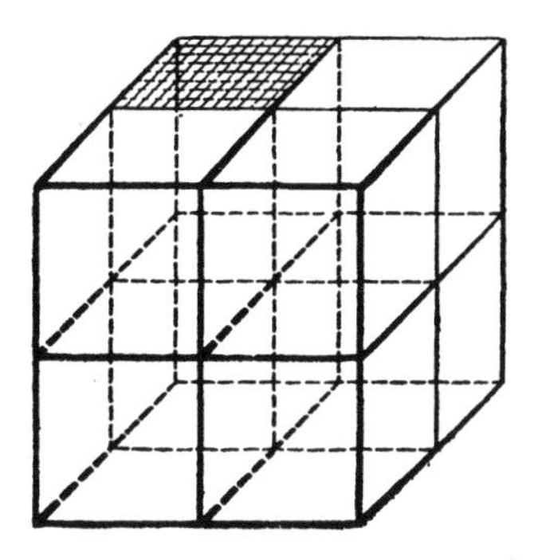

179. A JAR WITH LEAD SHOT

They poured the shot into the jug and then poured in water, which filled all the spaces between the pellets. Now the water volume plus the shot volume equaled the jar's volume.

Removing the shot from the jar, they measured the volume of water remaining, and subtracted it from the volume of the jar.

180. WHITHER THE SERGEANT?

As shown, he finished at his starting point.

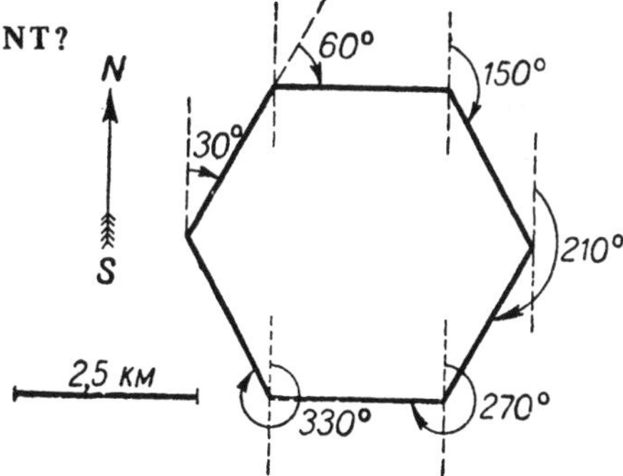

181. A LOG'S DIAMETER

The distance from one knothole to another cut of the same knothole is about two-thirds the width of the whole plywood sheet or 30 inches. The diameter of the log is $\frac{30}{\pi}$ = about 10 inches.

182. A CALIPER DIFFICULTY

He placed an object between one leg and its indentation, so that the calipers could be removed without opening the legs. He then subtracted the length of the object from the spread of the calipers.

183. WITHOUT A GAUGE

(A) Wind a number of coils tightly around a cylinder as shown in *a*. Twenty diameters make 2 inches, so 1 diameter is 0.1 inch.

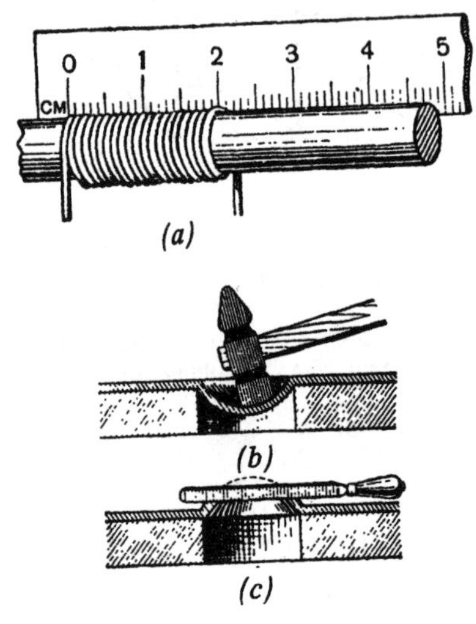

(a)

(b)

(c)

(B) Place the sheet on a support which has a round hole in it. Hammer the sheet to make a cuplike depression *(b)*. Turn over the sheet, file off the protuberance *(c)*, and a round hole remains.

184. CAN YOU GET 100 PERCENT SAVINGS?

No invention can save 100% on fuel, since energy cannot arise from nothing.

The correct calculation is not 30% + 45% + 25% = 100%, but:

$$\begin{aligned} &100\% - (100\% - 30\%)\,(100\% - 45\%)\,(100\% - 25\%) \\ &= 100\% - (70\% \times 55\% \times 75\%) \\ &= 71.125\%. \end{aligned}$$

This assumes the three inventions are independent in effect.

185. SPRING SCALES

Hang the bar from the hooks of 4 spring scales. Each hook will carry one-fourth the weight of the load. The sum of the readings on the 4 scales is the weight of the bar. In the illustration, the bar weighs 16 pounds.

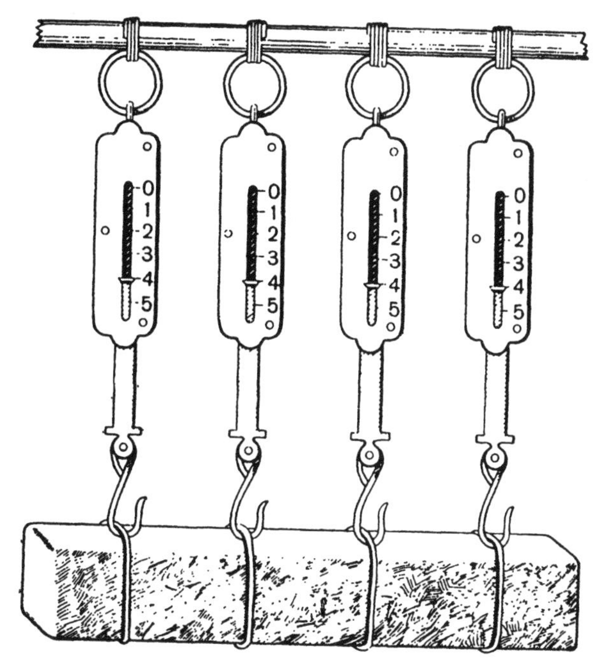

186. INGENUITY IN DESIGN

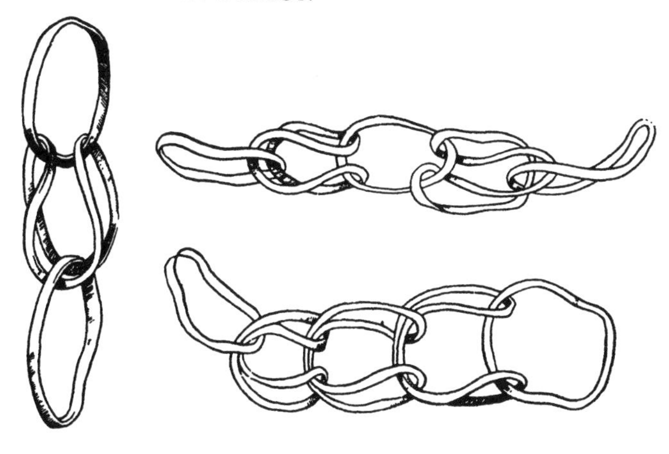

187. CUTTING A CUBE

(A) No. The plane would have to cut exactly 5 faces of the cube. But a cube's opposite faces are parallel, so a plane cannot cut an odd number of faces. (If it cuts a given face it must cut its opposite.)

(B) Yes. In *a*, the sides of triangle AD_1C, being diagonals of faces of the cube, are equal. In *b*, the hexagon's sides are equal because they are diagonals of squares with sides half those of the cube.

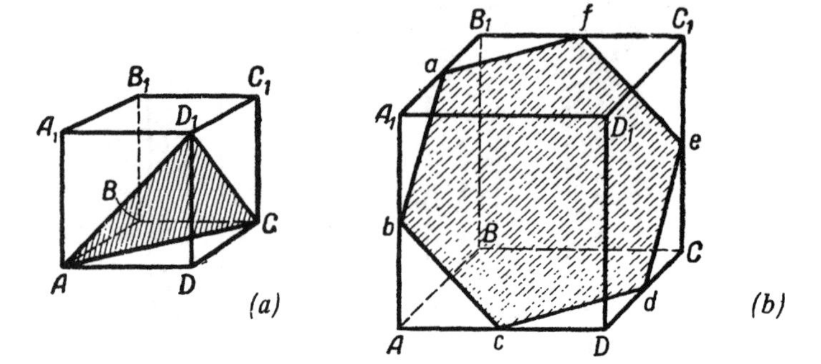

(C) No. A cube has only 6 faces, and a plane can only cut each one of them once.

188. TO FIND THE CENTER OF A CIRCLE

Place *C*, the right angle of the drafting triangle on the circumference, as shown. *D* and *E*, where the triangle's legs cross the circumference, are the endpoints of a diameter. Draw it, and get a second diameter the same way. Their intersection is the center of the circle.

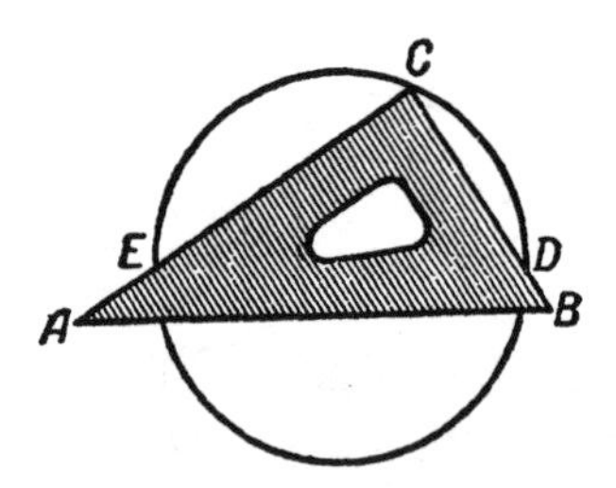

189. WHICH BOX IS HEAVIER?

One box has the balls arranged $3 \times 3 \times 3$, the other $4 \times 4 \times 4$. The diameter of a large ball, then, is 4/3 that of a smaller ball. Its volume, therefore weight, is 64/27 that of a smaller ball. There are as 27/64 as many large balls, so the 2 boxes weigh the same.

This would also be true for other pairs of cubical numbers.

190. A CABINETMAKER'S ART

In the diagram, right angles A and A_1, and B and B_1, coincide when the cube is assembled. The 2 parts can be disconnected only by sliding them along AB.

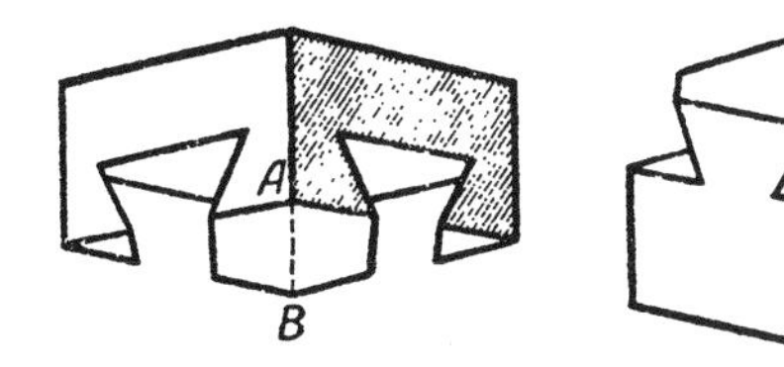

191. GEOMETRY ON A BALL

Put the point of the compasses at any point M on the ball, and draw on the ball a circle of any radius. Mark 3 random points on the circle *(a* in diagram). Using the compasses draw a triangle on the sheet of paper with vertices ABC the same distance apart as A, B, and C on the ball *(b* in diagram).

Draw a circle through the triangle's vertices and draw two perpendicular diameters PQ and GH. The circle equals the circle drawn on the ball, so $PQ = KL$.

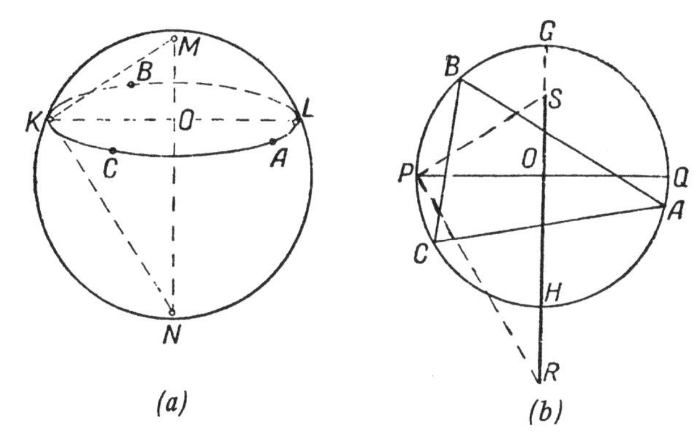

(a) (b)

Let P on the circle correspond to K on the ball's surface. With your compasses open to a distance of KM, put the compass point on P and locate S on GH, so that $PS = KM$. Draw PR perpendicular to PS, with R on the extension of GH. Line segment SR will equal the ball's diameter. This is easily proved by showing that triangle MKN on the ball is congruent to triangle SPR on the sheet of paper.

192. THE WOODEN BEAM

Cut the beam into 2 congruent, 4 steplike parts, as shown in diagram *a*, next page. Each step's height is 9 inches and its width 4 inches. By shifting the upper part to the step below, we get a new parallelepiped with edges of 12, 8, and 18 inches (see *b*).

The new solid is again cut into 2 congruent, steplike parts, in a direction perpendicular to the first one. Each step's height is 6 inches and its width 4 inches. Another shift of upper part to the step below forms the cube shown in *c*.

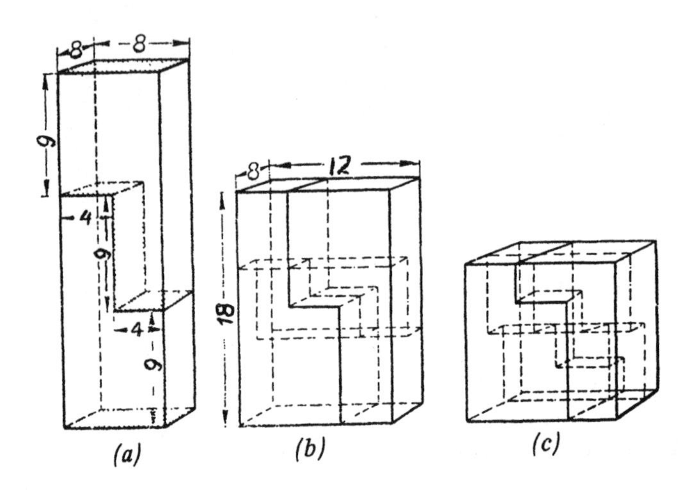

193. A BOTTLE'S VOLUME

The area of a circle, square, or rectangle can easily be calculated after measuring sides or diameter with a ruler. Call the area *s*.

With the bottle upright (see illustration), measure the height h_1 of the liquid. The full part of the bottle has the volume sh_1.

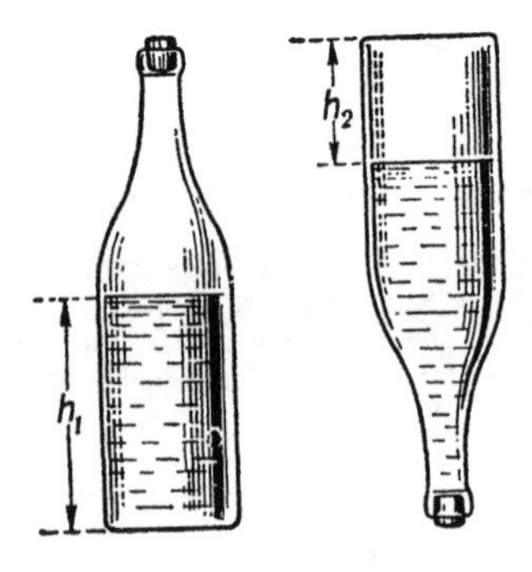

Turn the bottle upside down and measure the height h_2 of the air space. The empty part of the bottle has the volume sh_2. The whole bottle has the volume $s(h_1 + h_2)$.

194. LARGER POLYGONS

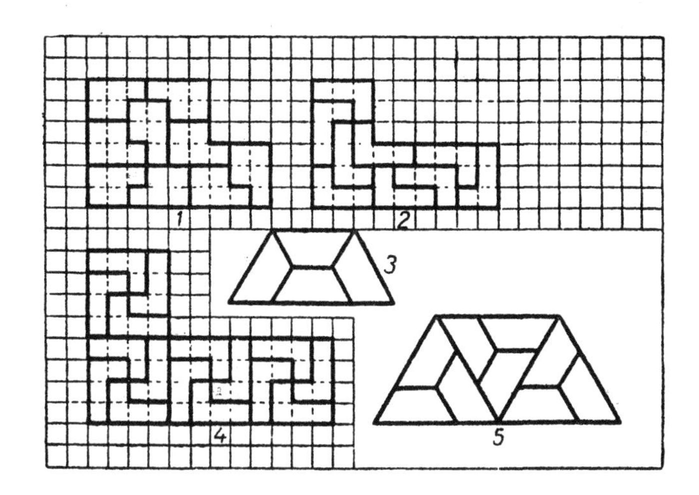

195. LARGER POLYGONS IN TWO STEPS

Several examples are given in *a, b,* and *c.*

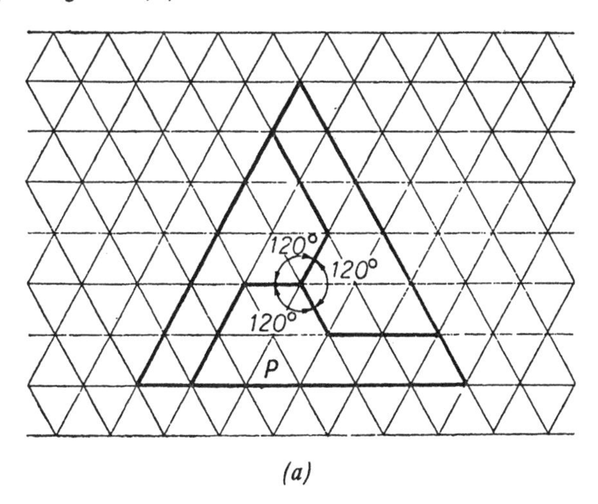

(*a*)

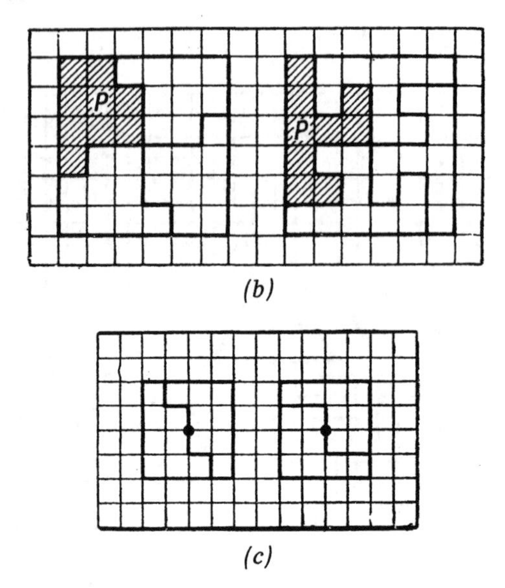

(b)

(c)

A good way to find unit polygons is to mark the center of a rectangle containing an even number of squares (marked with dots in *d)* and draw stair-step or notched lines through the center, dividing the rectangle into 2 congruent parts, as shown. Since you can always make a square out of rectangles, you can always make a square out of the unit polygons found.

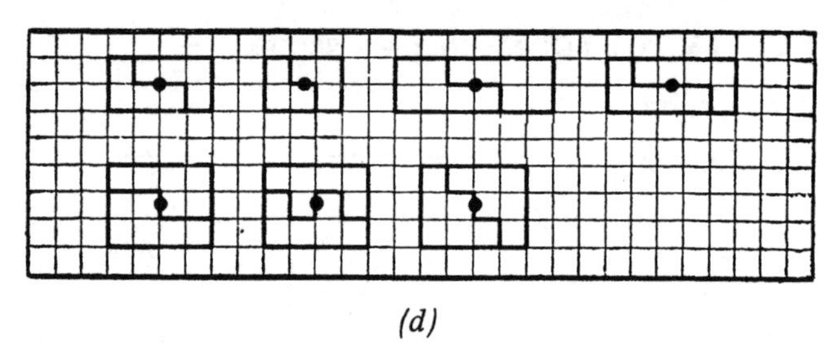

(d)

196. A HINGED MECHANISM FOR CONSTRUCTING REGULAR POLYGONS

The interior angles of a regular 9-gon and 10-gon equal 140° and 144° respectively. Construct both using the hinged mechanism, subtract the 140° angle from the 144° angle, and you have a 4° angle. Bisect it twice with straightedge and compasses, and you get a 1° angle.

VI. Dominoes and Dice

197. HOW MANY DOTS?

Five. Each value appears on an even number of squares (8). Inside the chain the values match in pairs, therefore a 5 at one end of the chain must be matched by another 5 at the other end.

198. A TRICK

The two values you concealed must appear at the ends, since you left an odd number of each value, and they can only appear in pairs on the inside of the chain.

199. A SECOND TRICK

The values of the 13 leftmost tiles are 12 through 0. Before transposing, the middle (thirteenth) tile is 0. If 1 tile is transposed, the middle tile becomes 1; if 2 tiles, the middle is 2, and so on.

200. WINNING THE GAME

The four unused tiles were 0-2, 1-2, 2-5, 6-2. Those laid down were 2-4, 3-4, 3-2, 2-2.

The following distribution of tiles is possible among players *B*, *C*, and *D*:
Player *B*: 0-1, 0-3, 0-6, 0-5, 3-6, 3-5;
Player *C*: 0-0, 1-1, 2-2, 3-3, 4-4, 3-4;
Player *D*: 6-6, 5-5, 6-5, 6-4, 5-4, 6-1.

201. HOLLOW SQUARES

(A) Top (left to right): 4-3, 3-3, 3-1, 1-1, 1-4, 4-6, 6-0. Right (top to bottom): 0-2, 2-4, 4-4, 4-5, 5-5, 5-1, 1-2. Bottom (right to left): 2-3, 3-5, 5-0, 0-3, 3-6, 6-2, 2-2. Left (bottom to top): 2-5, 5-6, 6-6, 6-1, 1-0, 0-0, 0-4. The joinings at the two top corners are shown in the small diagram.

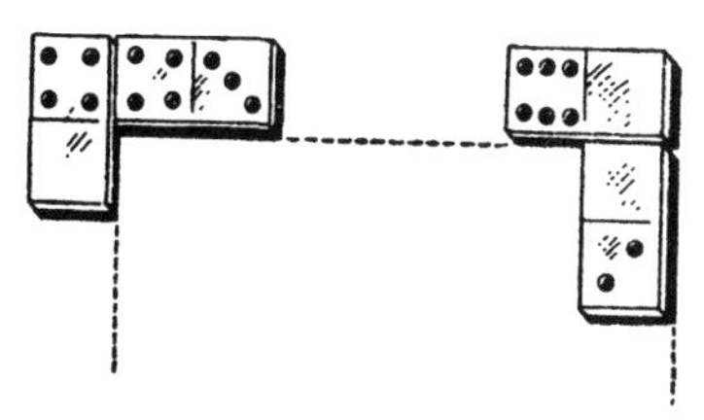

(B) Placing as many blanks as possible in the eight corners, each side can total 21. If the point sum of the eight corners is 8, each side can total 22 (see diagram on next page); if 16, 23; if 24, 24; if 32, 25; if 40, 26.

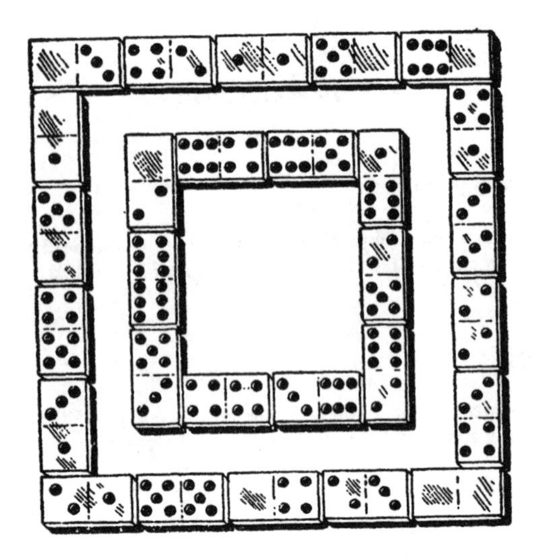

202. WINDOWS

203. MAGIC SQUARES OF DOMINO TILES

(A)

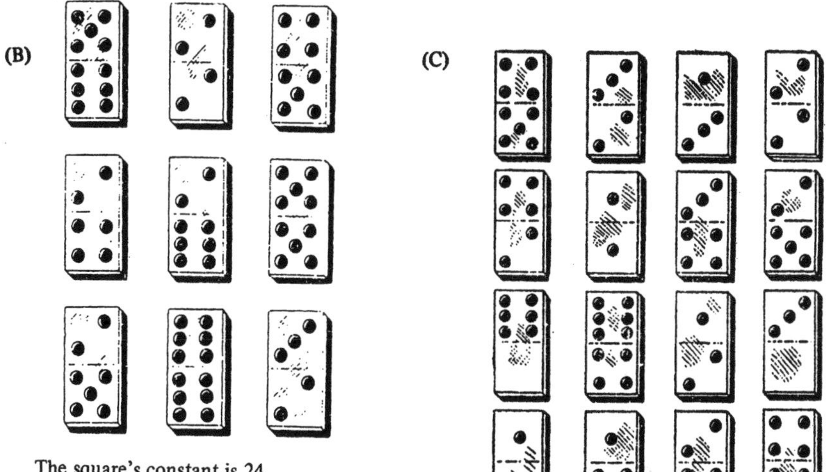

The square's constant is 24.

(D) The solution is:

5-3	0-3	0-6	2-2	1-5
1-1	3-2	1-6	4-5	0-4
6-2	4-6	0-0	1-2	2-4
0-1	1-3	2-5	3-6	3-3
4-4	1-4	3-4	0-2	0-5

204. A MAGIC SQUARE WITH A HOLE

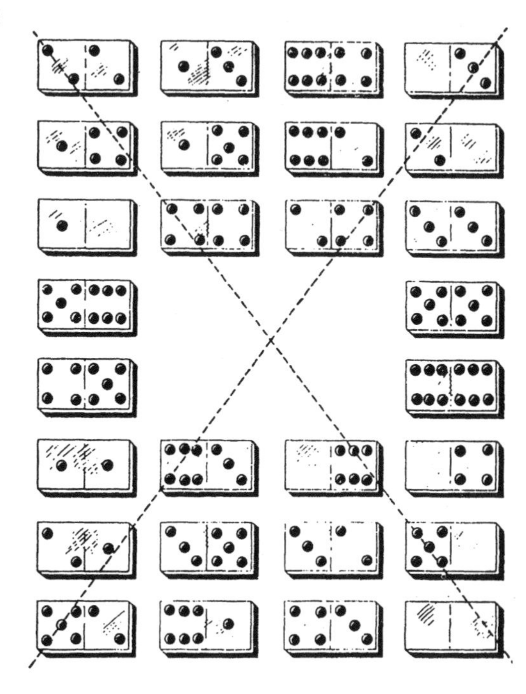

205. MULTIPLYING DOMINOES

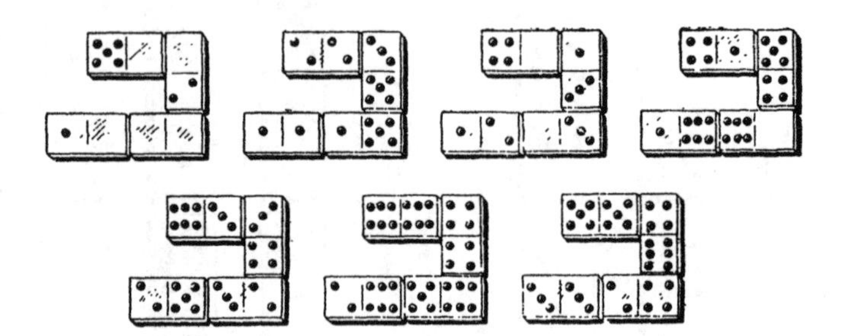

206. GUESS A TILE

Let the mentally noted tile be x-y, and let the first half used be x. His result is $(10x + 5m + y)$. You subtract $5m$ and get a two-digit number composed of x and y. If he starts with y, you get $10y + x$, which is just as good.

207. A TRICK WITH THREE DICE

The sum to be guessed is the number of points on the upper faces of the three dice in their final positions plus the sum on two opposite faces of a die. The latter is 7.

208. GUESSING THE SUM OF SPOTS ON HIDDEN FACES

The top and bottom of the middle die add to 7; the top and bottom of the bottom die add to 7 more. The bottom of the top die is 3 (7 minus the top of the top die).

There are two ways of numbering the faces of a die so that opposite sides add to 7. Each is a mirror image of the other (see figure). Modern dice are numbered as shown on the right, with 1, 2, and 3 reading counterclockwise around their common corner. Thus you need only see two adjacent faces of a die to know what the other four faces are. On your own, can you name the hidden faces on the three dice shown below?

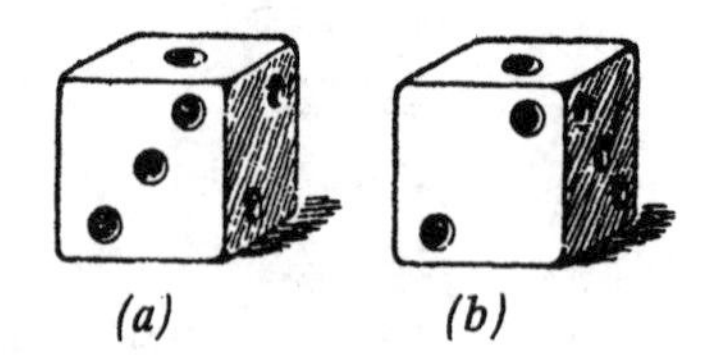

(a) (b)

209. IN WHAT ORDER ARE THE DICE ARRANGED?

Let A be the original three-digit number. The second number is $777 - A$, and the six-digit number is $1{,}000A + 777 - A = 999A + 777 = 111(9a + 7)$. Divide by 111, subtract 7, and divide by 9, and you get A again.

VII. Properties of Nine

210. WHICH DIGIT IS CROSSED OUT?

(A) The 9-remainder of your friend's number is the same as the 9-remainder of the sum of its digits. The latter is unchanged when a 9 is crossed out, so the former is unchanged too.

(B) The 9-remainder of the difference between two numbers composed of the same digits is 0. When he crosses out 1, 2, 3, . . . , 9, he leaves a 9-remainder in what is left of the difference, and in the digit sum he calls out, of 8, 7, 6, . . . , 0. When you subtract one of these numbers from 9, or subtract $9n$ plus the number from $(9n + 9)$, you get 1, 2, 3, . . . , 9: that is, the crossed-out digit.

(C) Add the digits of the number he calls out $(6 + 9 + 8 = 23)$. Subtract from the next higher 9 multiple $(27 - 23 = 4)$.

211. THE NUMBER 1,313

Suppose he subtracts 48 from 1,313 (1,265), appends 148 (1,265,148), calls out 125,148. The digit sum is 21, 6 less than the nearest higher multiple of 9. Then 6 was crossed out.

Explanation: Subtracting a number, then adding 100 plus the number just subtracted, adds 1 to the digit sum of 1,313. This sum (8) simplifies your work because $(8 + 1)$ has a 9-remainder of 0. The rest of the problem is the same as in Problem 210(B).

212. GUESS A MISSING NUMBER

(A) The sum of the numbers 1 through 9 is 45. The sum of the numbers of the tally line is 40. The number I didn't pick is 5.

Or, even quicker: the *digit* sum (the final digit after repeated summations leave only one digit) of the numbers 1 through 9 is $4 + 5 = 9$. The sum of the digits on the tally line is 13, which has a digit sum of $1 + 3 = 4$. This is 5 less than 9.

(B) 3. The digit sum of the 9 given two-digit numbers is 9. The digit sum on the tally line is 6, which is 3 less than 9. Therefore the missing digit is 3.

213. FROM ONE DIGIT

$99 \times 11 = 1{,}089$ $\quad$ $99 \times 66 = 6{,}534$
$99 \times 22 = 2{,}178$ $\quad$ $99 \times 77 = 7{,}623$
$99 \times 33 = 3{,}267$ $\quad$ $99 \times 88 = 8{,}712$
$99 \times 44 = 4{,}356$ $\quad$ $99 \times 99 = 9{,}801$
$99 \times 55 = 5{,}445$

By inspection, the product is 4,356. The fact that each column of digits in the products ascends or descends regularly by 1s suggests that we can find a general method to use without referring to the table:

1. Digits in positions 1 and 3 always add to 9. If the third digit is 5, the first digit must be 4.
2. The second digit is 1 less than the first, therefore the second digit is 3.

Digits in positions 2 and 4 always add to 9, therefore the fourth digit is 6. The product is 4,356.

214. GUESSING THE DIFFERENCE

The number and its reverse have the same middle digit. The number with the smaller first digit is subtracted from the number with the larger first digit; therefore, the latter has the larger last digit, and the difference always has 9 (not 0) for its middle digit.

One of the properties of 9 discussed is that the digit sum of this difference is 9. Its first and third digits must therefore add up to 9.

So if the called-out last digit of the difference is 5, the first digit is 4 and the second digit 9, giving 495.

215. THREE AGES

The difference between A's and B's ages is a multiple of 9 ranging from 0 to $91 - 19 = 72$. C is 0, $4\frac{1}{2}$, 9, $13\frac{1}{2}$, . . . 36. Since 10 times his age is a two-digit number, he is $4\frac{1}{2}$ or 9. But if he is 9, B is 90 and A is 09 or 9, which contradicts the conditions of the problem.

Thus C is $4\frac{1}{2}$, B is 45, and C is 54.

216. WHAT IS THE SECRET?

He picked numbers with an odd number of digits, choosing the central digit at random and adding the others mentally as he went along so they added to a multiple of 9.

VIII. With Algebra and without It

217. MUTUAL AID

The problem can be easily solved backward:

First		Second		Third	
24	+	24	+	24	= 72;
↓		↓			
12	+	12	+	48	= 72;
↓				↓	
6	+	42	+	24	= 72;
		↓		↓	
39	+	21	+	12	= 72.

The last line is the answer. Reading up and reversing the directions of the arrows, you can see that the conditions of the problem have been met.

218. THE IDLER AND THE DEVIL

We will also solve this problem backward, but verbally. Before the third crossing the idler had $12. Adding the $24 he gave the devil after the second crossing, he had $36, twice the $18 he had before the second crossing. Adding $24 again, he had $42, twice the $21 he started with.

219. A SMART BOY

Before the oldest gave half his apples to his brothers he had 16; the middle and youngest had 4 each. Before the middle brother divided his apples he had 8; this means the oldest had $16 - \frac{1}{2}(4) = 14$ and the youngest had 2. Before the youngest divided his apples he had 4; the middle brother had $8 - \frac{1}{2}(2) = 7$, and the oldest 13.

The youngest brother is 7, the middle brother 10, and the oldest 16.

220. HUNTERS

Let x = the answer.
$$x - (4 \times 3) = \frac{1}{3}x$$
$$\frac{2}{3}x = 12;$$
$$x = 18.$$

221. TRAINS MEETING

When the locomotives meet, the cabooses are 2/6 = 1/3 mile apart, and their net approaching speed is 120 miles per hour. It takes them 1/360 hour = 10 seconds to meet.

222. VERA TYPES A MANUSCRIPT

Her mother was right. To average 20 pages a day, Vera would have had to type the second half of the manuscript in *no* time.

223. A MUSHROOM INCIDENT

At the end, let each boy have x mushrooms. Marusya gave Kolya $(x - 2)$ mushrooms, Andryusha $\frac{1}{2}x$, Vanya $(x + 2)$, and Petya $2x$. The sum, $4\frac{1}{2}x = 45$ mushrooms. Then $x = 10$ and the four boys received 8, 5, 12, and 20 mushrooms respectively.

224. MORE OR LESS?

Did you say, "They will both take the same time"? Let us approach the problem both "with algebra and without it." Apparently A on the river will be slower, because the current helps him for less time (he is covering the same distance faster) than it hinders him. If the current is half as fast as his rowing, for example, it will take him as long to do half the trip upstream as it takes B to do the whole trip. And if the current is as fast as his rowing, he will never get upstream at all!

Let's nail it down with algebra. Rate times time equals distance, so distance divided by rate equals time. B's distance is $2x$. Let his rate be r; then B's time is: $2x/r$.

A goes distance x downstream at rate $(r + c)$, where c is the rate of the current, and distance x upstream at rate $(r - c)$. Then A's total time is:

$$\frac{x}{r+c} + \frac{x}{r-c} = \frac{2xr}{r^2 - c^2}\,.$$

Now divide A's time by B's time:

$$\frac{2xr}{r^2 - c^2} \div \frac{2x}{r} = \frac{r^2}{r^2 - c^2} .$$

But r^2 is greater than $(r^2 - c^2)$, so the fraction is greater than 1, and A takes more time than B does.

225. A SWIMMER AND A HAT

Consider the problem from the viewpoint of the hat. Then it is not the hat that floats downstream from bridge to bridge. Instead, the second bridge moves with the current's speed toward the hat, which rests in *still* water. Also in still water, the swimmer moves away from the hat for 10 minutes and swims toward it 10 minutes. At the moment he rejoins the hat, the second bridge has "reached" the hat. Thus the speed of the current is $1{,}000 \div 20 = 50$ yards per minute. The swimmer's speed is immaterial.

226. TWO DIESEL SHIPS

From the buoy's point of view (floating downstream) the ships move away from it at equal speeds in still water. Then they return at equal speeds in still water. Thus the two ships reach the buoy simultaneously.

227. HOW SHARP-WITTED ARE YOU?

At first meeting, the boats have traveled a combined distance equal to 1 length of the lake; at second meeting, 3 lengths. (See diagram.) Elapsed time and distance for each is three times as

great. Then at second meeting M has traveled $500 \times 3 = 1{,}500$ yards. Since this is 300 yards longer than the length of the lake, the latter is 1,200 yards.

The ratio of M's speed to N's equals the ratio of the distances they travel before their first meeting:

$$\frac{500}{1{,}200 - 500} = \frac{5}{7}.$$

228. YOUNG PIONEERS

Let x be the answer. Kiryusha's pledge is to plant $\frac{1}{2}x$ trees. Vitya's pledge is to plant $\frac{1}{3}x$ trees. The preceding brigades plant the remaining trees: $\frac{1}{6}x$. Since this is 40, x is 240.

229. HOW MANY TIMES AS LARGE?

Twice as large. Call half the smaller number m. The smaller number minus m is m; the larger number minus m is three times as large, or $3m$. Then the smaller number is $m + m = 2m$ and the larger number is $3m + m = 4m$.

230. A DIESEL SHIP AND A SEAPLANE

Perhaps you can spot without any algebra or extended calculation that the seaplane goes 200 miles while the ship goes another 20.

231. BICYCLE FIGURE RIDERS

In $\frac{1}{3}$ hour, they travel $\frac{1}{3}$ mile 6, 9, 12, and 15 times. The largest number these numbers are all divisible by is 3. They return to their original array 3 times in 20 minutes: after $6\frac{2}{3}$, $13\frac{1}{3}$, and 20 minutes.

232. SPEED OF WORK BY TURNER BYKOV

The old time is 14 times the new, so the new velocity, inversely, is 14 times the old:

$$\frac{v}{v - 1{,}690} = 14.$$

Then $v = 1{,}820$ inches per minute.

233. JACK LONDON'S JOURNEY

If London had covered 50 more miles at full speed he would have reached the camp 24 hours sooner; so if he had covered 100 more miles at full speed he would have reached the camp 2 days sooner, without being late at all. Then 100 more miles remained to the camp at the end of the first day. With 5 dogs he could have gone not 100 but $\frac{5}{3}(100) = 166\frac{2}{3}$ miles. Not going the extra $66\frac{2}{3}$ miles would have saved him the 48 hours the problem mentions. Full speed, then, was $33\frac{1}{3}$ miles per day. He traveled $33\frac{1}{3}$ miles (first day) plus 100 miles (after the dogs ran away), or a total of $133\frac{1}{3}$ miles.

234. FALSE ANALOGY

(A) 30% (from 1 to $\frac{13}{10}$).
(B) Not 30% but almost 43% (from 1 to $\frac{10}{7}$).
(C) Not 10% + 8% = 18%. If 90% of the original price is 108% of what a book costs the store, then 100% of the original price is $108\% \times \frac{100}{90} = 120\%$ of what a book costs the store, and the answer is 20%.
(D) Not by $p\%$. He used to make 1 part of 1 time unit, but now it takes:

$$1 - \frac{p}{100} \text{ time unit.}$$

In 1 time unit he makes, not $1 + \frac{p}{100}$ parts, but:

$$\frac{1}{1 - \frac{p}{100}} = \frac{100}{100 - p} \quad \text{parts,}$$

an increase of:

$$100\left(\frac{100}{100-p} - 1\right)\% = \frac{100p}{100-p}\ \%.$$

For example, in (B) of this problem, p was 30% and the increase was:

$$\frac{100(30)}{100-30} = \frac{3{,}000}{70} = \text{almost } 43\%.$$

235. A LEGAL TANGLE

There is no one "right" answer. The Roman jurist, Salvian Julian, proposed:
The father's intent is clearly that the daughter should receive half as much as her mother, and the son twice as much. The inheritance should be divided into 7 equal parts: 2 for the mother, 4 for the son, and 1 for the daughter.
An opposing view:
The father wished the mother to inherit *at least* 1/3 of the estate, but Salvian Julian would give her only 2/7. Why not give her 1/3 and divide the rest between son and daughter according to the obviously intended ratio, 4 to 1? Divide the estate, then, into 15 equal parts: 5 for the mother, 8 for the son, and 2 for the daughter.
We received another opinion from Azimbai Asarov of the Kazakh SSR:
One of the twins must have been born first. If it was the boy, he is entitled to 2/3 of the estate, the remainder being divided between the daughter, 1/9 and the mother (twice as much), 2/9. But if it was the girl, she should get 1/3 of the estate, the remainder being shared by the son, 4/9, and the mother (half as much), 2/9.

236. TWO CHILDREN

(A) For 2 children in general there are 4 equally likely events: boy – boy, boy – girl, girl – boy, and girl – girl. Since boy – boy is ruled out, the probability of 2 girls is 1/3.

(B) Let each of the same 4 equally likely events be in older – younger order. Since girl – boy and girl – girl are ruled out, the probability of 2 boys is 1/2.

237. WHO RODE THE HORSE?

This problem will be solved with and then without algebra. Let the city be x miles from the village: that is the y miles the older man rode and the $(x - y)$ miles he had left. If he had ridden $3y$ miles, he would have had $(x - 3y)$ miles left. Or he would have had "half as far" left: $\frac{1}{2}(x - y)$ miles. Then:

$$x - 3y = \frac{1}{2}(x - y);$$
$$y = \frac{1}{5}x.$$

The young man rode z miles and had $(x - z)$ miles left. If he had ridden $\frac{1}{2}z$ miles, he would have had $(x - \frac{1}{2}z)$ miles left. Or he would have had "three times as far" left: $3(x - z)$ miles. Then:

$$x - \frac{1}{2}z = 3(x - z);$$
$$z = \frac{4}{5}x.$$

The young man rode four times as far as the older man, so it was the older man who rode the horse.

Second solution: Eighth-grader Lyalya Grechko used a geometric construction.

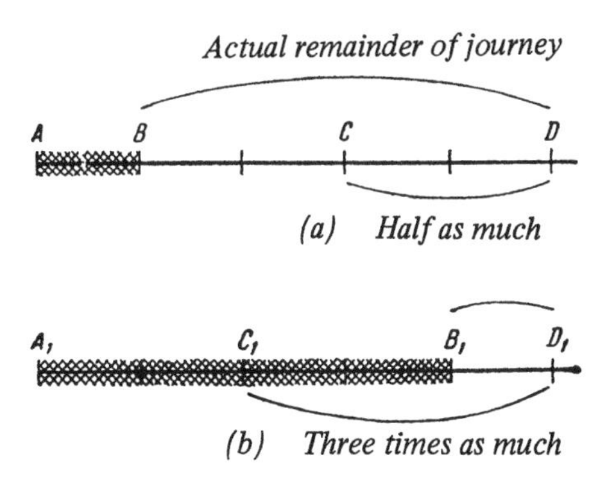

(a) Half as much

(b) Three times as much

In a, let a random segment AB represent how far the older man rode. Mark two more segments equal to AB. If the older man had ridden three times as far as he had, he would have ridden to C. Now mark D so CD is half as far to ride as BD, the actual remainder of the journey.

Note the distance to the city has 5 equal parts, as shown.

In b, we make a similar construction for the young man. Let a random segment A_1B_1 represent how far he rode. Divide it in two at C_1 to show how far he would have ridden if he had ridden half as far as he had. Mark D_1 so C_1D_1 is three times as far to ride as B_1D_1, the actual remainder of the journey.

• Again the line has 5 equal parts. Since $AD = A_1D_1$, the young man rode four times as far as the older one, as the shading shows. So the older man rode the horse.

238. TWO MOTORCYCLISTS

Suppose the first motocyclist rode x hours and rested $\frac{1}{3}y$, and the second rode y hours and rested $\frac{1}{2}x$.
Then:

$$x + \tfrac{1}{3}y = y + \tfrac{1}{2}x;$$
$$x = \tfrac{4}{3}y.$$

Since the first motorcyclist rode more hours to go the same distance, the second one is faster.

239. IN WHICH PLANE DID VOLODYA'S FATHER FLY?

Algebra is unwieldy for such a simple problem. Just make a table:

Plane	*To right*	*To left*	*Product*
1	8	0	0
2	7	1	7
3	6	2	12
4	5	3	15
5	4	4	16
6	3	5	15
7	2	6	12
8	1	7	7
9	0	8	0

His father flew in the third plane. (Twelve is 3 less than 15.)

240. EQUATIONS TO SOLVE IN YOUR HEAD

Adding and subtracting the equations we see that the numbers become 10,000, 10,000, and 50,000; and 3,502, – 3.502, and 3,502. Dividing by 10,000, and by 3,502, we obtain:

$$x + y = 5;$$
$$x - y = 1.$$

Anyone can solve such simple equations in his head.

241. TWO CANDLES

Let x be the original length of the long candle, and y of the short candle. After 2 hours $2 \div 3\frac{1}{2} = \frac{4}{7}x$ has burned, and $\frac{2}{5}y$ has burned, leaving the equal lengths of $\frac{3}{7}x$ and $\frac{3}{5}y$. Then the short candle had $\frac{5}{7}$ the height of the long one.

242. WONDERFUL SAGACITY

Any four-digit number may be written as:

$$1{,}000a + 100b + 10c + d.$$

Transposing the first digit to the end, we get:

$$1{,}000b + 100c + 10d + a.$$

The sum is:

$$1{,}001a + 1{,}100b + 110c + 11d.$$

Obviously, the sum is divisible by 11, and only Tolya's number was divisible by 11. (See Problem 314 for a quick way to verify this.)

243. CORRECT TIMES

No, the 2-minute differences do not cancel out by subtraction, and the wristwatch does not keep correct time.

In 1 hour the wall clock ticks away 58 minutes.

In 1 hour by the wall clock the table clock ticks away 62 minutes. In 58 minutes by the wall clock (1 actual hour) the table clock ticks away:

$$58 \times \frac{62}{60} \text{ minutes.}$$

And in this number of minutes by the table clock (1 actual hour) the alarm clock ticks away:

$$58 \times \frac{62}{60} \times \frac{58}{60} \text{ minutes.}$$

And in this number of minutes by the alarm clock (1 actual hour) the wristwatch ticks away:

$$58 \times \frac{62}{60} \times \frac{58}{60} \times \frac{62}{60} = 59.86 \text{ minutes.}$$

The wristwatch loses 0.14 minute per real hour, or 0.98 minute in 7 hours. At 7:00 p.m. the wristwatch will show, to the nearest minute, 6:59.

244. WATCHES

The watches will show the same time again when the gain of mine plus the lag of Vasya's equals 12 hours (43,200 seconds). In x hours my watch will be x seconds fast and Vasya's watch will be $\frac{3}{2}x$ seconds slow. Then:

$$x + \tfrac{3}{2}x = 43{,}200;$$
$$x = 17{,}280 \text{ hours} = 720 \text{ days.}$$

To show the same correct time will take even longer—until my watch is a multiple of 12 hours fast and Vasya's is a multiple of 12 hours slow. This will happen to my watch every 43,200 hours (1,800 days), and to Vasya's watch every 1,200 days. The lowest common multiple of 1,800 and 1,200 days is 3,600 days (almost 10 years), which is the second answer.

245. WHEN?

(A) During lunch hour the clock hands move a total of 360°—a full circle. The minute hand, moving twelve times as fast as the hour hand, covers $\frac{12}{13}$ of the circle, and the hour hand $\frac{1}{13}$. The craftsman was absent $\frac{12}{13}$ hour, or 55 $\frac{5}{13}$ minutes.

From noon till x minutes later when the craftsman goes to lunch, the minute hand moves x minutes past 12 and the hour hand $\frac{1}{12}x$ minutes past 12. As he leaves, the hands are $\frac{11}{12}x$ apart. But this distance has been shown to be $\frac{1}{13} \times 60$ minutes. Then:

$$\frac{11}{12}x = \frac{1}{13} \times 60;$$
$$x = 5\,\frac{5}{143} \text{ minutes.}$$

The craftsman left for lunch at 12:05$\frac{5}{143}$p.m., was absent 55$\frac{5}{13}$ minutes, and returned at 1:00$\frac{60}{143}$p.m.

(B) Two hours after I leave, the minute hand will be where it is as I leave, and the hour hand will have covered $\frac{2}{12}$ of a full circle. For the hands to move until their original positions are exchanged, the sum of their distances from their original positions must increase, after 2 hours, by $\frac{10}{12}$ of a circle, or 50 minutes. The minute hand moves twelve times as fast as the hour hand, so the distance still to be covered by it is $\frac{12}{13} \times 50 = 46\frac{2}{13}$ minutes. The walk lasted 46$\frac{2}{13}$ minutes more than 2 hours.

(C) At 4:00 p.m. the hour hand is at 20 minutes. It moves $\frac{1}{12}x$ minutes as the minute hand moves x minutes to meet it. Then:

$$20\,x \times \frac{1}{12}x = x;$$
$$x = 21\frac{9}{11} \text{ minutes.}$$

The boy starts the problem at 4:21$\frac{9}{11}$p.m.

Again starting at 4:00 p.m. the hour hand moves $\frac{1}{12}y$ minutes from 20 as the minute hand moves y minutes to pass the hour hand by 30 minutes (half a full circle). Then:

$$20 + \frac{1}{12}y + 30 = y;$$
$$y = 54\frac{6}{11} \text{ minutes.}$$

The boy finishes the problem at 4.54 $\frac{6}{11}$ p.m. It takes him 32$\frac{8}{11}$ minutes to solve it.

246. WHEN DOES THE CONFERENCE START AND END?

The hour and minute hands in the illustration show about when the conference starts. The

minute hand has moved y minutes since 6:00 p.m., and the hour hand $(x - 30)$ minutes. Since the minute hand moves twelve times as fast as the hour hand:

$$y = 12(x - 30).$$

Now imagine that the hands have exchanged places. The hour hand (the *long* hand in the illustration) has moved $(y - 45)$ minutes since 9:00 p.m., and the minute hand (the *short* hand) has moved x minutes. Then:

$$x = 12\,(y - 45).$$

Substituting from the first equation:

$$x = 12\,[12(x - 30) - 45] = 144x - 4{,}860;$$
$$x = 33\frac{141}{143} \text{ minutes;}$$
$$y = 12(x - 30) = 47\frac{119}{143} \text{ minutes.}$$

The conference starts at 6:47 $\frac{119}{143}$ p.m. and ends at 9:33 $\frac{141}{143}$ p.m.

247. A SERGEANT TEACHES HIS SCOUTS

The first scout gets there first whether they run first or walk first.

Say they run first. They both run half the distance. The second scout starts walking. But the first scout still has some running to do, because he runs farther in half the time than he walks in half the time. So he takes the lead. When he changes to walking the second scout is also walking, so the first scout doesn't lose the lead and he finishes first.

Or say they walk first. They both walk half the *time* and—but it isn't necessary to finish the proof this way. It is exactly the same race only done backward, so again the first scout finishes first.

248. TWO DISPATCHES

Let x be the train's length and y its speed. It passes the observer—that is, it covers a distance equal to its own length in t_1 seconds:

$$y = \frac{x}{t_1} .$$

It crosses the bridge—that is, it covers a distance equal to the sum of its own length and a yards—in T_2 seconds:

$$y = \frac{x + a}{t_2} .$$

Thus:

$$x = \frac{at_1}{t_2 - t_1} ;$$

$$y = \frac{a}{t_2 - t_1} .$$

249. NEW STATIONS

If y stations are added to x already existing stations each new station will require $(x + y - 1)$ sets of tickets: for y new stations this is $y(x + y - 1)$ sets. Each old station needs y sets so:

$$y(x + y - 1) + xy = 46, \text{ or}$$
$$y(2x + y - 1) = 46.$$

Thus y must be a positive integer which is a factor of 46: it is 1, 2, 23, or 46. "Some new stations" cannot be 1 station, and 23 and 46 give a minus number of old stations, so $y = 2$ and $x = 11$.

250. SELECT FOUR WORDS

School, oak, overcoat, mathematical. Multiply the left and right sides of the given equations: $a^3d = b^3\,dc$, or $a^3 = b^3c$.

It follows that c must be a cube of an integer. The only cube from 2 through 15 is 8, so $c = 8$, and $a^3 = 8b^3$, or $a = 2b$.

Substituting in the first given equation, $4b^2 = bd$, or $4b = d$. In the given range of numbers, b must equal 2 or 3, and d must equal 8 or 12. Since c already is 8, d is 12 and b is 3. Then $a = 2b = 6$.

251. FAULTY SCALES

When the pans balance, then whether or not the scales are even armed this equation holds:

$$a(p + m) = b(q + m),$$

where a and b are the lengths of the arms, p and q are being weighed, and m is the weight of each pan (see diagram).

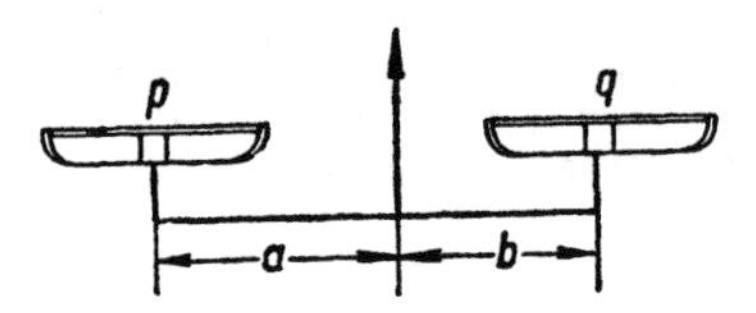

Say x pounds of sugar and then y pounds of sugar are balanced against 1 pound of sugar, as described in the problem. Then:

$$a(1 + m) = b(x + m), \text{ and } a(y + m) = b(1 + m).$$

Therefore:

$$x = \frac{a + am - bm}{b}, \quad y = \frac{b + bm - am}{a}$$

The sugar adds up to:

$$x + y = \frac{a}{b} + \frac{b}{a} + m \quad \frac{a}{b} + \frac{b}{a} \quad - 2m.$$

We will now show that the sum of any positive fraction a/b (where a is not equal to b) and its inverse, b/a is greater than 2. Since $(a - b)^2$ is positive:

$$a^2 - 2ab + b^2 > 0, \text{ or}$$
$$a^2 + b^2 > 2ab.$$

Dividing each side by ab (which is positive):

$$\frac{a}{b} + \frac{b}{a} > 2.$$

Multiply by m (which is positive):

$$m\left(\frac{a}{b} + \frac{b}{a}\right) > 2m.$$

It follows that:

$$m\left(\frac{a}{b} + \frac{b}{a}\right) - 2m > 0.$$

and since:

$$\frac{a}{b} + \frac{b}{a} > 2,$$

it is easy to see, in the equation for the total weight of the sugar, that $(x + y)$ must be greater than 2.

To get proper weighing from faulty scales the sugar and weights must be on the same side of the scales! Well, not exactly, but you can put the 1-pound weight on the left pan and balance it with lead shot on the right pan. Then remove the 1-pound weight and add sugar till it balances the lead shot. Do it twice and you have 2 pounds of sugar.

252. AN ELEPHANT AND A MOSQUITO

The wrong square root of $(y - v)^2$ was used. According to the conditions of the problem, it should have been $-(y - v)$; not $(y - v)$:

$$x - v = -(y - v);$$
$$x + y = 2v.$$

Note that $(x - v)$ (an elephant minus a half-elephant, half-mosquito) is positive, while $(y - v)$ is negative. If numbers had been used, you would have seen the fallacy, for example, of:

$$81 = 81;$$
$$9 = -9.$$

253. A FIVE-DIGIT NUMBER

A with 1 after it is $10A + 1$. With a 1 before it, it is $100{,}000 + A$. Then $10A + 1 = 3(100{,}000 + A)$, and $A = 42{,}857$.

254. LIVE TO ONE HUNDRED WITHOUT AGING

If my age is AB and yours is CD, then segment KB shows how long ago my age equaled your present age. But your age then was less by $ND = KB$ and equaled CN, which is half AB.

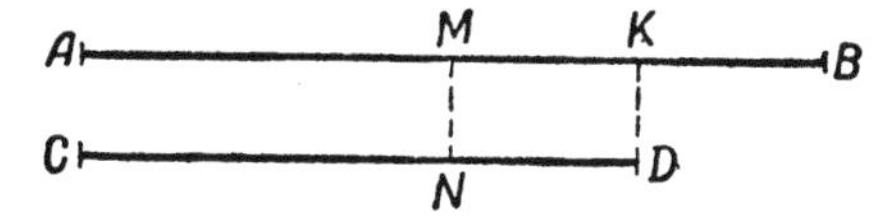

Since $ND = MK = KB$, then $MB = 2KB$, $AB = 4KB$, and $CD = 3KB$.

When you are as old as I am now, your age will be represented by a segment equal to AB, which contains KB four times. By then my age will increase by KB and will be represented by a segment containing KB five times. Since $4KB = 5KB = 63$, segment KB represents 7 years. You are 21 and I am 28. Seven years ago you were 14, half my present age.

255. THE LUCAS PROBLEM

If you answered 7, bearing in mind the ships that haven't started yet, you forgot about the ships already en route. A convincing solution is shown in the diagram.

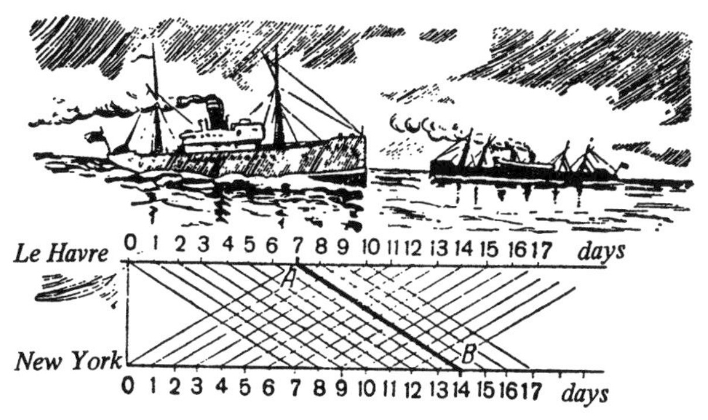

AB is the ship leaving Le Havre today. It will meet 13 ships at sea and 1 in each harbor, a total of 15. The meetings are daily, at noon and midnight.

256. A SINGULAR TRIP

We do not know when or how often the bicycle changes hands. In problems with such fluid conditions, a graph is often helpful. The vertical axis shows miles and the horizontal axis, hours.

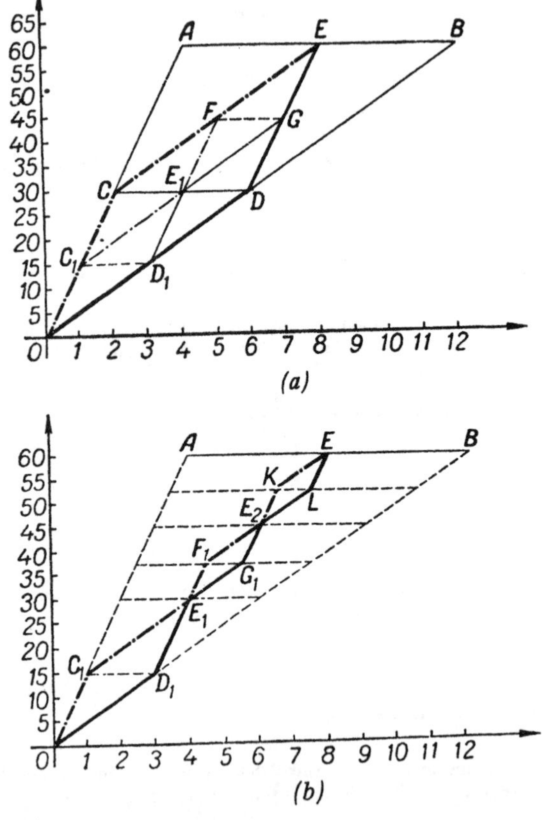

One boy bicycling all the way (at 15 miles per hour) is represented by *OA*, where *A* is at 60 miles and 4 hours. One boy walking all the way (5 miles per hour) is *OB*, where *B* is at 60 miles and 12 hours. The boys' actual paths lie between *OA* and *OB*, and they intersect on *AB* (since they finish their trip simultaneously).

Suppose the bicycle changes hands once. Their paths form a parallelogram (see *s*). (Obviously, to finish simultaneously, they must switch halfway, at 30 miles.) One boy bicycles to *C*, then walks to *E*. Note the bend in his path *OCE*. His walk on *CE* is parallel to the second boy's walk from *O* to *D*. The second boy picks up the bicycle, which was left there 4 hours ago by the first, and rides from *O* to *C*, parallel to *OA*.

D is at the same height as *C* because the bicycle is picked up where it was left. *CDAE* is a parallelogram and $AE = CD = 4$ hours. Then E is at 8 hours, which is how long their trip took after the breakdown with one bicycle change.

With three changes (see *a*), the second is shown at E_1, the midpoint of CD. Here the second boy, who walked OD, and rode D_1E_1, catches up with the first boy, who rode OC_1 and walked C_1E_1. Their total paths are OC_1E_1FE (first boy) and OD_1E_1GE (second boy). The last changeover occurs 15 miles from the destination (on FG).

For five changes, see *b*. After their meeting at E_1 their paths are $E_1F_1E_2KE$ and $E_1G_1E_2LE$, with the last changeover 7½ miles from the destination (on KL).

No matter how many times they change, they will always end at E. (Total bicycling distance and time is always the same.) So the answer is, they will always reach the destination simultaneously, no matter where the bicycle is left behind for the last time.

257. A CHARACTERISTIC OF SIMPLE FRACTIONS

For this kind of general problem with rigid conditions, try algebra.

Let
$$\frac{a_1}{b_1}, \frac{a_2}{b_2}, \frac{a_3}{b_3}, \ldots, \frac{a_n}{b_n}$$
be fractions whose numerators and denominators are positive integers. They have been arranged in ascending order, the smallest fraction being $\frac{a_1}{b_1}$ and the largest $\frac{a_n}{b_n}$. We must prove that:

$$\frac{a_1}{b_1} < \frac{a_1 + a_2 + a_3 + \ldots + a_n}{b_1 + b_2 + b_3 + \ldots + b_n} < \frac{a_n}{b_n}.$$

We have:

$$\frac{a_2}{b_2} > \frac{a_1}{b_1} \text{ or } a_2 > b_2\frac{a_1}{b_1};$$

$$\frac{a_3}{b_3} > \frac{a_1}{b_1} \text{ or } a_3 > b_3\frac{a_1}{b_1};$$

$$\ldots\ldots\ldots\ldots\ldots$$

$$\frac{a_n}{b_n} > \frac{a_1}{b_1} \text{ or } a_n > b_n\frac{a_1}{b_1}.$$

Therefore:

$$a_2 + a_3 + \ldots + a_n > (b_2 + b_3 + \ldots + b_n)\frac{a_1}{b_1}.$$

To the left side we add a_1, and to the right side we add $\frac{b_1 \; a_1}{b_1}$:

$$a_1 + a_2 + a_3 + \ldots + a_n > (b_1 + b_2 + b_3 + \ldots + b_n)\frac{a_1}{b_1}.$$

Therefore:

$$\frac{a_1 + a_2 + a_3 + \ldots + a_n}{b_1 + b_2 + b_3 + \ldots + b_n} > \frac{a_1}{b_1}.$$

The second part of the theorem is proved similarly.

IX. Mathematics with Almost No Calculations

258. SHOES AND SOCKS

4 shoes, 3 socks. Of 4 shoes, 2 must be of the same brand; of 3 socks, 2 must be the same color.

259. APPLES

4 apples; 7 apples.

260. A WEATHER FORECAST

No—it will be midnight again.

261. ARBOR DAY

Ten. The sixth-graders overfilled their assignment by 5 trees; the fourth-graders underfilled theirs by 5.

262. MATCH NAMES AND AGE

1. Of Burov's two grandfathers, one is named Burov and one Serov. Petya's grandfather is named Mokrasov. Petya is not Burov.
2. Kolya is not Burov.
3. Burov is Grisha.
4. Petya is not Gridnev, so he must be Klimenko.
5. By elimination, Kolya is Gridnev.
6. If Petya was 7 when he started first grade, he is starting sixth grade at 12.
7. Gridnev and Grisha are 13.

Summary:
Grisha Burov is 13.
Kolya Gridnev is 13.
Petya Klimenko is 12.

263. A SHOOTING MATCH

Tabulate the results and satisfy yourself that there is only one way to partition the 18 shots into 3 equal sets of 6:

25, 20, 20, 3, 2, 1;
25, 20, 10, 10, 5, 1;
50, 10, 5, 3, 2, 1.

The first row is Andryusha's, since it is the only row that contains two numbers whose sum is 22.

There are 3s in the first and third row. Then the third row is Volodya's, and he hit the bull's-eye.

264. A PURCHASE

Four cents, 20 cents, 8 notebooks, and 12 sheets of paper are all divisible by 4, but 170 cents is not.

265. PASSENGERS IN A RAILROAD COMPARTMENT

Numbers in the table indicate that certain matchings of city and initial are impossible. In *A*'s column, for example, "1" refers to statement 1 in the problem, which implies that *A* is not from Moscow, and "1-2" refers to statements 1 and 2, which imply that *A*, a physician, cannot be the Leningrader, who is a teacher.

	A	*B*	*C*	*D*	*E*	*F*
Moscow	1	7	7-8 1-3	–	1-2	*
Leningrad	1-2	*	2-3	–	2	–
Kiev	–	–	*	–	–	–
Tula	1-3	4	3	*	2-3	4
Odessa	*	–	6	–	–	–
Kharkov	5	7-8	8	–	*	–

After entering all possible numbers in the table, we eliminated all cities for *C* except Kiev. We put a star in the *C*–Kiev box, and put dashes in all the other Kiev boxes, to show that they are eliminated. The dash in *A*'s column eliminated all cities for *A* except Odessa, and so on until every empty box had a star or a dash.

Statements 1-3 link 6 initials or cities to a profession. The matchings we have just marked with stars link the other 6 initials or cities to a profession. We have: *A*, Odessa, physician; *B*, Leningrad, teacher; *C*, Kiev, engineer; *D*, Tula, Engineer; *E*, Kharkov, teacher; *F*, Moscow, physician.

The facts given are sufficient, but not all are necessary. There are two entries in the table showing that *C* is not from Moscow.

Fifteen facts are necessary to match initials and cities: 5 of 6 initials must be eliminated to match the first city, then 4 of 5 to match the second city, 3 of 4, 2 of 3, and 1 of 2.

266. A CHESS TOURNAMENT

	Inf	*Fly*	*Tank*	*Artil*	*Car*	*Mort*	*Sap*	*Com*
Col	9-10	1-2	7-12	10	1	1-13	11-12	*
Maj	3-4	–	7-8	*	–	–	–	–
Cap	5-9	*	5-7	5-10	5-13	5-13	5-11	–
Lieut	9	–	7-11	9-10	*	–	11	–
SS	3-4	–	7-12	10-12	1-12	*	11-12	–
JS	6-9	–	6-7	6-10	–	–	*	–
Corp	3	–	*	–	–	–	–	–
Priv	*	–	–	–	–	–	–	–

Again, the stars show matchings. There are 28 facts in the table, which is just enough (7 + 6 + 5 + 4 + 3 + 2 + 1 = 28).

267. VOLUNTEERS

Sawing 2-yard logs produces a number of ½-yard logs divisible by 4; sawing 1½-yard logs, divisible by 3; 1-yard logs, divisible by 2. Pastukhov's team's 27 logs is not divisible by 2 or 4, so it is the 1½-yard-log team of Petya and Kostya. The team leader is Petya Galkin, so Pastukhov's first name is Kostya.

268. WHAT IS THE ENGINEER'S LAST NAME?

The passenger who lives nearest to the conductor is not Petrov (4-5). He does not live in Moscow or Leningrad, since at best these are only tied for nearest to the conductor (2), so he is not Ivanov (1). By elimination, he is Sidorov.

Since the passenger from Leningrad is not Ivanov (1), by elimination he is Petrov, and the conductor's name is Petrov (3). Since Sidorov is not the fireman (6), by elimination he is the engineer.

269. A CRIME STORY

Theo is innocent because he says so twice. Then (9) is a lie. Since (9) is a lie, (8) is true. Since (8) is true, (15) is a lie. Since (15) is a lie, (14) is true. Judy is the thief.

270. HERB GATHERERS

(A) The first digit of the sum is 1, since no two one-digit numbers add to 20 or more. The second digit is 7, as in the divisor of (B). The addends are 9 and 8, the only one-digit numbers to add to 17. The 9 comes first because the first group collected more than the second group.

(B) The divisor is 17, from (A).

The dividend equals the sum of the products in (C) and (D). Its first digit is 1, since no two

two-digit numbers add to 200 or more. The quotient times a number with first digit 1 is a number with first digit 1, so the quotient has first digit 1.

The line under the dividend is 1 × 17 = 17. The second digit of the dividend is 8 or 9; if it was 7, there would be no star two lines under it. And it is not 9, because two lines under the dividend would start with 2, and a two-digit number starting with 2 cannot be divided by 17 without a remainder. Then it is 8, two lines under the dividend is 17, and the third digit of the dividend is 7.

The division is 187 ÷ 17 = 11.

(C) The first group received 11 × 9 = 99 cents.

(D) The second group received 11 × 8 = 88 cents.

271. A HIDDEN DIVISION

The key to a quick solution is that the quotient has five digits, but only three products are shown. The bringing down of two numbers for the second and last products indicates that the second and fourth digits of the quotient are 0s. The first and last digits of the quotient times the two-digit divisor are three-digit numbers; 8 only produces a two-digit number; the first and last digits are 9.

The quotient is 90,809. The divisor is 12, the only number that multiplied by 8 gives a two-digit number and multiplied by 9 gives a three-digit number. The dividend is 1,089,709.

272. CODED OPERATIONS

(A) Consider the product of A and ABC. A is 1, 2, or 3; if it was more, the product would have four digits. A is not 1, or the product would end on C, not A. If A is 3, C is 1 (1 × 3 = 3), but C cannot be 1, or C × ABC would have three digits. Then A is 2. Again, C is not 1, so it is 6.

Now consider the product of B and ABC. B equals 4 or 8, since the last digit of B × 6 is B. But if B is 4 the product has three digits (4 × 246 = 984). So B is 8. Since we now know that ABC = 286 and BAC = 826, the other stars are decided by simple multiplication.

(B) 1. The result of multiplying a three-digit number by 2 is a four-digit number, but the other two products are three-digit numbers. Then both stars in the second line are 1s and the multiplier is 121.

2. Since the third product results from multiplying by 1, the 8 in that product is repeated in the first line and the first product:

```
        * 8 *
        1 2 1
        -----
        * 8 *
    * * * *
    * 8 *
-----------
  * * 9 * 2 *
```

3. The first digit of the first line is 5 or more, or the fourt line could not be a four-digit number: the first digit of the fourth line must be 1. The last digit of the fourth line is 4, the only digit that will produce the 2 in the bottom line:

```
        * 8 *
        1 2 1
        -----
        * 8 *
    1 * * 4
    * 8 *
-----------
  * * 9 * 2 *
```

4. The first digit of the bottom line must be 1. The first digit of the fifth line (and third and first lines) is either 8 or 9, or the product would be a five-digit number.

5. Since the last digit of the fourth line is 4, the last digit of the first, third, and fifth lines is 2 or 7.

6. The third digit of the fourth line is either 6 or 7, since it is the last digit of the product of 2 and 8, possibly increased by 1. The second digit of the fourth line is either 7 or 9, depending on whether the first digit of the first line is 8 or 9. If the second digit of the fourth line is 7, its column (7 + 8) has to have 4 carried over to produce the 9 below the column. But the sum of three numbers in the third column cannot produce a carry digit of 4. Therefore, the second digit of the fourth line is 9.

7. The first digit of the first line, and therefore also the first digit of lines 3 and 5, is not 8 (paragraph 6), so it is 9 (paragraph 4);

```
        9 8 *
        1 2 1
        -----
        9 8 *
    1 9 * 4
    9 8 *
  -----------
  1 * 9 * 2 *
```

8. The 9 in the product is the result of carrying 2 from the next column. If the third digit of the top line is 2, the column will be 9 + 6 + 2 + 1 carried over = 18, which has 1 to be carried over, not 2. Then (paragraph 5) the third digit of the top line is 7 (9 + 7 + 7 + 1 = 24). The other digits "fill themselves in," and:

$$987 \times 121 = 119{,}427.$$

(C) Hint: If the third product has only six digits, what does that mean about the first digit of the divisor? Proceed by narrowing down the choices of digits, especially in the third and fourth products versus the divisor, and with careful work you will reach the answer:

$$7{,}375{,}428{,}413 \div 125{,}473 = 58{,}781.$$

(D)

$$1{,}337{,}174 \div 943 = 1{,}418;$$
$$1{,}202{,}464 \div 848 = 1{,}418;$$
$$1{,}343{,}784 \div 949 = 1{,}416;$$
$$1{,}200{,}474 \div 846 = 1{,}419.$$

(E) Hint: In the second sum, what is I? In the third, can S in SOL be greater than 2? If it is 2, what are R and L, and is this consistent with the first sum?

DOREMIFASOL is 34569072148 or 23679048135.

(F)

$$1{,}091{,}889{,}708 \div 12 = 90{,}990{,}809.$$

(G) Hint: What is the last digit of $M \times M$? What four digits have this property? Which two can be immediately eliminated? Then consider $OM \times OM$, and so on.

Can you prove, on your own, that $ATOM = 9{,}376$ is the only possible solution?

273. A PRIME CRYPTARITHM

Hint: If a, b, and c are the last digits of lines 1, 2, and 3, can b be 2? Can a? Can c? What does $(a \times b)$ end in?

```
   775
    33
  ----
  2325
 2325
 -----
 25575
```

274. THE MOTORCYCLIST AND THE HORSEMAN

The motorcyclist would have taken 20 minutes to go from where he met the horseman to the airport and back. Thus he was 10 minutes from the airport when he met the horseman. These 10 minutes plus the 30 minutes the horseman had been riding before they met makes 40 minutes the plane was ahead of schedule.

275. ON FOOT AND BY CAR

The car was scheduled to reach the station at 8:30 a.m. When it met the engineer, it saved 10 minutes – 5 to get to the station and 5 to come back to the meeting point. Therefore, the engineer met the car at 8:25 a.m.

276. PROOF BY CONTRADICTION

(A) Suppose neither of the integers is greater than 8. Then either both integers are 8, or one is 8 and the other less than 8, or both are less than 8. In each case their product is less than 75, which is impossible. Therefore, at least one integer is greater than 8.

(B) Suppose the first digit of the multiplicand differs from 1. Then it is not less than 2, and the multiplicand is not less than 20. But $20 \times 5 = 100$ and any larger number times 5 is over 100. But the product is a two-digit number. Then the first digit of the multiplicand is 1.

277. TO FIND A FALSE COIN

(A) Weighing 1: Weigh 3 coins against 3. If a pan goes up, one of the 3 coins on it is counterfeit. If the pans balance, one of the 3 coins not on the balance is counterfeit.

Weighing 2: Of the 3 coins that include the counterfeit, weigh 1 against 1. If a pan goes up, the coin on it is counterfeit. If the pans balance, the coin just set aside is counterfeit.

(B) Once you see that 2 coins can be left off the balance as easily as 3, there is no problem. Weighing 1: 3 coins against 3. If a pan goes up, continue as in (A). If the pans balance, one of the 2 coins not on the balance is counterfeit.

Weighing 2 if the pans balance in weighing 1: Weigh one of the 2 that were left off the balance against the other. The pan that goes up contains the counterfeit.

(C) We number the coins from 1 through 12.

Weighing 1: Weigh 1, 2, 3, 4 against 5, 6, 7, 8. If the pans balance, one of the 4 coins not on the balance is counterfeit.

Weighing 2, if the pans balance in weighing 1: Weigh 1, 2, 3 against 9, 10, 11.

If the pans balance, 12 is counterfeit. Weigh it against 1 to see if it is lighter or heavier. If the first pan goes down, one of 9, 10, 11 is lighter (since 1, 2, 3 were proved good coins in weighing 1). You can find out which in one weighing, as in (A)'s second weighing above. The procedure if the first pan goes up is similar.

But if, in weighing 1, the pan with 1, 2, 3, 4 goes down, proceed as in the diagram. (If the pan goes down, the procedure is similar.)

O — *Coin.*
— *Coin on heavy pan in first weighing.*
— *Good coin.*
— *Possible counterfeit.*
— *Weighed against.*

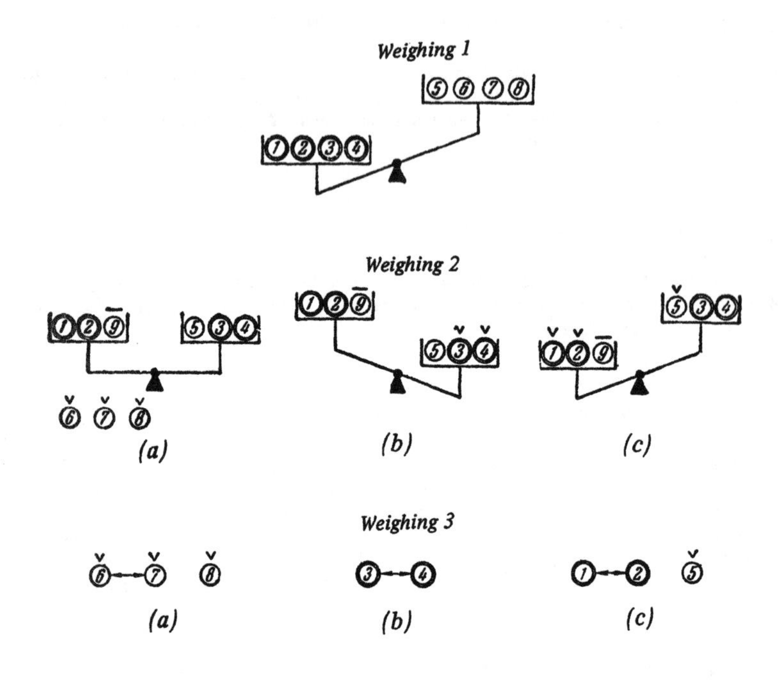

278. A LOGICAL DRAW

A reasoned: "My friends have white; suppose I have black. Then *B* would say to himself, '*A*'s paper has black and *C* has white. If I have black *C* would see two pieces of black paper and immediately announce he has white. But *C* is silent. Therefore, I announce I have white.' But *B* is silent. Therefore, I announce I have white."

B and *C* reasoned the same way. But *A* could also have reasoned: "In a fair contest, we must all have the same problem to solve. If I see two white pieces of paper, so do they."

279. THREE SAGES

A reasoned: "*B* is confident his own face is not smeared. If he saw mine unsmeared he would be surprised at *C*'s laughing, because *C* would have no smeared face to laugh at. But *B* is not surprised. Therefore, my face is smeared."

280. FIVE QUESTIONS

(A) 1. Two. 2. Acute.
(B) 1. Chord. 2. Triangle. 3. Diameter. 4. Equilateral triangle. 5. Concentric circles.
(C) Altitude, median, perpendicular bisector, bisector of vertex angle, axis of symmetry.

(D) Geometric figure, plane figure, polygon, convex quadrilateral, parallelogram, rhombus, square.

(E) No convex polygon can have more than 3 obtuse exterior angles. Therefore, no convex polygon can have more than 3 acute interior angles.

281. REASONING WITHOUT EQUATIONS

(A) The only even multiple of 9 from 10 through 22 is 18. Check:

$$18 \times 4\tfrac{1}{2} = 81.$$

(B) The answer is

$$6 \times 7 \times 8 \times 9 = 3{,}024.$$

282. A CHILD'S AGE

Squares which can be considered 3 years older than a child's age are 4, 9, 16. Of these, only 9 gives its square root when you subtract 3 and again 3. The child's age is 6.

With differences other than 3, we might have:

$$2 + 1 = \text{age } 3; \quad 3 + 1 = 2 \times 2;$$
$$4 + 6 = \text{age } 10; \quad 10 + 6 = 4 \times 4;$$
$$5 + 10 = \text{age } 15; \quad 15 + 10 = 5 \times 5.$$

283. YES OR NO?

The key to the solution is that 2 to the tenth power is 1,024 (that is, over 1,000). With each question you knock out half the remaining numbers, and after ten questions only the "thought" number is left. Say the number is 860. The ten questions are:

1. "Is your number greater than 500?" "Yes." Add 250.
2. "Greater than 750?" "Yes." Add 125.
3. "Greater than 875?" "No." Subtract (not 62½ but the nearest even number) 62.
4. "Greater than 813?" "Yes." Add 31.
5. "Greater than 844?" "Yes." Add (not 15½ but) 16.
6. "Greater than 860?" "No." Subtract 8.
7. "Greater than 852?" "Yes." Add 4.
8. "Greater than 856?" "Yes." Add 2.
9. "Greater than 858?" "Yes." Add 1.
10. "Greater than 859?" "Yes."

The number is 860.

X. Mathematical Games and Tricks

284. ELEVEN MATCHES

(A) Yes. Start from the end. To leave 1 match on your last play is to win. The play before, 5 is the number to leave. He picks up 1, 2, or 3; you pick up 3, 2, or 1, and leave him 1.

Before 5, you should leave him 9. Whether he picks up 1, 2, or 3, you can leave him 5, and so on.

So, on the first play, pick up 2 matches, leaving 9.

(B) Yes. Extend the series by 4s: 1, 5, 9, 13, 17, 21, 25, 29. On the first play, pick up 1 match, leaving him 29. Then leave him 25, 21, and so on.

(C) No. You may have to start with a number of matches you would like to leave the other player, and unless he makes a mistake he can win just as you would.

You want to leave him 1, $p + 2$, $2p + 3$, $3p + 4$ and so on—working backward, of course, to the number in the series, of those not more than n, that is largest—call it N. If N is not n, play $(n - N)$ matches the first time, and win. But if N is n, the second player can win.

285. WINNER PICKS UP THE LAST MATCH

Again working backward, you win if you leave him 7 matches. If he picks up 1, 2, 3, 4, 5, 6 matches, you pick up all the rest.

Before that, leave him 14 matches; before that, 21; and, before that, 28. Pick up 2 matches the first time.

286. AN EVEN NUMBER OF WINS

The strategy is more complicated than before. Take 2 matches, and then:

If your opponent takes an even number of matches, leave him a number that is larger by 1 than a multiple of 6 (19, 13, 7).

If he takes an odd number, leave him a number that is smaller by 1 than a multiple of 6 (23, 17, 11, 5).

If this is impossible, leave him a number that is a multiple of 6 (24, 18, 12, 6). For example, you take 2 and he takes 3, leaving 22. You can't take 5 (leaving 17), so you take 4 (leaving 18).

See if you can prove on your own that this strategy wins.

287. WYTHOFF'S GAME

Yes: (3, 5), (4, 7), (6, 10), (8, 13), (9, 15), (11, 18), (12, 20), . . .

[This series of number pairs is closely related to Fibonacci numbers and the golden ratio. The first pair differ by 1, the second pair by 2, and the nth by n. Every positive integer appears once and only once in the series of pairs. For a full discussion see H.S.M. Coxeter, "The Golden Section, Phyllotaxis, and Wythoff's Game," *Scripta Mathematica,* 19 (1953): 139 f.–*M.G.*]

290. WHO WILL BE FIRST TO REACH 100?

To call 100, call 89; to call 89, call 78; call 67, 56, 45, 34, 23, 12, and first of all 1. There is no way for *B* to disrupt the series.

291. THE GAME OF SQUARES

A draws any side of the central square, such as *v* (see *a* in the diagram).

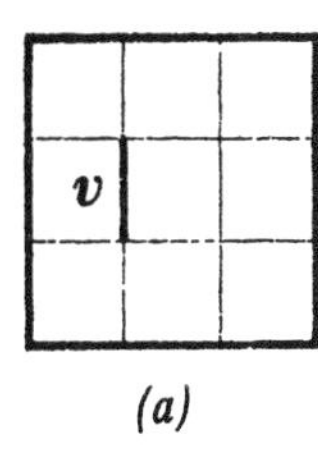

(a)

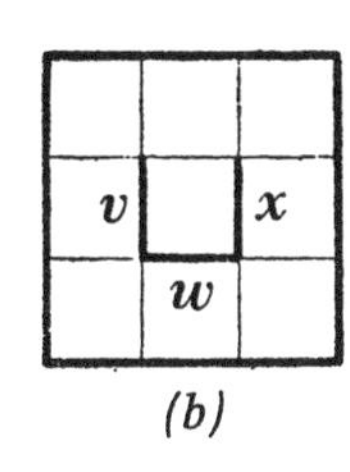

(b)

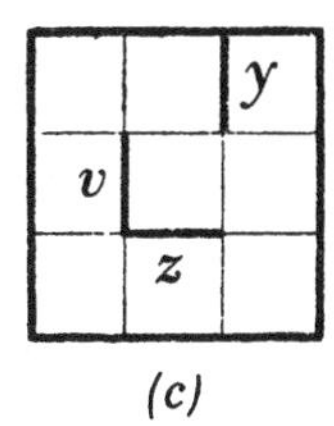

(c)

If *B* draws a line in the left column, *A* wins the 3 squares of the column and opens the 3-by-2 rectangle as in *c* on page 123, winning all 9 squares.

If *B* plays correctly, drawing another side *w* of the central square (as in *b*), *A* draws *x*. B can finish the central square but *A* wins the other 8. (If *B* makes any other move, *A* wins all 9.)

And if *B* draws a line *y* in the right column, *A* counters with *z* (see *c*). Any move *B* makes again loses 8 or 9 squares.

295. NUMBER CROSSWORD PUZZLES

(A) Continuing from the second diagram in the problem, 5 Down is 543 or 567. If it is 543, 6 Across starts with 34 and is the product of 77 and a two-digit ending in 3. The fact that 77 × 43 = 3,311 shows there is no such number. So 5 Down is 567, 8 Across is 47, and 6 Across is 3,619.

The only tricky entry remaining is 7 Down. The prime factors of 3,087 are 7 (cubed) and 3 (squared); of 77, 7 and 11. The only product of any of these that is a two-digit number starting with 9 is 11 × 3 × 3 = 99. The complete answer is shown in the diagram below.

[1] 3	0	[2] 8	[3] 7
[4] 4	[5] 5	6	7
[6] 3	6	1	[7] 9
[8] 4	7	[9] 3	9

(B) Since 3 Down has only three digits, the first digits of 1 and 8 Across differ by 1 (they cannot be the same). Then the middle digit of 1 Down is 1—also the first digit of 10 Across.

Since the largest two-digit factor of 3 Down starts with 1 (5 Across), so does the smallest (11 Down). The first two digits of 2 Down are 17, so the last digit of 3 Down is 7. The last two digits of 9 Across are 11, so the last two digits of 1 and 8 Across add to 99; in fact, the last digits and penultimate digits each add to 9. Then the fourth digit of 1 Across is 2, as is the last digit of 7 Across.

In 4 Down, the last digit y of 8 Across equals twice the last digit x of 1 Across. But we have shown that $x + y = p$. Then $x = 3, y = 6$.

3 Down is the difference between 1 and 8 Across. Looking at their last digits, since 3 Down ends in 7, 1 Across is greater than 8 Across. The middle digit of 3 Down must be 4. Of the five digits missing in 1 and 8 Across, 1 Across gets 50 and 8 Across gets 498. (See diagram *a* below.)

5	0	1	2	3
1		7	4	2
4	9	8	7	6
			1	1

(a)

5	0	1	2	3
1	9	7	4	2
4	9	8	7	6
1	1	1	1	1
7	4	2	9	3

(b)

9 Across is 11,111. 3 Down's prime factors 19 and 13 go in 5 Across (and 10 Down) and in 11 Down.

Since 6 Down starts with 9, its reverse is the product of 3 Down and a two-digit prime ending in 7—and beginning with 1 (since 9 Down does). We find $247 \times 17 = 4{,}199$; 6 Down is 9,914, and 9 Down is 17.

For 12 Across, products of two factors of 6 Down are 221, 247, 323. These times two-digit primes ending in 9, 7, and 3 respectively (since 12 Across ends in 9) are to equal 12 Across. The only products that look close are $221 \times 29 = 6{,}409$ and $323 \times 23 = 7{,}429$; the latter is 12 Across. (See diagram *b* for complete solution.)

(C) A table of squares shows that the only symmetrical 6-digit square is 698,896 (10 Across); its square root is 836 (11 Down), which reversed is 638 (10 Down). The prime factors of these numbers are:

$$638 = 29 \times 11 \times 2;$$
$$836 = 19 \times 11 \times 2 \times 2.$$

5	3	2	9		1
1	1		7	2	9
	7	3		4	
6	9	8	8	9	6
3	9		3	6	5
8		1	6	0	0

The highest common factor is 11 × 2 = 22, and half of it is 11 (5 Across). 4 Down must be 11 or 19; to be half a number ending in 8 (9 Down) it must be 19, and 9 Down is 38. Then 13 Across is 39 and 15 Across is 1,600 (the square of 40).

For 1 Down, the lowest common multiple of 2 through 6 is 60, and 1 less (leaving remainders of 1 less than 2 through 6, respectively) is 59. Subtracting 8 gives 51 (1 Down).

The only four-digit squares that begin with 5 are 5,041 and 5,329. We reject the first because 2 Down cannot begin with 0. So 1 Across is 5,329 and 8 Across is 73. Checking, the digit sum of 2 Down is 29. 3 Down is 97, the only two-digit prime beginning with 9. Completing the solution is easy (see last diagram).

296. GUESSING "THOUGHT" NUMBERS

(C) There are four cases.

Case 1: The "thought" number has the form $4n$. Then:

$$4n + 2n = 6n;\ 6n + 3n = 9n;\ 9n \div 9 = n$$

There is no remainder. The "thought" number is $4n$.

Case 2: Form $(4n + 1)$. Its larger part is $(2n + 1)$:

$$(4n + 1) + (2n + 1) = 6n + 2;\ \ (6n + 2) + (3n + 1) = 9n + 3;$$
$$(9n + 3) \div 9 = n, \text{ with a remainder of } 3.$$

The remainder is less than 5. The "thought" number is $(4n + 1)$.

Case 3: Form $(4n + 2)$:

$$(4n + 2) + (2n + 1) = 6n + 3.$$

Adding the larger part $(3n + 2)$:

$$(6n + 3) + (3n + 2) = 9n + 5;\ (9n + 5) \div 9 = n, \text{ with a remainder of } 5.$$

The "thought" number is $(4n + 2)$.

Case 4: Form $(4n + 3)$. Its larger part is $(2n + 2)$:

$$(4n + 3) + (2n + 2) = 6n + 5.$$

Adding the larger part $(3n + 3)$:

$$(6n + 5) + (3n + 3) = 9n + 8;\ (9n + 8) \div 9 = n, \text{ with a remainder of } 8.$$

The remainder is greater than 5. The "thought" number is $(4n + 3)$.

(D) If the "thought" number has the form $4n$, no larger part is used. The answer will be $4n + 2n + 3n = 9n$, which is a multiple of 9. The sum of the digits of $9n$ are divisible by 9, so, in order for the unnamed digits and the concealed digit to add to a multiple of 9, nothing need be done. (That is, "add 0.")

For numbers of the form $(4n + 1)$, $(4n + 2)$, and $(4n + 3)$, a larger part is used in the first step only, second step only, and both steps respectively. As in (C), the answers are $(9n + 3)$, $(9n + 5)$, and $(9n + 8)$. Their digit sums become multiples of 9 when we add 6, 4, and 1 respectively. When the new digit sum is subtracted from the next higher multiple of 9, the difference equals the concealed digit.

(E) Let the "thought" number be x, and the added number y. Then:

$$(x + y)^2 - x^2 = 2xy + y^2 = 2y\left(x + \frac{y}{2}\right) = z;$$
$$x = \tfrac{1}{2}z - y - \tfrac{1}{2}y.$$

(F) Let the "thought" number be x, the quotients of the division of x by 3, 4, and 5 be a, b, and c respectively, and the remainders be r_3, r_4, and r_5 respectively. Obviously:

$$x = 3a + r_3;$$
$$x = 4b + r_4;$$
$$x = 5c + r_5.$$

Hence:

$$r_3 = x - 3a;\ r_4 = x - 4b;\ r_5 = x - 5c.$$

We calculate:

$$\begin{aligned} &40r_3 + 45r_4 + 36r_5 \\ &= 40(x - 3a) + 45(x - 4b) + 36(x - 5c) \\ &= 121x - 120a - 180b - 180c. \end{aligned}$$

The remainder when divided by 60 is x.

297. WITHOUT ASKING QUESTIONS

(A) Let his "thought" number be n, and the three numbers you supply be a, b, and c. First he gets $\frac{na + b}{c}$. When he subtracts $\frac{na}{c}$, the answer is $\frac{b}{c}$. The answer doesn't include x, so no questions are necessary.

(B) Let the "thought" number be y and your number sealed in an envelope be x. First the spectator gets $y + 99 - x$, which is a number from 100 through 198. Then crossing out the first digit and adding it means subtracting 99. When the spectator subtracts $y - x$ from his "thought" number y, he gets your sealed number x.

Variant: The spectators pick numbers from 201 through 1,000: The sealed number is from 100 through 200; in the calculations 999 is used instead of 99.

298. I FOUND OUT WHO TOOK HOW MANY

	A	B	C
Initially	$4n$	$7n$	$13n$
After first step	$8n$	$14n$	$2n$
After second step	$16n$	$4n$	$4n$
After third step	$8n$	$8n$	$8n$

After the third step, each has twice as many as A had initially. The rest is clear.

299. ONE, TWO, THREE ATTEMPTS

Let the "thought" numbers be a and b. Then:

$$(a + b) + ab + 1 = a + 1 + b(a + 1) = (a + 1)(b + 1).$$

Tricks can be based on adding or subtracting the difference and product of two numbers.

300. WHO TOOK THE ERASER, WHO TOOK THE PENCIL?

A is a prime number, and B is a composite number not divisible by A. Two other numbers y and a are mutually prime, and y is a factor of B. $Ay + Bx$ is divisible by y and the boy with y took the pencil. $Ac + By$ is not divisible by y, and the boy with y took the eraser.

301. GUESSING THREE CONSECUTIVE NUMBERS

The sum of three consecutive numbers and a multiple of 3 is:

$$a + (a + 1) + (a + 2) + 3k = 3(a + k + 1).$$

Multiplying by 67, we obtain:

$$201(a + k + 1).$$

We know that $a<59$ and $3k<100$, or $k<34$. Therefore, $(a + k + 1)$ is no larger than a two-digit number, and the last two digits of $201(a + k + 1)$ are $(a + k + 1)$. By subtracting $(k + 1)$ from this, you get the first "thought" number.

Furthermore, the two or three digits that precede the last two digits of $201(a + k + 1)$ are $2(a + k + 1)$.

302. GUESSING SEVERAL "THOUGHT" NUMBERS

If there are two "thought" numbers, a and b: $5(2a + 5) + 10 = 10a + 35$; $10a + 35 + b = 10a + b + 35$.

Subtracting 35, we obtain a two-digit number $(10a + B)$ composed of the "thought" numbers.

The proof is similar for three or more "thought" numbers.

303. HOW OLD ARE YOU?

With age x, the answer is $10x - 9k$, where k is a one-digit number. Let us transform the difference:

$$10x - 9k = 10x - 10k + k = 10(x - k) + k.$$

Now x is over 9 and k cannot exceed 9; therefore $(x - k)$ is positive. Then $10(x - k) + k$ has k as its last digit. If we discard k, then $(x - k)$ is left. Adding k gives x.

304. GUESS HIS AGE

His age is x, and:

$$(2x + 5) \times 5 = 10x + 25 = 10(x + 2) + 5.$$

Thus 5 is the last digit. Discarding it leaves the number $(x + 2)$, and subtracting 2 leaves x.

305. A GEOMETRICAL "VANISH"

No line vanished. The 13 lines were replaced by 12 lines that are each one-twelfth longer than the old ones—if you draw the lines long enough, you can measure the difference with a ruler.

The effect is enhanced in the figure below. Copy the left half and cut all the way around the circle. When you turn the circle a little counterclockwise, a line appears to vanish (right).

[Here Kordemsky gives the mathematical basis of Sam Loyd's famous paradox of the vanishing Chinese warrior. For this paradox, and closely related paradoxes involving vanishing pictures, see chapter 7 of my *Mathematics, Magic and Mystery* (New York: Dover Publications, 1956).–*M.G.*]

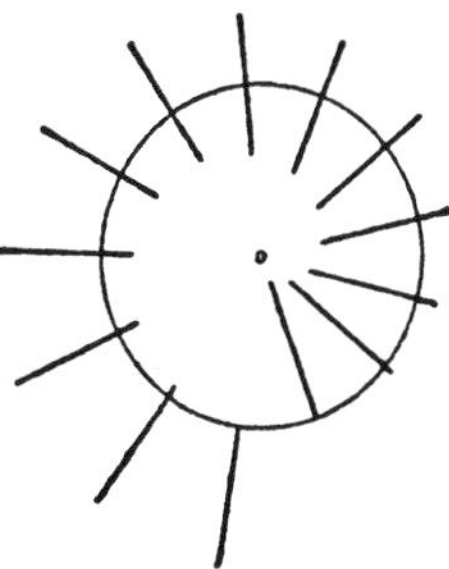

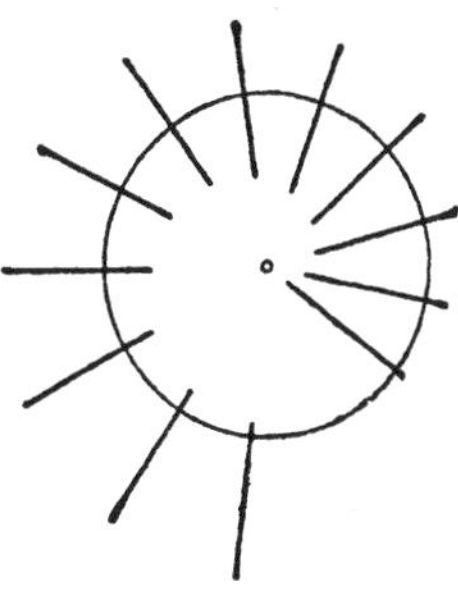

XI. Divisibility

306. THE NUMBER ON THE TOMB

The lowest common multiple (LCM) of a set of numbers is the product of their distinct prime factors, each taken the greatest number of times that it occurs in any one number. For numbers 1 through 10 the LCM is the product of:

$$2 \times 2 \times 2 \times 3 \times 3 \times 5 \times 7 = 2{,}520.$$

The LCM of 1 through 10 equals the LCM of 6 through 10. In general, the LCM of 1 through $2n$ equals the LCM of $(n + 1)$ through $2n$.

307. NEW YEAR'S GIFTS

With 1 more orange, their number would have been divisible by 10, 9, 8, ... As we just learned, such a number is a multiple of 2,520.

We had 2,519 oranges, or $2{,}519 + 2{,}520n$, where n is any positive integer.

308. IS THERE SUCH A NUMBER?

There is an infinity of such numbers. The difference between divisor and remainder is always 2. Then 2 plus the desired number is a multiple of the divisors given. The lowest common multiple of 3, 4, 5, and 6 is 60, and $60 - 2 = 58$, the smallest answer.

309. A BASKET OF EGGS

The lowest common multiple of 2, 3, 4, 5, and 6 is 60. We have to find a multiple of 7 that is larger by 1 than a multiple of 60. Now:

$$60n + 1 = (7 \times 8n) + 4n + 1.$$

The number $(60_n + 1)$ is divisible by 7 if $(4n + 1)$ is divisible by 7. The lowest value of n that satisfies this condition is 5.

Therefore, there were 301 eggs in the basket.

310. A THREE-DIGIT NUMBER

The LCM of 7, 8, and 9 is 504. This is the answer, since no multiples of it have three digits.

311. FOUR DIESEL SHIPS

The LCM of 4, 8, 12, and 16 is 48. The ships met again after 48 weeks, on December 4, 1953.

312. THE CASHIER'S ERROR

The prices of lard and soap are multiples of 3. The number of packages of sugar and of pastries are also multiples of 3. Then the total cost in cents should have been, but was not, a multiple of 3.

313. A NUMBER PUZZLE

The left side of the equation is divisible by 9, so the right side is too. It follows that the sum of the digits on the right side is divisible by 9, so a is 8, which gives t a value of 4.

314. A TEST OF DIVISIBILITY BY 11

(A) $7 + 1 + 2 + 1 - (3 + a + 0 + 0) = 0$; $a = 8$.

(B) The left side of the equation is divisible by 11, so b is again 8. Since $61^2 = 3{,}721$ and $62^2 = 3{,}844$, the expression in brackets is approximately 6,150, and x is approximately 68. On testing this and nearby values, x is found to be 67.

315. A TEST OF DIVISIBILITY BY 7, 11, AND 13

Take for example, 31,218,001,416:

$$31{,}218{,}001{,}416 = 416 + (1 \times 10^3) + (218 \times 10^6) + (31 \times 10^9) = 416 + 1 \, (10^3 + 1 - 1) + 218 \, (10^6 - 1 + 1) + 31 \, (10^9 + 1 - 1) = (416 - 1 + 218 - 31) + [(10^3 + 1) + 218\,(10^6 - 1) + 31\,(10^9 + 1)].$$

The expression in brackets is divisible by 7, 11, and 13. Therefore, divisibility of the entire number by 7, 11, and 13 depends only on the divisibility of the expression $(416 - 1 + 218 - 31)$, which is between the sums of even- and odd-positioned groups. (In the example the difference is 602, which, like the entire number, is divisible by 7 but not by 11 or 13.

316. SHORTENING THE SHORTCUT FOR DIVISIBILITY BY 8

We must prove that the three-digit number $(100x + 10y + z)$ is divisible by 8 if $(10x + y + z/2)$ is divisible by 4.

Let $10x + y + z/2 = 4k$, where k is any positive integer. Then:

$$20x + 2y + z = 8k; \quad z = 8k - 20x - 2y;$$

$$100x + 10y + z = 100x + 10y + 8k - 20x - 2y = 80x + 8y + 8k.$$

The last expression is obviously divisible by 8.

Show on your own that if $10x + y = z/2 = 4k + 1$ or $4k + 2$ or $4k + 3$, where k is any positive integer, then $(100x + 10y + 2)$ is not divisible by 8.

317. A REMARKABLE MEMORY

Let n be the nine-digit number, and write it as:

$$N = 10^6a + 10^3b + c,$$

where a, b, and c are the three triplets of digits. We know that $a + b + c$ is divisible by 37. Let $a + b + c = 37k$. We can now write:

$$N = 10^6a + 10^3b + 37k - a - b$$
$$= a(10^6 - 1) + b(10^3 - 1) + 37k.$$

Since each of these three terms is divisible by 37, N is divisible by 37.

318. A TEST OF DIVISIBILITY BY 3, 7, and 19

Remove the last two digits from the number to be tested. Add, to what remains of the number, 4 times the two-digit number removed. Repeat, until you get a number small enough to test easily.

Example: 138,264. Add 1,382 to 64×4 to get 1,638. Repeating the process, $16 + 152 = 168$.

There is no need to continue: 168 is divisible by 3 and 7, but not by 19, so 138,264 is divisible by 3 and 7, but not by 19.

XII. Cross Sums and Magic Squares

324. A STAR

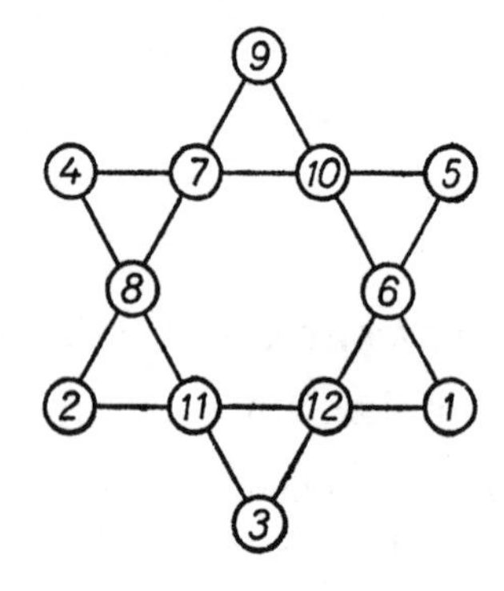

325. A CRYSTAL

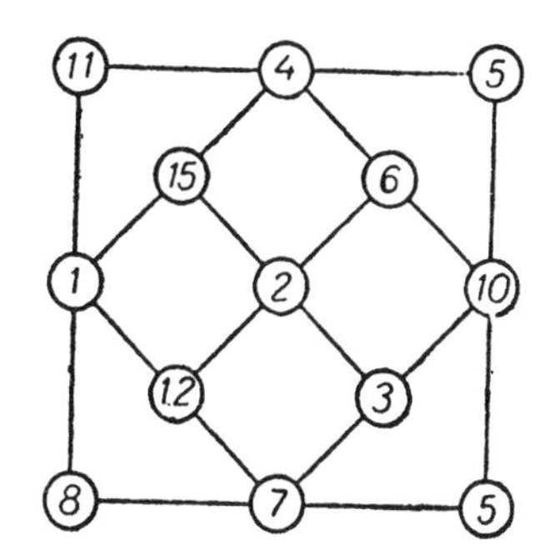

326. AN ORNAMENT FOR A WINDOW

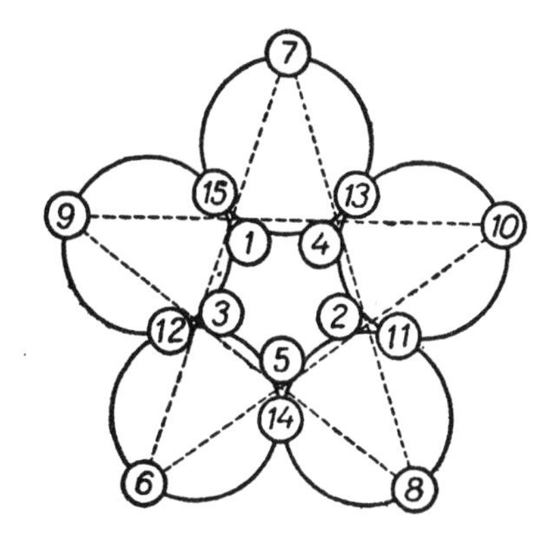

327. THE HEXAGON

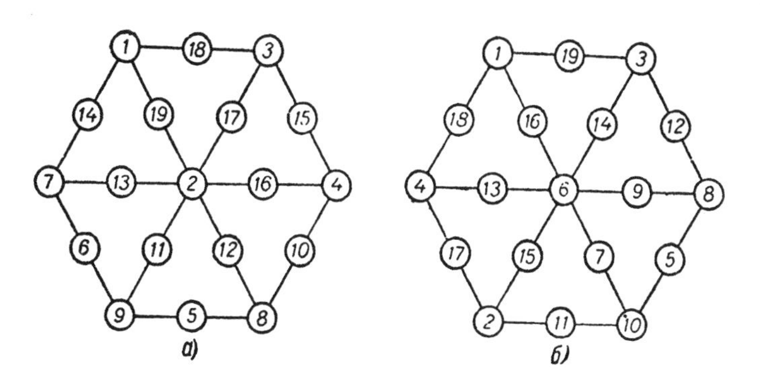

328. A "PLANETARIUM"

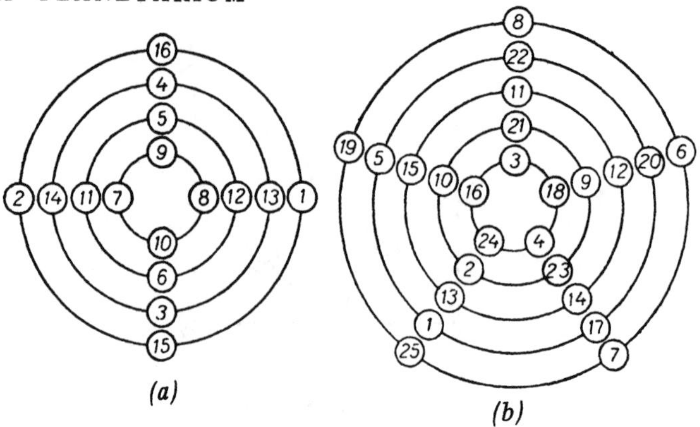

329. OVERLAPPING TRIANGLES

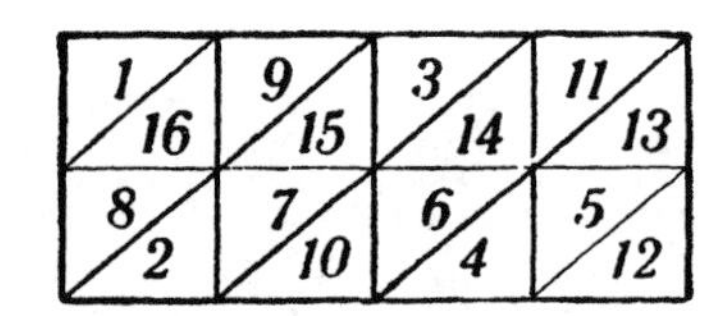

330. INTERESTING GROUPINGS

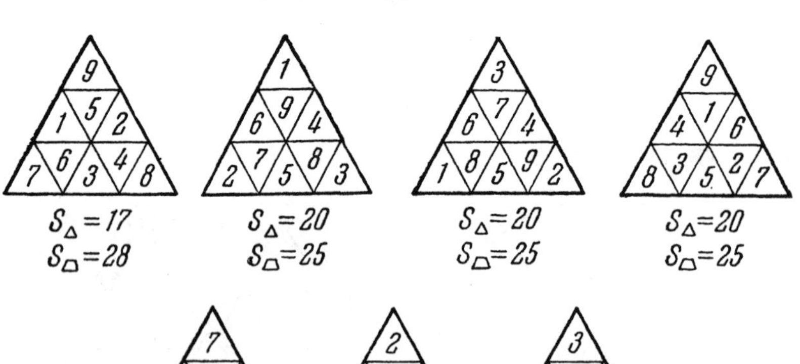

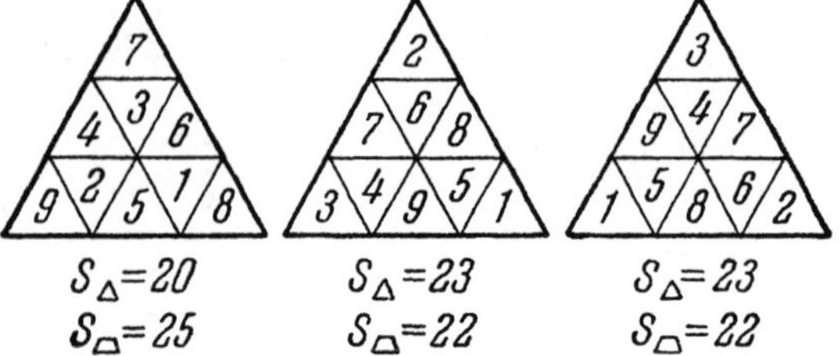

331. WAYFARERS FROM CHINA AND INDIA

In *a*, the rows and columns of the magic square from India are numbered. We want the numbers discussed in the text along the main diagonals: 12, 14, 3, 5; and 15, 9, 8, 2. To get them there, we move row II to first place and row IV to second place, with row I in third place, and row III in fourth place; then exchange columns 2 and 3. The square in *b* has the desired characteristics.

	1	2	3	4
I	1	14	15	4
II	12	7	6	9
III	8	11	10	5
IV	13	2	3	16

(a)

12	6	7	9
13	3	2	16
1	15	14	4
8	10	11	15

(b)

332. HOW TO MAKE A MAGIC SQUARE

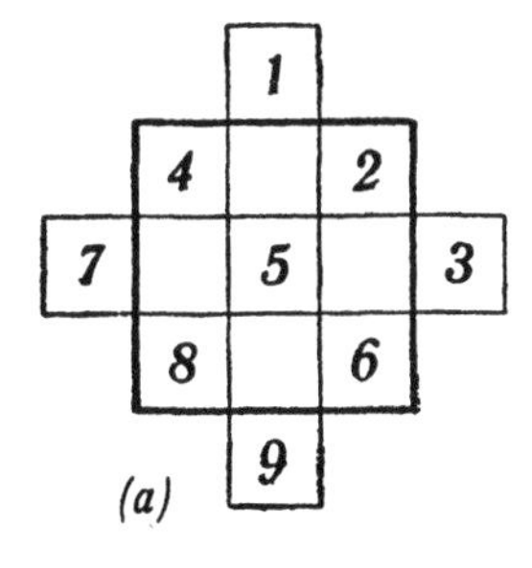

(a)

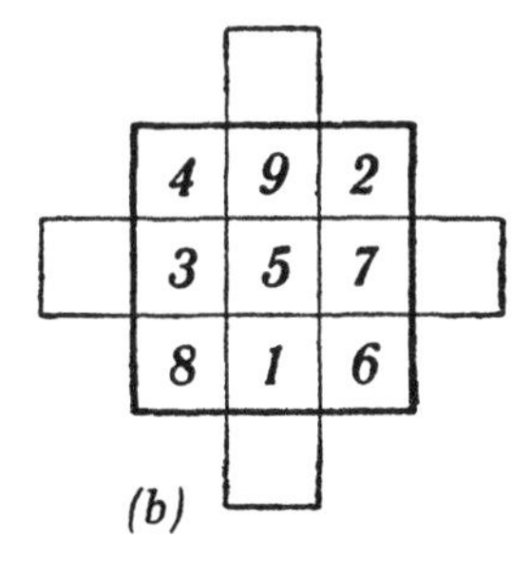

(b)

333. TESTING YOUR WITS

The 7-by-7 square contains four horizontal and four vertical rows of four cells each, and three horizontal and three vertical rows of three cells each. It can be formed by putting a 9-cell magic square "inside" a 16-cell one.

The first diagram constructs the order-4 square, and the second, the order-3 square. Each has a magic constant of 150. The last diagram shows how the squares are combined to solve the problem.

1

30	31	32	33
34	35	36	37
38	39	40	41
42	43	44	45

→

30	31	32	33
41	40	39	38
34	35	36	37
45	44	43	42

30	44	43	33
41	35	36	38
34	40	39	37
45	31	32	42

→

30	44	43	33
41	35	36	38
37	39	40	34
42	32	31	45

2

		46		
	49		47	
52		50		48
	53		51	
		54		

49	54	47
48	50	52
53	46	51

3

30		44		43		33
	49		54		47	
41		35		36		38
	48		50		52	
37		39		40		34
	53		46		51	
42		32		31		45

334. THE MAGIC GAME OF "15"

Move these blocks in order: 12, 8, 4, 3, 2, 6, 10, 9, 13, 15, 14, 12, 8, 4, 7, 10, 9, 14, 12, 8, 4, 7, 10, 9, 6, 2, 3, 10, 9, 6, 5, 1, 2, 3, 6, 5, 3, 2, 1, 13, 14, 3, 2, 1, 13, 14, 3, 12, 15, 3.

Exactly 50 moves! While moving the blocks you may reach a different magic square, but I do not know a solution shorter than 50 moves.

13	1	6	10
14	2	5	9
	12	11	7
3	15	8	4

1	2	3	4
13	5	6	10
14	12	11	9
	15	8	7

335. UNORTHODOX MAGIC SQUARES

(A) See the diagram below. The conditions laid down suggest placing the pairs of equal numbers in the relation of a knight's move at chess.

1	7	2	8
4	6	3	5
7	1	8	2
6	4	5	3

(B) Form a magic square by the method of Problem 332 (diagram *a*). Exchange the first two rows, then the first two columns to obtain the desired magic square (diagram *b*).

31	3	5	25
9	21	19	15
17	13	11	23
7	27	29	1

(a)

21	9	19	15
3	31	5	25
13	17	11	23
27	7	29	1

(b)

(C) Turn the square upside down. It is still magic and has the same constant.

336. THE CENTRAL CELL

$$a_1 + a_4 + a_7 = S; \qquad a_3 + a_6 + a_9 = S;$$
$$a_1 + a_5 + a_9 = S; \qquad a_3 + a_5 + a_7 = S.$$

$$a_4 + a_7 = S - a_1; \qquad a_6 + a_9 = S - a_3;$$
$$a_5 + a_9 = S - a_1; \qquad a_5 + a_7 = S - a_3.$$

Therefore, $a_4 + a_7 = a_5 + a_9$, and $a_6 + a_9 = a_5 + a_7$. Adding these equations:

$$a_4 + a_7 + a_6 + a_9 = 2a_5 + a_9 + a_7 \text{ or } a_4 + a_6 = 2a_5.$$

To each side of this equation add a_5:

$$a_4 + a_5 + a_6 = 3a_5.$$

But $a_4 + a_5 + a_6 = S$, so $S = 3a_5$. When $S = 15, a_5 = 5$.

XIII. Numbers Curious and Serious

340. TEN DIGITS

(A) Take a digit, say 1. It can occupy any of the ten places in a ten-digit number. Each of the ten choices has nine empty places for a second digit, say 2; eight places for 3; seven for 4, and so on up to 1 place for 0. This gives 10 × 9 × 8 × 7 × 6 × 5 × 4 × 3 × 2 × 1 = 3,628,800 ways. But wait. A number cannot start with 0, and one-tenth of these ways start with 0. Subtract 362,880 to get the right answer: 3,265,920.

(B) 4,938,271,605 ÷ 9 = 548,696,845.

(C) Each product of *a* or *b* and a number of the first group has no repeating digits; each product of *a* or *b* and a number of the second group contains repeating digits. Numbers of the first group do not have factors (except 1) in common with *a* and *b*, but numbers of the second group do.

(D) Because 12,345,679 × 9 = 111,111,111.

341. MORE NUMBER ODDITIES

(A) 2,025 = 45^2. Simply examine a table of squares from 32 through 99 and do the needed calculations. One must not slight such direct methods of solution.

342. REPEAT THE OPERATION

(A) We consider (a, b, c, d) to be the same row as its transpositions (d, a, b, c), (c, d, a, b), and (b, c, d, a). Then there are six ways to arrange even and odd numbers in a quartet:

$$(e, e, e, e); \quad (e, e, o, o); \quad (e, o, o, o);$$
$$(e, e, e, o); \quad (e, o, e, o); \quad (o, o, o, o).$$

The difference of two even or two odd numbers is even; the difference of an odd and an even number is odd. What are the fourth differences of the six arrangements?

For (e, e, e, e) all differences are (e, e, e, e).

For (e, e, e, o):

$$A_1 = (e, e, o, o); \quad A_3 = (o, o, o, o);$$
$$A_2 = (e, o, e, o); \quad A_4 = (e, e, e, e).$$

The third, fourth, and sixth arrangements are A_1, A_2, and A_3 in this series, so they will permanently reach (e, e, e, e) by the fourth difference. On your own, show that the fourth difference of (e, o, o, o) is (e, e, e, e).

Thus the fourth difference of any row is composed of even numbers.

Now we temporarily replace the numbers of A_4 by their halves. How does this row differ from the real A_5?

Its numbers are halves of A_5. For example, if $A = (4, 6, 12, 22)$, $A_5 = (2, 6, 10, 18)$. The row of halves is (2, 3, 6, 11), and its first row of differences (1, 3, 5, 9) is composed of halves of the numbers of A_5.

The numbers of the fourth row of differences of the halves are still half those of A_8. But the fourth row of differences of the halves consists of even numbers, so A_8 consists of multiples of 4. Similarly, A_{12} consists of multiples of 8, and A_{4n} consists of multiples of 2^n.

Every row has a largest number x. Since no number less than 0 will ever be subtracted from x, no number in any row of differences will exceed x. Say the first power of 2 greater than x is 2^y. Then A_{4y} consists of multiples of 2^y but there are no numbers as great as 2^y, in any row of differences. Therefore, A_{4y} is (0, 0, 0, 0).

(B) Pick a number from the first column of the table and subtract the number on the same line in the second column. Pick the same or another number from the first column and subtract the number on the same line in the third column. Repeat, using the first and fourth columns. Optionally, keep repeating, using the first and as many more columns as desired. The last matching must not be from the first line (or the chosen integer would begin with 0).

x^2	a	$10b$	$100c$	$1{,}000d$	. . .
0	0	0	0	0	. . .
1	1	10	100	1,000	. . .
4	2	20	200	2,000	. . .
9	3	30	300	3,000	. . .
16	4	40	400	4,000	. . .
25	5	50	500	5,000	. . .
36	6	60	600	6,000	. . .
49	7	70	700	7,000	. . .
64	8	80	800	8,000	. . .
81	9	90	900	9,000	. . .

We want the maximum sum of these subtractions. Clearly we should begin by selecting (81 – 9) on the last line. The last matching (which is with at least the fourth column) should be from the second line, where we lose the least ground. And the digits in between should be 0, giving 0 differences (all other differences are negative).

Then the chosen integer has the form $1Z9$, where Z represents one or more 0s. Of such integers, we should choose 109, since (1-100) loses less ground than (1-1,000), (1-10,000), and so on. But:

$$1^2 + 0^2 + 9^2 = 82,$$

which is less than 109. Therefore, the sum of the squares of the digits of any number of three or more digits is less than the number, and if we repeat the operation enough times, we will reach a number of less than three digits.

346. PATTERNS OF DIGITS

(E)

$$49 = 7^2;$$
$$4{,}489 = 67^2;$$
$$444{,}889 = 667^2;$$
$$44{,}448{,}889 = 6{,}667^2.$$

(F)

$$81 = 9^2;$$
$$9{,}801 = 99^2;$$
$$998{,}001 = 999^2;$$
$$99{,}980{,}001 = 9{,}999^2.$$

347. ONE FOR ALL AND ALL FOR ONE

(A)

$$11 = 22 \div 2 + 2 - 2;$$
$$12 = 2 \times 2 \times 2 + 2 + 2;$$
$$13 = (22 + 2 + 2) \div 2;$$
$$14 = 2 \times 2 \times 2 \times 2 - 2;$$
$$15 = 22 \div 2 + 2 + 2;$$
$$16 = (2 \times 2 + 2 + 2) \times 2;$$
$$17 = (2 \times 2)^2 + \tfrac{2}{2};$$
$$18 = 2 \times 2 \times 2 \times 2 + 2;$$
$$19 = 22 - 2 - \tfrac{2}{2};$$
$$20 = 22 + 2 - 2 - 2;$$
$$21 = 22 - 2 + \tfrac{2}{2};$$
$$22 = 22 \times 2 - 22;$$
$$23 = 22 + 2 - \tfrac{2}{2};$$
$$24 = 22 - 2 + 2 + 2;$$
$$25 = 22 + 2 + \tfrac{2}{2};$$
$$26 = 2 \times (\tfrac{22}{2} + 2).$$

(B)

$$1 = (4 \div 4) \times (4 \div 4);$$
$$2 = (4 \div 4) + (4 \div 4);$$
$$3 = (4 + 4 + 4) \div 4;$$
$$4 = 4 + (4 - 4) \times 4;$$
$$5 = (4 \times 4 + 4) \div 4;$$
$$6 = 4 + (4 + 4) \div 4$$
$$7 = 4 + 4 - 4 \div 4;$$
$$8 = 4 + 4 + 4 - 4;$$
$$9 = 4 + 4 + 4 \div 4;$$
$$10 = (44 - 4) \div 4.$$

(C)

$$3 = \frac{17{,}469}{5{,}823}; \quad 5 = \frac{13{,}485}{2{,}697}; \quad 6 = \frac{17{,}658}{2{,}943}; \quad 7 = \frac{16{,}758}{2{,}394};$$

$$8 = \frac{25{,}496}{3{,}187}; \quad 9 = \frac{57{,}429}{6{,}381}.$$

(D)

$$9 = \frac{95{,}742}{10{,}638} = \frac{75{,}249}{08{,}361} = \frac{58{,}239}{06{,}471}.$$

348. EVEN NUMBERS CAN BE ODD

(B)

$$14 \times 82 = 41 \times 28; \quad 34 \times 86 = 43 \times 68;$$
$$23 \times 64 = 32 \times 46; \quad 13 \times 93 = 31 \times 39.$$

(I)

$$1{,}466 - 1 = 1 + 24 + 720 + 720;$$
$$81{,}368 - 1 = 40{,}320 + 1 + 6 + 720 + 40{,}320;$$
$$372{,}970 - 1 = 6 + 5{,}040 + 2 + 367{,}880 + 5{,}040 + 1;$$
$$372{,}973 + 1 = 6 + 5{,}040 + 2 + 362{,}880 + 5{,}040 + 6.$$

(J) Let n be a three-digit number for which $(n^2 - n)$ ends in three 0s. Consider the expression $(n^k - n)$, where k is any positive integer:

$$n^k - n = n(n^{k-1} - 1).$$

Thus the expression is divisible by both n, and $(n - 1)$. Hence $(n^k - n)$ is divisible by $n(n - 1) = (n^2 - n)$. But $(n^2 - n)$ ends in three 0s, so $(n^k - n)$ also ends in three 0s and n^k ends in the same three digits that n does. We only need to establish then, that only 376 and 625 have squares that end in the same three digits.

Such numbers must be of the form n or $(n - 1)$, where $n(n - 1)$ is a multiple of 1,000. Now n and $(n - 1)$, being adjacent integers, have no common prime factor. Then one of them is divisible by $2 \times 2 \times 2 = 8$, and the other by $5 \times 5 \times 5 = 125$ (and not by 2). There are four of the latter: 125, 375, 625, and 875, with neighbors 124 and 126, 374 and 376, 624 and 626, and 874 and 876. Of the neighbors, only 376 and 624 are divisible by 8.

The only possible three-digit numbers are 375, 376, 624, and 625. But $375^2 = 140{,}625$ and $624^2 = 389{,}376$. This concludes the proof.

349. A ROW OF POSITIVE INTEGERS

(D) No. No. Yes. The only solution in positive integers to $n^2 + (n + 1)^2 = (n + 2)^2$ is $n = 3$, and the only one to $n^2 + (n + 1)^2 + (n + 2)^2 = (n + 3)^2 + (n + 4)^2$ is $n = 10$. But there are equations with four, five, . . . terms on the left:

$$21^2 + 22^2 + 23^2 + 24^2 = 25^2 + 26^2 + 27^2;$$
$$36^2 + 37^2 + 38^2 + 39^2\ 40^2 = 41^2 + 42^2 + 43^2 + 44^2.$$

Prove, on your own, that if n is the number of integers on the right side, the first term of the equation is $n(2n + 1)$.

(F)

$$\left(\frac{n(n+1)}{2}\right)^2$$

350. A PERSISTENT DIFFERENCE

Let a, b, c, d be the digits of a number, with a equal to or greater than b, c equal to or greater than d, and a greater than d. $M = abcd$ and $m = dcba$. To find $(M - m)$:

1. If $b > c$:

$$\begin{array}{rcccc} & [a] & [b] & [c] & [d] \\ - & [d] & [c] & [b] & [a] \\ \hline & [a-d] & [b-1-c] & [10+c-1-b] & [10+d-a] \end{array}$$

2. If $b = c$:

$$\begin{array}{rcccc} & [a] & [b] & [c] & [d] \\ - & [d] & [c] & [b] & [a] \\ \hline & [a-1-d] & [10+b-1-c] & [10+c-1-b] & [10+d-a] \end{array}.$$

In the first (1) the sum of the end digits of the difference is 10, and the sum of the middle digits is 8. In (2) the sums are 9 and 18, and both middle digits are 9s.

This is true for subsequent subtractions. Then we need only test the twenty-five numbers like the difference in (1), and the five like the difference in (2) (order of digits can be ignored):

9,801	8,802	7,803	6,804	5,805
9,711	8,712	7,713	6,714	5,715
9,621	8,622	7,623	6,624	5,625
9,531	8,532	7,533	6,534	5,535
9,441	8,442	7,443	6,444	5,445

and 9,990, 8,991, 7,992, 6,993, 5,994.

The diagram below shows the thirty numbers (with digits in descending order) in boxes with arrows leading (usually through others of the thirty numbers) to the circled 6,174. From any four-digit number not more than seven steps are needed.

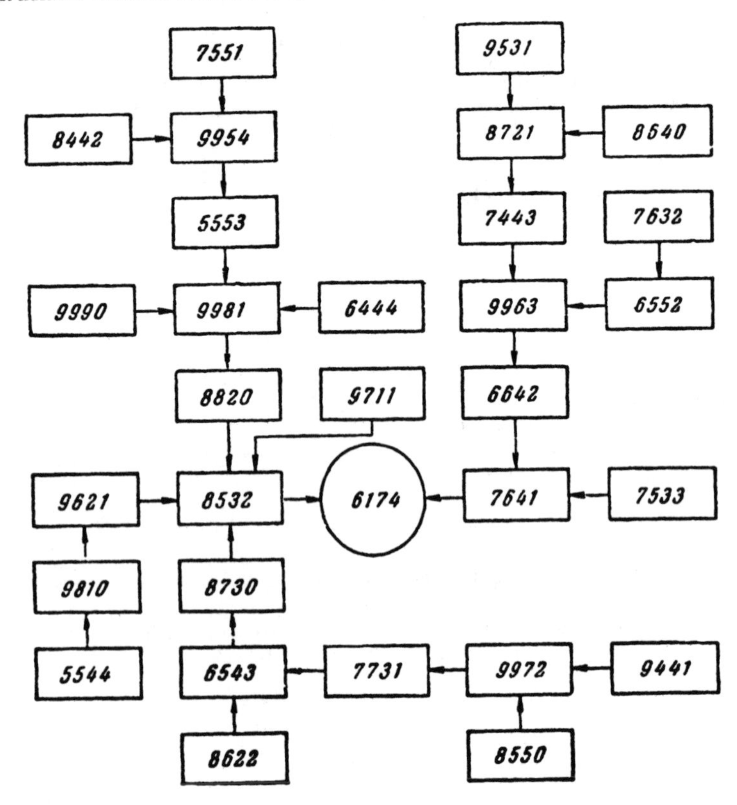

We may call the final difference the pole. The pole of all three-digit numbers is 495. Two-digit numbers do not have a pole; instead, they give a repeating cycle of differences:

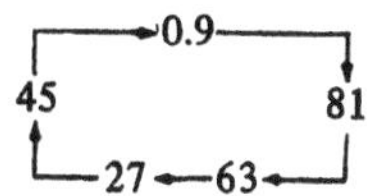

For five-digit numbers the central digit of each difference is 9, and the remaining four digits have the same structure as they have for four-digit numbers (see for yourself). The investigation of five-digit numbers is reduced, therefore, to testing the same thirty sets tested for four-digit numbers. These numbers form three separate cycles:

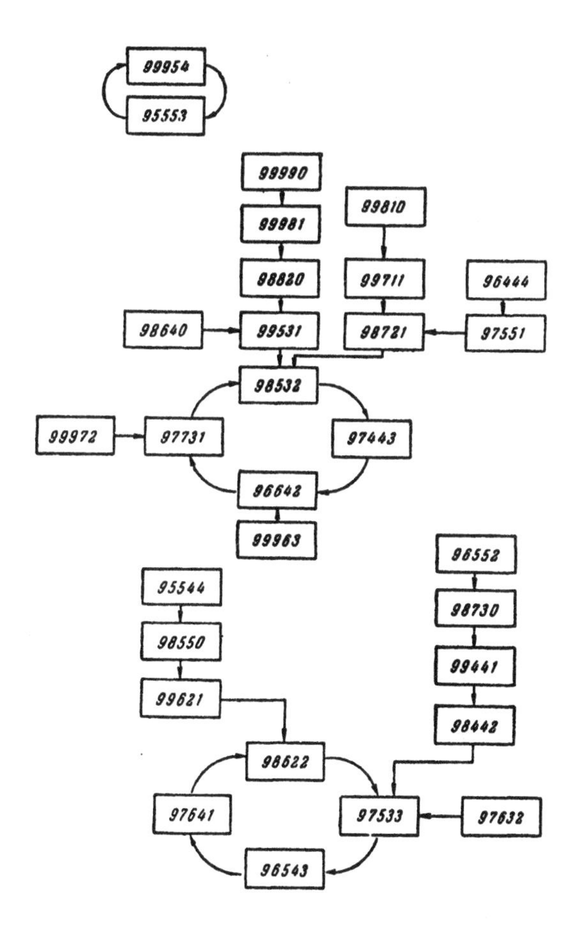

V. A. Orlov, a Moscow engineer, has found an amusing peculiarity of poles. Consider 495, the pole of three-digit numbers. Separate its digits into three groups and insert the digits 5, 9, 4, as shown:

5 9 4
↓4↓9↓5.

Starting with the resulting number, 549,945, we get 995,554 − 445,599 = 549,945, which, therefore, is a pole of six-digit numbers.

Let's continue:

5 9 4
↓54↓99↓45.

Since 999,555,444 − 444,555,999 = 554,999,445, we have a pole of nine-digit numbers.

Now let's take the pole of four-digit numbers, 6,174. Separate its digits into three groups and insert the digits 3 and 6:

3 6
6↓17↓4.

The result is 631,764, another pole of six-digit numbers. Again:

3 6
63↓17↓64.

We have a pole of eight-digit numbers. These procedures can be continued endlessly.

Moreover, if we take a cycle of six-digit numbers (six-digit numbers give one cycle with branches, and two poles, one of which is isolated—see for yourself) and insert 3s and 6s, we get a cycle of eight-digit numbers:

→ 87 ↓6 64 ↓3 20 → 87 ↓6 54 ↓3 21 → 87 ↓6 54 ↓3 30 —

— 66 ↓6 54 ↓3 42 → 86 ↓6 63 ↓3 22 ← 88 ↓6 63 ↓3 20 → 88 ↓6 54 ↓3 20 ←

→ 87,664,320 → 87,654,321 → 87,654,330 → 88,654,320 —

— 66,654,432 → 86,663,322 → 88,663,320 ←

The insertion can be continued endlessly.

XIV. *Numbers Ancient but Eternally Young*

357. A PARADOX

The rectangle's area, $x(2x + y)$, minus the square's, $(x + y)^2 = x^2 - xy - y^2$, which as we know from Problem 356 equals 1 or −1 when x and y are Fibonacci numbers, as here. The rectangle's parts overlap slightly or do not quite meet, which accounts for the differences in area. The diagram shows the "hole" *KHEF* in the 13-by-5 rectangle, a thin quadrilateral with an area of 1.

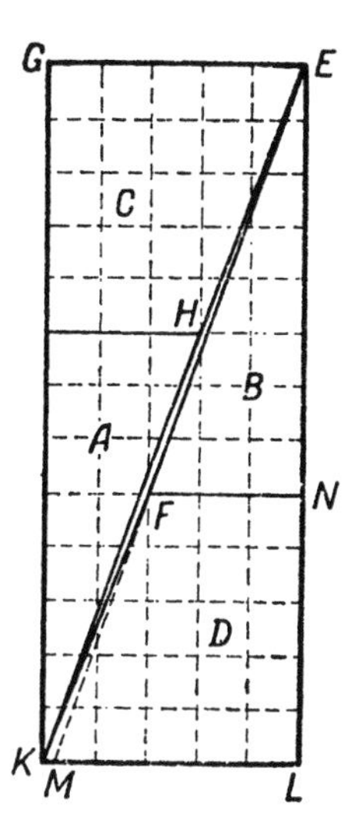

To prove it, let M be the point where the extension of side EF of triangle EFN crosses KL. (If EFK is a straight line, M coincides with K.)

Since triangles EFN and EML are similar, we have these ratios:

$$\frac{ML}{FN} = \frac{EL}{EN}, \text{ or } \frac{ML}{3} = \frac{13}{8}.$$

Then $ML = 4.875$. Since $KL = 5$, M does not coincide with K. Therefore, EFK and EHK are not straight lines, and there is a hole as shown.

To cut the square into a true rectangle, we let $x^2 - xy - y^2 = 0$ instead of 1 or -1. Solving, and accepting only the positive solution:

$$x = \frac{1 + \sqrt{5}}{2} y.$$

This is the ratio of the "golden section," long valued by artists and architects.

Index